Prof P R Eisenstadt
M.E. Dept
Union College
Schenectady N.Y.

**THE COMPONENT ELEMENT
METHOD IN DYNAMICS**
with application to earthquake
and vehicle engineering

**McGRAW-HILL
BOOK COMPANY**
New York
St. Louis
San Francisco
Auckland
Düsseldorf
Johannesburg
Kuala Lumpur
London
Mexico
Montreal
New Delhi
Panama
Paris
São Paulo
Singapore
Sydney
Tokyo
Toronto

SAMUEL LEVY
General Electric Company

and

JOHN P. D. WILKINSON
General Electric Company

The Component Element Method in Dynamics

WITH APPLICATION TO EARTHQUAKE AND VEHICLE ENGINEERING

Library of Congress Cataloging in Publication Data

Levy, Samuel, 1912–
    The component element method in dynamics.

    Includes bibliographical references and index.
    1. Structural dynamics—Data processing.   2.   Finite
element method.   I.   Wilkinson, John P. D., joint
author.   II.   Title.
TA654.L47   624'.171   76-16551

ISBN 0–07–037398–1

**THE COMPONENT ELEMENT
METHOD IN DYNAMICS**
WITH APPLICATION TO EARTHQUAKE
AND VEHICLE ENGINEERING

5 4 3 2 1   7 9 8 7 6

CONTENTS

PREFACE

Things are in continuous motion. Gusts of wind shake trees and buildings, machinery vibrates on a factory floor, vehicles bounce over bumps in the roadway, bridges oscillate as trucks run across them. These events are normal occurrences, and structures are designed to safely accommodate these dynamic forces. To produce such a design requires a knowledge of the forces and stresses occurring during the dynamic excitation.

The objective of this book is to set forth useful and general methods for making calculations to determine the dynamic response of a variety of structures. The methods discussed are oriented toward the use of a digital computing machine. The use of the computer widely extends the range of structural situations that the design analyst can study.

The component element method uses simple conceptual elements for dynamic systems. The assembly of the elements into a complex dynamic system is done automatically in a computer program. The elements themselves can be either linear or nonlinear with no added computational difficulty. This method is particularly advantageous for systems having nonlinearities or having forces which cannot be readily integrated in the dynamic equations of motion. Therefore, it uses numerical integration techniques to solve the dynamic equations. Excitation can be by external forces and by displacement of supports. A key feature of the component element method is that each element is described by coordinates in such a way that it is uncoupled from other elements. The coupling takes place through a coupling matrix in a simple manner by the computer program logic.

The book is written for practicing engineers and designers, as well as for graduate and undergraduate students of structural dynamics. All of the essential tools of structural dynamics pertinent to the methods described are developed in a self-contained manner within the book. We describe three computer programs basic to many of the examples discussed in the text. Complete listings are given for each of the programs. The first program is applicable to a single mass structural system restrained by nonlinear force elements, like springs, stops, and frictional elements. The second program is intended for use in two mass or three mass systems. The third is somewhat more complex, having been written for structures with up to 65 degrees of freedom. The first two programs are suitable for use on a time-sharing computer, and thus allow immediate interaction between the analyst and the computer.

The computer programs described in this book can be obtained by the reader in the form of cards by writing to:

Technology Marketing Operation
General Electric Company
1 River Road
Schenectady, New York 12345, USA.

Price and shipping information will be supplied on request.

The reader is expected to be familiar with the FORTRAN computer language. The computer programs are written in this language. The serious reader is expected not only to read the book, but also to actually load the computer programs into a computer, and to use them to run the examples discussed in the text. Of course, the casual reader is also welcome to peruse the text for the flavor of what is involved in a structural dynamic analysis, and the features and details of structural modeling.

Chapter 1 presents some building blocks necessary for the development of the computing methods. The blocks include some basic modeling elements such as nonlinear springs, dampers, stops, and frictional elements, as well as the finite difference approximations for the equations of motion. Chapters 2 and 3 describe the basic computing methods. The resulting computer programs are applied to simple examples for both single mass nonlinear systems and systems with many degrees of freedom. Chapter 4 contains some case studies from the transportation field. We discuss how automobiles, locomotives, and tracked air cushion vehicles can be modeled dynamically and how their motions can be calculated.

The remaining five chapters concern the dynamics of structures appropriately modeled by a continuum. Chapter 5 presents the basics of the finite element method of structural analysis, followed by a discussion in chapter 6 on how the mode shapes and natural frequencies of continuous structures (such as turbine blades) can be determined by using the stiffness matrices deduced by the finite element method. The final three chapters are devoted to special applications. Chapter 7 describes a study of the dynamics of an aircraft engine fan blade, chapter 8 discusses computations on buildings subject to earthquake excitations, and chapter 9 discusses some aspects of the vibrations of structures submerged in water.

The methods and approaches discussed in this book have been used as a basis for a summer course in structural dynamics taught by one of the authors (S. Levy) at Union College, Schenectady, New York, over the past several years. Many of the illustrative examples discussed in the book have been derived from the personal experiences of the authors in industrial situations. Their inclusion is intended to illustrate the application of structural dynamic analysis to current technological needs and challenges.

We wish to acknowledge the help and encouragement of many of our colleagues at the General Electric Company. In particular, we wish to thank Norman J. Lipstein, who critically read the initial manuscript, and who has materially assisted us in preparing the book. Robert A. Rand has worked with us on the finite element method and produced some of the computer-generated drawings used in chapter 7. Charles T. Salemme and Steven Yokel assisted significantly in the developments associated with the aircraft engine blade analysis of chapter 7, Howard A. Eagle and Frank R. Krieger made the animated movie of the blade, and Robert Zirin kindly provided his analysis of the gas turbine bucket that is discussed in chapter 6. In the seismic studies, we wish to acknowledge useful discussions with Dr Lun K. Liu, Dr Gerald Mok, and Dr Peter T. K. Wu. Gene Martin provided some useful criticisms of the study on locomotive dynamics described in chapter 4. Melody Slater typed the first draft and numerous revisions of the text; to her we are deeply grateful.

S. LEVY
J. P. D. WILKINSON

INTRODUCTION

There is an increasing need for detailed knowledge of how structures behave under applied loads. Two factors appear to account for this need. Technological demands require more exact understanding of structural behavior under load, so that economical and reliable designs can be synthesized. At the same time, social pressures have forced a more detailed assessment of the safety of structures and machines. One area of assessment concerns how the structure behaves under dynamic loads. Some of these loads may be routinely encountered in service—e.g., a vehicle should ride smoothly or quietly. Other loads are encountered only in extreme circumstances, or, in all probability, not at all—e.g., a nuclear power plant must be designed to withstand earthquake tremors and survive them safely. This wide-ranging demand has introduced a challenge to the structural dynamicist. Can one develop a unified set of approaches and tools that are applicable alike to vehicles, to civil engineering structures, to machinery, to biological systems, and to other as yet unencountered structures? The objective of this book is to formulate one such framework, whereby rather general, yet very simple, concepts are developed and applied to a variety of situations.

In this book we advocate the use of three fundamental strategies in structural dynamics. They are: the use of the component element method as the basic description of all parts of a system and their assembly; the use of finite difference approximations in time so that the numerical integration procedures can be used in determining structural response; and the use of the digital computer to carry out the computations in a logical and general manner. These three fundamental aspects of computation allow the solution of a wide variety of problems in structural dynamics. Examples in the text are taken from a range of areas of current technological interest. They include discussions of automotive dynamics, railroad locomotive dynamics, aircraft landing shock, air cushion vehicles, the response of buildings to seismic excitations, and some specialized problems concerning impacts of foreign objects on aircraft engine fan blades.

The type of problem that can be tackled numerically by means of the three basic tools includes those where the major features are nonlinear. Thus, situations where friction or nonlinear springs are of design importance can be treated as routinely as any others. Such nonlinear elements abound in the transportation field, particularly in vehicle suspensions. Through the use of the component element method, a mixture of modeling aids can be used that is of a wider variety than has traditionally been the case. Thus, for example, in studying the response of a water tower situated on a soil foundation, the dynamic behavior of the soil and water can be described in terms of how they vibrate in their normal modes of free vibration. The normal modes can be determined by the finite element method, which is described in the text. These normal modes, since they satisfy orthogonality, can be taken as component elements to describe the response of the soil and water. The structural portion of the tower, on the other hand, can be modeled by lumped mass elements connected by lumped springs or by beams. The coordinates describing the motion are merely the physical displacements of the components.

At the heart of the book are three computer programs which are provided as an essential feature of the main text. One is a simple program for the solution of single degree-of-freedom nonlinear systems. The second is a program for two and three degree-of-freedom nonlinear systems. The third is a more complicated nonlinear program that is dimensioned for a total of 65 degrees of freedom. The text is developed around these programs. First, the means of computing dynamic responses are presented, then simple illustrative examples are inspected by means of the programs for single and multiple degree-of-freedom systems, and, finally, more general situations of practical interest are discussed in the context of these computer programs.

1

THE BUILDING BLOCKS

1.1 INTRODUCTION

This chapter brings together some elementary concepts that lead to the component element method of dynamic analysis. The aim of the chapter is to introduce the basic differential equation of motion for a single mass constrained by a spring. The differential equation, which is obtained through the application of Newton's second law, is written in an approximate manner so that it remains valid at discrete time intervals. The introduction in this way of finite difference equations leads naturally to the observation that differential equations of motion can be written in a finite difference form so that the prediction of movement at one time increment depends solely on the position of the body at one or two instants prior to that time, and on the external forces. This process is known as the time-stepping numerical integration process. The convergence of the numerical process is discussed by means of a simple example as well as by analytical evaluation.

Another set of building blocks involves the behavior of certain force component elements. These elements may derive from real springs of one form or another, from stops, or from the friction phenomenon. We show how the elements develop restraining forces that are used in the time-stepping numerical integration scheme.

1.2 NEWTON'S SECOND LAW OF MOTION

When a mass is acted on by a force, the rate of change of the momentum of the mass in the direction of the force is equal to the force. Thus,

$$\frac{\mathrm{d}}{\mathrm{d}t}(mv) = F \tag{1.1}$$

In all systems discussed in this book, the mass m will remain constant in time, although this is not a requirement of the component element method. Thus, Newton's second law can be written as

$$m\frac{\mathrm{d}v}{\mathrm{d}t} = m\frac{\mathrm{d}^2y}{\mathrm{d}t^2} = F \tag{1.2}$$

This equation states that the sum of the forces acting on a mass gives rise to a change in acceleration of the mass. This equation is the basic differential equation of motion. In the example given here, only a mass has been considered. In practice, the differential equations that describe the motion of a structure are far more complex than this equation. Ordinarily, a significant amount of engineering and physical judgement has to be used to set up the differential equations of motion. Part of the aim of this book is to describe how component elements of structures may be assembled as a system so that a computer program automatically sets up the equations. Part of the aim is also to show how these equations, once set up, can be solved numerically to obtain a fairly detailed description of the dynamic behavior of the structure.

1.3 COMPONENT MODELING ELEMENTS—MASS, SPRING, DAMPER

Consider a structure that is made up of a single mass m, supported by a spring, as shown in Fig. 1.1. The spring has a stiffness k, which means that a force k will extend the spring by a unit length. The mass is also connected to the ground by a damper which exerts a force proportional to the velocity of the mass. The quantity c is known as the damping constant. For unit velocity of the mass, the damper develops a resisting force c. This particular damper is known as a viscous damper, because the property of proportionality to velocity is observed in viscous fluids. Damping in physical systems is rarely as simple as purely viscous damping.

   The three basic elements of mass, spring stiffness, and damping are frequently used in structural modeling to simulate a physical system in simple terms. Because these elements are characterized by simplicity and general usefulness, many basic patterns of behavior in more complex systems can often be deduced from a study of the behavior of a model made up of elements of this type.

   If the mass–spring–damper system under consideration is acted on by a force F, as shown in Fig. 1.1, it will tend to move in the direction of F in a manner that is governed by Newton's second law. The spring will push back with a force ky, the damper will push back with a force $c\dot{y}$, and the mass inertia will push back with a force $m\ddot{y}$, as shown. The motion is defined by a differential equation that can be

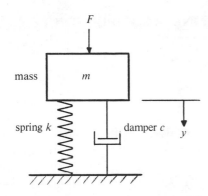

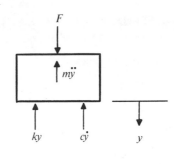

(a) Simple mass–spring–damper system   (b) Free body diagram of forces acting on the system

FIGURE 1.1
A simple dynamic system

obtained by adding up all the forces acting on the mass and equating them to the product of mass and acceleration, which is also known as the inertia force. Thus, following the free body diagram of Fig. 1.1(b), the equation of motion can be written as

$$F - ky - c\dot{y} = m\ddot{y} \tag{1.3}$$

or

$$m\ddot{y} + c\dot{y} + ky = F \tag{1.4}$$

For a single mass system, this equation can be solved in a variety of ways. If the force F varies with time in a simple enough manner, then the solutions can be obtained through the classical means of integrating the equation analytically. Closed form solutions are then obtained that can be readily understood, and which provide an overall view of the behavior of the system. Many excellent books on structural dynamics contain discussions on this aspect of structural dynamics. Such solutions provide considerable insight into the behavior of simple systems. However, when we consider complex structures which contain an assemblage of many such component masses, springs, and dampers, then an analytical solution becomes so complex as to defeat its purposes. It is in this situation that other means are sought to gain insight into the detailed structural dynamic behavior. One such method is the component element method which describes the structural system as an assemblage of elementary components and integrates the dynamic equations by the use of finite differences. This method will be used as a basic tool in this book because it is amenable for use with digital computers, and because it can also be used in situations where the basic elements in the structural model are more complicated than the three discussed thus far. Before entering into a description of this approach, however, we will first discuss, by means of the classical closed form solutions, some of the basic properties of the single mass–spring–damper system. These properties will serve as a guide in the study of more complex systems that will be encountered subsequently.

1.4   SOME BASIC CHARACTERISTICS OF LINEAR DAMPED SINGLE MASS SYSTEMS

Natural frequencies

The differential equation

$$m\ddot{y} + c\dot{y} + ky = 0 \tag{1.5}$$

readily yields a solution if we assume that $y = A\,e^{rt}$, where both A and r may be complex numbers. Substitution of this expression into eq. (1.5) gives a quadratic equation in r:

$$r^2 + \frac{c}{m}r + \frac{k}{m} = 0 \tag{1.6}$$

This equation has two roots, r_1 and r_2:

$$\begin{Bmatrix} r_1 \\ r_2 \end{Bmatrix} = -\frac{c}{2m} \pm \sqrt{\frac{c^2}{4m^2} - \frac{k}{m}} \tag{1.7}$$

The final solution of eq. (1.5) is, therefore,

$$y = A_1\,e^{r_1 t} + A_2\,e^{r_2 t} \tag{1.8}$$

A_1 and A_2 are constants that depend on the initial conditions. The character of these roots, and thus of the solution itself, is strongly dependent on the amount of viscous damping present in the system.

If no damping is present, then

$$\begin{Bmatrix} r_1 \\ r_2 \end{Bmatrix} = \pm i\,\sqrt{\frac{k}{m}} \tag{1.9}$$

and the solution becomes

$$y = A_1\,e^{r_1 t} + A_2\,e^{-r_1 t} \tag{1.10}$$

Because of the relationship between exponential and trigonometric functions,

$$e^{i\alpha} = \cos\alpha + i\sin\alpha \tag{1.11}$$

The solution in eq. (1.10) can also be written as:

$$y = B_1\cos\omega_0 t + B_2\sin\omega_0 t \tag{1.12}$$

Here, $\omega_0 = \sqrt{k/m}$ is known as the *natural frequency* of the undamped system. The system will oscillate in time, always with the same frequency ω_0. The units of ω_0 are in radians per unit time, generally rad/sec.

In the presence of damping, there are two physically distinct behaviors that depend on the relative magnitude of c. If $c > 2\sqrt{km}$, then the roots of eq. (1.6) are both real. If $c < 2\sqrt{km}$, the roots are complex conjugates. In each regime of c, the physical behavior of the mass is quite different. On the one hand, when $c > 2\sqrt{km}$, eq. (1.6) tells us that both roots are real and negative. There,

$$y = A_1\, e^{r_1 t} + A_2\, e^{r_2 t} \tag{1.13}$$

and we observe that no oscillations occur; the motion will decay exponentially with time. In this case, the system is said to be overdamped.

On the other hand, when $c < 2\sqrt{km}$, the roots are complex conjugates:

$$r = -\frac{c}{2m} \pm i\sqrt{\frac{k}{m} - \frac{c^2}{4m^2}} \tag{1.14}$$

In this case, the solution can be recast into a form also containing trigonometric functions. First note that

$$y = e^{-ct/2m}\left[A_1\, e^{i\omega t} + A_2\, e^{-i\omega t}\right] \tag{1.15}$$

where

$$\omega = \sqrt{\frac{k}{m} - \frac{c^2}{4m^2}} \tag{1.16}$$

Then, using the trigonometric identity eq. (1.11), this solution can be written as

$$y = e^{-ct/2m}\left[(A_1 + A_2)\cos \omega t + i(A_1 - A_2)\sin \omega t\right] \tag{1.17}$$

Since A_1 and A_2 are arbitrary constants dependent on initial conditions, B_1 and B_2 can be defined so as to finally write:

$$y = e^{-ct/2m}\left[B_1 \cos \omega t + B_2 \sin \omega t\right] \tag{1.18}$$

This solution will decay exponentially according to the exponential, $e^{-ct/2m}$. However, in contrast to the previous case, it also contains oscillatory terms governed by the trigonometric functions, which have a frequency ω. In this case, the system is said to be underdamped.

The transition between the two solution types occurs when

$$c = c_c = 2\sqrt{mk} = 2m\omega_0 \tag{1.19}$$

This value of the damping coefficient c_c is known as *critical damping*. A structure is said to be critically damped when, upon release from a small initial displacement with no velocity, it just manages to return to rest without oscillatory motion. We also note here the relationship between the critical damping coefficient and the undamped natural frequency ω_0. When the damping is less than critical damping, there is damped oscillatory motion of a frequency:

$$\omega = \sqrt{\frac{k}{m} - \frac{c^2}{4m^2}} = \omega_0\sqrt{1 - \left(\frac{c}{c_c}\right)^2} \tag{1.20}$$

Thus, the damped natural frequency of a system is always less than the undamped natural frequency. The damped natural frequency, in fact, approaches zero as the damping coefficient c approaches the critical damping c_c.

Initial conditions

When the system is given an initial velocity $\dot{y}_0$ and an initial displacement y_0 at time $t = 0$, it is possible to evaluate the constants in the solution. For example, consider the case when the system is underdamped and $c < 2\sqrt{km}$. Then the general solution has been obtained in terms of the constants B_1 and B_2 as

$$y = e^{-ct/2m}[B_1 \cos \omega t + B_2 \sin \omega t] \tag{1.21}$$

By simple substitution, we find that, at time $t = 0$,

$$y_0 = B_1$$

$$\dot{y}_0 = \omega B_2 - \frac{c}{2m}B_1$$

Thus,

$$B_1 = y_0$$

$$B_2 = \frac{1}{\omega}\left(\dot{y}_0 - \frac{c}{2m}y_0\right)$$

and the full solution is

$$y = e^{-ct/2m}\left[y_0 \cos \omega t + \frac{\dot{y}_0 - (c/2m)y_0}{\omega} \sin \omega t\right] \tag{1.22}$$

This solution is also a decaying oscillation. Its importance here is to illustrate that even in the absence of external forces, an initial velocity or displacement will cause a dynamic response in the system. Thus, the initial conditions of a dynamic situation are of the greatest consequence.

Magnification factor and resonance curves

If the single mass system is under a sinusoidal excitation $F \sin \omega t$, then a solution can be obtained for the times at which the transient initial conditions have ceased to matter. For that case, then, the equation of motion is

$$m\ddot{y} + c\dot{y} + ky = F \sin \omega t \tag{1.23}$$

The solution for y is obtained by letting

$$y = C_1 \sin \omega t + C_2 \cos \omega t \tag{1.24}$$

Substituting this assumed solution into eq. (1.23) and collecting together the coefficients of $\sin \omega t$ and $\cos \omega t$ give two equations in C_1 and C_2:

$$\left.\begin{array}{l} (k - m\omega^2)C_1 - c\omega C_2 = F \\ c\omega C_1 + C_2(k - m\omega^2) = 0 \end{array}\right\} \tag{1.25}$$

Simultaneous solution of these two equations yields

$$C_1 = \frac{(k - m\omega^2)F}{(k - m\omega^2)^2 + c^2\omega^2}$$

$$C_2 = -\frac{c\omega F}{(k - m\omega^2)^2 + c^2\omega^2}$$

(1.26)

Rearranging the solution into the form

$$y = C_3 \sin(\omega t - \psi)$$

we get

$$y = \frac{F}{[(k - m\omega^2)^2 + c^2\omega^2]^{1/2}} \sin(\omega t - \psi)$$

(1.27)

$$\tan \psi = \frac{c\omega}{k - m\omega^2}$$

Again, this solution can be rearranged by recalling the definitions of ω_0, the undamped natural frequency of the system, and c_c, the critical damping of the system:

$$\omega_0 = \sqrt{\frac{k}{m}}$$

$$c_c = 2m\omega_0$$

Then,

$$y = \frac{(F/k)\sin(\omega t - \psi)}{\left[\left(1 - \frac{\omega^2}{\omega_0^2}\right)^2 + 4\frac{\omega^2}{\omega_0^2}\frac{c^2}{c_c^2}\right]^{1/2}}$$

(1.28)

$$\tan \psi = \frac{2\dfrac{\omega}{\omega_0}\dfrac{c}{c_c}}{1 - \dfrac{\omega^2}{\omega_0^2}}$$

This solution can now be plotted to show how $|y|$ varies with the frequency of excitation. Such a plot is shown in Fig. 1.2, where the magnification factor $|y|/(F/k)$ is plotted. Such a curve is known as a *resonance curve*. The curves show that the magnification factor has a maximum at a frequency just below $\omega = \omega_0$. In fact, it may be shown that the maximum occurs at

$$\frac{\omega}{\omega_0} = \left(1 - \frac{2c^2}{c_c^2}\right)^{1/2}$$

(1.29)

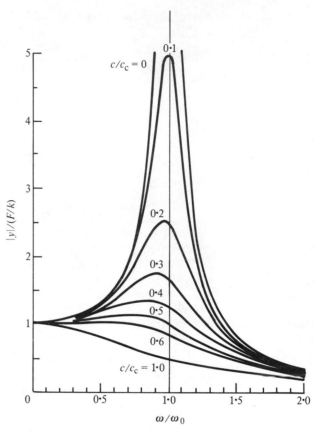

FIGURE 1.2
Magnification factor

and that the magnification factor is then

$$\frac{|y|}{(F/k)} = \frac{1}{2(c/c_c)} = \frac{1}{c}\sqrt{km} \tag{1.30}$$

Response curves such as the ones in Fig. 1.2 are often used to illustrate the response of a system to sinusoidal forces. They will be encountered in later sections when we discuss the response of systems with nonlinear spring elements.

1.5 FINITE DIFFERENCES

The derivative of y with respect to t is defined as the limit of the ratio of the increment in y to the increment in t, as the increments approach zero. Thus,

$$\frac{dy}{dt} = \lim_{\Delta t \to 0} \left(\frac{\Delta y}{\Delta t}\right) \tag{1.31}$$

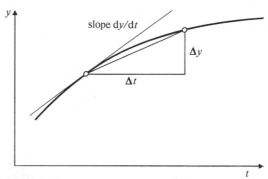

FIGURE 1.3
Approximation involved in replacing dy/dt with $\Delta y/\Delta t$

This relationship is shown in Fig. 1.3. It is evident from this figure that even when the increment Δt is quite large, the error involved in approximating dy/dt by the ratio $\Delta y/\Delta t$ is quite small as long as the curve has a slowly varying slope.

Consider now the curve shown in Fig. 1.4, which crosses the three coordinate points (y_{-1}, t_{-1}), (y_0, t_0), (y_1, t_1). The following approximation can be made for the slope of the curve:

$$\left(\frac{dy}{dt}\right)_{t=t_0} = \frac{\Delta y}{\Delta t} = \frac{y_1 - y_{-1}}{t_1 - t_{-1}} = \frac{y_1 - y_{-1}}{2\Delta t} \tag{1.32}$$

Likewise, a similar approximation can be made for the second derivative (acceleration):

$$\left(\frac{d^2 y}{dt^2}\right)_{t=t_0} = \frac{\left[\left(\dfrac{dy}{dt}\right)_{t=t_0+\Delta t/2} - \left(\dfrac{dy}{dt}\right)_{t=t_0-\Delta t/2}\right]}{\Delta t}$$

$$= \frac{\left[\dfrac{(y_1 - y_0)}{\Delta t} - \dfrac{(y_0 - y_{-1})}{\Delta t}\right]}{\Delta t}$$

$$= \frac{y_1 - 2y_0 + y_{-1}}{(\Delta t)^2} \tag{1.33}$$

By introducing these approximations, the derivatives in the equations of motion can be replaced by the differences between successive positions taken by the mass at successive increments in time. These differences are known as *finite differences* precisely because they are separated by finite time increments.

1.6 DIFFERENCE EQUATIONS

In the basic differential equation (1.4), the derivatives will be replaced by the approximations given by eqs. (1.32) and (1.33). Then

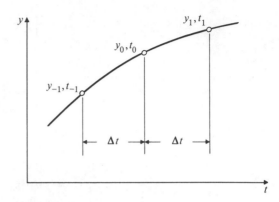

FIGURE 1.4
Positions along a curve at intervals Δt

$$m\left(\frac{y_1 - 2y_0 + y_{-1}}{(\Delta t)^2}\right) + c\left(\frac{y_1 - y_{-1}}{2\Delta t}\right) + ky_0 = F_0 \qquad (1.34)$$

Solving this equation for y_1 gives

$$y_1 = \left[\frac{1}{\dfrac{m}{(\Delta t)^2} + \dfrac{c}{2\Delta t}}\right]\left[\left(\frac{2m}{(\Delta t)^2} - k\right)y_0 + \left(\frac{c}{2\Delta t} - \frac{m}{(\Delta t)^2}\right)y_{-1} + F_0\right] \qquad (1.35)$$

Hence, it is possible to calculate the displacement of the mass y_1 if the previous history of displacements at y_0 and y_{-1}, as well as the present external force F_0, are known.

Having thus obtained the value of the displacement y_1, the same process can be repeated by changing the indices. Then the value of y_2 can be found in terms of y_1, y_0, and F_1. This process of finding the new displacement based on the knowledge of two previous displacements is known as the step-by-step process of integration. It is extraordinarily simple in concept, but can, with repetitive application, yield the complete time history of the behavior of a system. This approach is not only applicable to simple single mass systems described here; it is also applicable to systems made up of many masses and springs, and to systems where the relationships between the elements may be nonlinear. Its power derives from its very simplicity. However, a certain sacrifice is made: in order to make repetitive steps in time, a digital computer must be used, particularly if the system of equations is at all complicated. As a result, certain general conclusions obtainable from analytical closed form solutions may be hidden because the numerical solution yields specific results for a single situation only.

Example

Consider eq. (1.4) with $m = 1$, $k = 1$, and F and c set equal to zero. Initially, let y be 0 and $\dot{y}$ be 1. In that case, eq. (1.4) becomes

$$\ddot{y} + y = 0$$

with $y = 0$, $\dot{y} = 1$, at $t = 0$. It can readily be shown that $y = \sin t$ satisfies this equation and the initial conditions.

The difference equation (1.35) can also be used to solve this equation. First, assume $\Delta t = 1$. Then eq. (1.35) becomes

$$y_1 = y_0 - y_{-1}$$

$$y_{n+1} = y_n - y_{n-1}$$

Take $y_0 = 0$ and $y_{-1} = y_0 - \dot{y}_0 \Delta t = 0 - 1 = -1$. Then, with $y_0 = 0$ and $y_{-1} = -1$, it is seen that $y_1 = 1$, $y_2 = 1$, $y_3 = 0$, $y_4 = -1$, $y_5 = -1$, and so on, as given in the second column of Table 1.1. The results are plotted in Fig. 1.5.

Table 1.1 SOLUTIONS OF $\ddot{y} + y = 0$ WITH RELATIVELY LARGE TIME STEPS

t	$\Delta t = 1$	$\Delta t = 0.5$	Exact solution $\sin t$
-1.0	-1		
-0.5		-0.5	
0	0	0	0
0.5		0.5	0.479
1.0	1	0.875	0.841
1.5		1.030	0.997
2.0	1	0.926	0.909
2.5		0.590	0.598
3.0	0	0.106	0.141
3.5		-0.405	-0.350
4.0	-1	-0.815	-0.757
4.5		-1.022	-0.978
5.0	-1	-0.975	-0.959

FIGURE 1.5
Comparison of finite difference solution with exact solution of $\ddot{y} + y = 0$

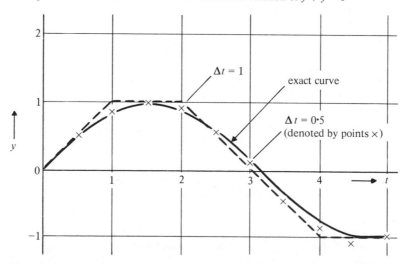

Now, halve the time increment so that $\Delta t = 0\cdot5$. Then $y_0 = 0$, and $y_1 = y_0 - \dot{y}_0 \Delta t = 0 - 0\cdot5 = -0\cdot5$. Thus, eq. (1.35) becomes

$$y_1 = 1\cdot75y_0 - y_1$$

$$y_{n+1} = 1\cdot75y_n - y_{n-1}$$

Hence, by sequential steps, the results in the third column of Table 1.1 are obtained. The results are also plotted in Fig. 1.5.

Several points are brought out by this example. First, the smaller the time step, the nearer the results come to the exact solution. Remarkably good accuracy is obtainable, however, even with quite a coarse time interval. Second, the choice of the size of the time step is an important ingredient to the success of the numerical scheme. In fact, a criterion has been proposed by Levy and Kroll[1] for choosing the size of the time increments. This criterion influences the convergence of such numerical solutions toward the exact solution of the given differential equation.

1.7   A STABILITY CRITERION

We will reproduce here the discussion of Levy and Kroll[1] which leads to a stability criterion suitable for numerical integration.

The basic problem considered in this section is shown in Fig. 1.6. The mass, if initially disturbed, should vibrate without damping or amplitude build-up at a natural frequency

$$\omega = \sqrt{k/m}$$

For such a system, the equation of motion is

$$m\frac{d^2y}{dt^2} + ky = 0 \tag{1.36}$$

In applying numerical integration methods to the solution of this equation, the following notation is used:

Δt = time increment between successive steps in the numerical integration process

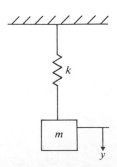

FIGURE 1.6
Basic problem considered in developing
the stability criterion

n = number of steps taken from $t = 0$ to $t = n\Delta t$; as a subscript, it indicates the value when $t = n\Delta t$

y_n = displacement when $t = n\Delta t$

Substituting into eq. (1.36):

$$\left(\frac{d^2 y}{dt^2}\right)_n = \frac{1}{(\Delta t)^2}(y_{n+1} - 2y_n + y_{n-1}) \tag{1.37}$$

gives

$$\frac{m}{(\Delta t)^2}(y_{n+1} - 2y_n + y_{n-1}) + ky_n = 0 \tag{1.38}$$

which reduces to

$$y_{n+1} + \left[\frac{k(\Delta t)^2}{m} - 2\right]y_n + y_{n-1} = 0 \tag{1.39}$$

To solve eq. (1.39) by the calculus of finite differences, we make the substitution

$$y_n = A\beta^n \tag{1.40}$$

where A is an arbitrary constant to be determined from initial conditions, and β is a number to be so chosen that eq. (1.39) is satisfied. Substituting from eq. (1.40) into eq. (1.39) gives

$$A\beta^{n+1} + \left[\frac{k(\Delta t)^2}{m} - 2\right]A\beta^n + A\beta^{n-1} = 0 \tag{1.41}$$

Dividing through by $A\beta^{n-1}$, eq. (1.41) reduces to

$$\beta^2 + \left[\frac{k(\Delta t)^2}{m} - 2\right]\beta + 1 = 0 \tag{1.42}$$

The following two cases are of interest.

Case 1: $0 < \Delta t < 2\sqrt{m/k}$
We will consider $\Delta t = \sqrt{2}\sqrt{m/k}$; that is, $\Delta t = 0{\cdot}225(2\pi)\sqrt{m/k}$ where $(2\pi)\sqrt{m/k}$ is the natural period of the system. In this case, eq. (1.42) reduces to

$$\beta^2 + 1 = 0 \tag{1.43}$$

from which

$$\beta = \pm\sqrt{-1} = \pm i = e^{\pm i\pi/2} \tag{1.44}$$

Substituting this value of β into eq. (1.40):

$$y_n = A\,e^{i\pi n/2} + B\,e^{-i\pi n/2}$$

$$= A'\sin\frac{\pi n}{2} + B'\cos\frac{\pi n}{2} \tag{1.45}$$

where A, B, A', B' are arbitrary constants determined from the initial conditions. Since

$$n = t/\Delta t = t\sqrt{k/m}/\sqrt{2}$$

eq. (1.45) can be written

$$y = A' \sin 1{\cdot}11\, t\sqrt{k/m} + B' \cos 1{\cdot}11\, t\sqrt{k/m} \qquad (1.46)$$

In this case, the effect of the numerical integration method is to increase the effective natural frequency from $\sqrt{k/m}$ to $1{\cdot}11\sqrt{k/m}$ without introducing damping or build-up of the response.

As the value of the finite time increment Δt is varied from 0 to $2\sqrt{m/k}$, the factor $1{\cdot}11$ in eq. (1.46) varies from 1 to $\pi/2$, but otherwise the form of eq. (1.46) does not change.

Case 2: $2\sqrt{m/k} < \Delta t$

We will consider $\Delta t = 3\sqrt{m/k}$. Equation (1.42) becomes

$$\beta^2 + 7\beta + 1 = 0 \qquad (1.47)$$

giving

$$\beta = -0{\cdot}1459 = -e^{-1{\cdot}925}$$

$$\beta = -6{\cdot}8541 = -e^{1{\cdot}925}$$

Substituting these values of β into eq. (1.40):

$$y_n = A(-1)^n e^{-1{\cdot}925n} + B(-1)^n e^{1{\cdot}925n}$$

$$= A' \cos n\pi \, \text{sh } 1{\cdot}925n + B' \cos n\pi \, \text{ch } 1{\cdot}925n \qquad (1.48)$$

Substituting

$$n = t/\Delta t = (t\sqrt{k/m})/3$$

into eq. (1.48) gives

$$y = \left[A' \, \text{sh } 0{\cdot}642t\sqrt{k/m} + B' \, \text{ch } 0{\cdot}624t\sqrt{k/m} \right]$$

$$\text{cos } 1{\cdot}047t\sqrt{k/m} \qquad (1.49)$$

In this case, the primary effect of the numerical integration method is to introduce hyperbolic functions of time in the answer. Since these functions increase indefinitely with time, the result is divergent. Similar results are obtained whenever $\Delta t > 2\sqrt{m/k}$.

We conclude, therefore, that the solutions are convergent, or stable, if $\Delta t < 2\sqrt{m/k}$. In practice, in order to adequately follow the response, and to retain good accuracy during the calculation, Δt should be made equal to about 1/20 the period of the oscillation, or $\Delta t = (2\pi/20)\sqrt{m/k} \approx 0{\cdot}31\sqrt{m/k}$. When systems are made up of several masses and springs, then the time interval should be chosen to be within $0{\cdot}31/\omega$, where ω is the highest natural frequency in rad/sec.

1.8   OTHER APPROXIMATIONS TO TIME DERIVATIVES

There is a large body of literature[2,3] on the subject of how derivatives can be approximated by finite differences. The approximations used in section 1.5 are probably the most simple. In this section, we present a discussion of a number of other difference schemes that have been found useful in dynamic calculations. The purpose of the section is to give the flavor of the possibilities that exist in differencing schemes, rather than to present an exhaustive compilation of all methods. Thus, while we present in some detail the methods developed by Houbolt,[4] by Newmark,[5] and by Wilson,[6] many other approaches to be found in the literature are not discussed.

Polynomial approximations

The process of fitting difference approximations to a curve and its derivatives can be viewed as fitting a polynomial to the points on the curve at the difference intervals in the immediate vicinity of the point of interest, as illustrated in Fig. 1.7. Consider how a curve $y(t)$ can be fitted at the three equally spaced difference points y_1, y_0, y_{-1} by a second-order polynomial

$$y_p = at^2 + bt + c \tag{1.50}$$

If the spacing is h, then

$$\left.\begin{array}{l} y_{-1} = ah^2 - bh + c \\ y_0 = c \\ y_1 = ah^2 + bh + c \end{array}\right\} \tag{1.51}$$

Solving for a, we get

$$2ah^2 = y_1 - 2y_0 + y_{-1} \tag{1.52}$$

However, since the second derivative of y_p is itself equal to $2a$, we have the approximation

$$\ddot{y}_0 = \frac{1}{h^2}(y_1 - 2y_0 + y_{-1}) \tag{1.53}$$

which is the same expression that was obtained in section 1.5.

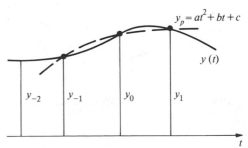

FIGURE 1.7
Approximation to the curve $y(t)$ by a polynomial y_p

Similarly, if we were to fit a cubic polynomial

$$y_p = at^3 + bt^2 + ct + d \tag{1.54}$$

through the points y_{-1}, y_0, y_1, y_2, we would demand that

$$\left.\begin{array}{l} y_{-1} = -ah^3 + bh^2 - ch + d \\ y_0 = d \\ y_1 = ah^3 + bh^2 + ch + d \\ y_2 = 8ah^3 + 4bh^2 + 2ch + d \end{array}\right\} \tag{1.55}$$

Solving for $6a$, which is also the third derivative of y_p, we find directly that

$$\dddot{y}_0 = \frac{1}{h^3}(y_2 - 3y_1 + 3y_0 - y_{-1}) \tag{1.56}$$

Other approximations can be derived in the same way. They may be of a wide variety and could be expressed in terms of other values of y. For example,

$$\ddot{y} = f(y_{-2}, y_{-1}, y_0, y_1), \text{ etc.}$$

Obviously, an approximation is involved. It is intuitively evident that some approximations may be more accurate than others. The formulas are not exact. There is an *error* involved in each term. To investigate the size of this error, we use Taylor's series, wherein the function $y(t + h)$ can be expanded as a series in its derivatives:

$$y(t + h) = y(t) + h\dot{y}(t) + \frac{h^2}{2!}\ddot{y}(t) + \frac{h^3}{3!}\dddot{y}_0 + \cdots \tag{1.57}$$

Each point y_i can therefore be written in terms of a series as, for example,

$$y_1 = y_0 + h\dot{y}_0 + \frac{h^2}{2!}\ddot{y}_0 + \frac{h^3}{3!}\dddot{y}_0 + \cdots$$

$$y_{-1} = y_0 - h\dot{y}_0 + \frac{h^2}{2!}\ddot{y}_0 - \frac{h^3}{3!}\dddot{y}_0 + \cdots \tag{1.58}$$

Upon eliminating y_0, we find that

$$y_1 - y_{-1} = 2h\dot{y}_0 - \frac{2h^3}{3!}\dddot{y}_0 + \cdots \tag{1.59}$$

so that

$$\dot{y}_0 = \frac{(y_1 - y_{-1})}{2h} + \frac{h^2}{3!}\dddot{y}_0 + \cdots \tag{1.60}$$

Thus, the error in the approximation

$$\dot{y}_0 = \frac{(y_1 - y_{-1})}{2h} \tag{1.61}$$

consists of the remaining terms in the series, which are all multiplied by small quantities h^2 or by even higher powers of h. This is the same expression obtained in section 1.5. The error in the first derivative for this particular approximation is said to be of order h^2. In a similar manner, we can eliminate $\dot{y}_0$ by addition, and obtain

$$y_1 + y_{-1} = 2y_0 + \frac{2h^2}{2!}\ddot{y}_0 + \text{terms of order } h^4 + \cdots$$

so that

$$\ddot{y}_0 = \frac{(y_1 - 2y_0 + y_{-1})}{h^2} + \text{terms of order } h^2 + \cdots \tag{1.62}$$

Here, also, the approximation is of order h^2.

Forward, backward, and central differences

In general, then, the derivatives can be expressed with any desired accuracy of order h^n, and in terms of a variety of points y_i. If we select the approximation of the sort

$$y_0^{(n)} = f(y_0, y_1, y_2, \ldots) \tag{1.63}$$

these approximations are said to be *forward differences,* because they are given in terms of points ahead of the point of interest t_0. In our calculation scheme, these approximations are not useful, since the values ahead of the point are yet to be calculated. If, on the other hand, we select

$$y_0^{(n)} = f(y_0, y_{-1}, y_{-2}, \ldots) \tag{1.64}$$

these approximations are said to be *backward differences,* and are useful to us because the derivatives are given in terms of points behind the point of interest t_0. In our calculation scheme, this approximation is useful where the derivative will not be used to obtain y_1. *Central differences* are useful to us as well. In fact, the approximations used in section 1.5 are central differences. They are so called because the zero point is central to the points y_1 and y_{-1}. In our calculation scheme, these approximations are useful to obtain y_1. Thus, the approximations used in section 1.5 are central differences, and have errors of order h^2 in velocity and acceleration.

Houbolt's method

Houbolt[4] has used a time step method based on a backward difference scheme that has an error of order h^3. To develop his differencing scheme, we start by deriving the backward differences appropriate to this approximation.

Given the points $y_0, y_{-1}, y_{-2}, y_{-3}, \ldots$, define a first backward difference

$$\nabla y_0 = y_0 - y_{-1} \tag{1.65}$$

The second backward difference is then

$$\dot{\nabla}(\nabla y_0) = \nabla^2 y_0 = (y_0 - y_{-1}) - (y_{-1} - y_{-2})$$
$$= y_0 - 2y_{-1} + y_{-2} \tag{1.66}$$

Similarly, the nth backward difference is

$$\nabla^n y_0 = \nabla(\nabla^{n-1} y_0) \tag{1.67}$$

The coefficients in the difference expressions are actually the coefficients in the binomial expansion $(a - b)^n$. Thus,

$$\nabla^3 y_0 = y_0 - 3y_{-1} + 3y_{-2} - y_{-3} \tag{1.68}$$
$$\nabla^4 y_0 = y_0 - 4y_{-1} + 6y_{-2} - 4y_{-3} + y_{-4} \tag{1.69}$$

Now Taylor's series gives

$$y(t + h) = y(t) + h\dot{y}(t) + \frac{h^2}{2!}\ddot{y}(t) + \frac{h^3}{3!}\dddot{y}(t) + \cdots$$

or

$$y(t + h) = y(t) + hDy(t) + \frac{h^2}{2!}D^2 y(t) + \frac{h^3}{3!}D^3 y(t) + \cdots$$
$$= \left(1 + hD + \frac{h^2}{2!}D^2 + \frac{h^3}{3!}D^3 + \cdots\right) y(t) \tag{1.70}$$

where $D = d/dt$. Thus,

$$y_1 = y(0 + h) = \left(1 + hD + \frac{h^2}{2!}D^2 + \frac{h^3}{3!}D^3 + \cdots\right) y_0$$
$$y_{-1} = y(0 - h) = \left(1 - hD + \frac{h^2}{2!}D^2 - \frac{h^3}{3!}D^3 + \cdots\right) y_0 \tag{1.71}$$

The first backward difference is

$$\nabla y_0 = y_0 - y_{-1}$$
$$= \left(hD - \frac{h^2 D^2}{2} + \frac{h^3 D^3}{6} - \frac{h^4 D^4}{24} + \cdots\right) y_0 \tag{1.72}$$

Similarly,

$$\nabla^2 y_0 = \left(h^2 D^2 - h^3 D^3 + \frac{7}{12}h^4 D^4 - \cdots\right) y_0 \tag{1.73}$$

$$\nabla^3 y_0 = \left(h^3 D^3 - \frac{3}{2}h^4 D^4 + \frac{5}{4}h^5 D^5 - \cdots\right) y_0 \tag{1.74}$$

Now these three equations can be rearranged so that

$$\left.\begin{aligned}
D &= \frac{\nabla}{h} + \frac{hD^2}{2} - \frac{h^2 D^3}{6} + \frac{h^3 D^4}{24} - \cdots \\[2mm]
D^2 &= \frac{\nabla^2}{h^2} + hD^3 - \frac{7h^2 D^4}{12} + \cdots \\[2mm]
D^3 &= \frac{\nabla^3}{h^3} + \frac{3hD^4}{2} - \frac{5h^2 D^5}{4} + \cdots
\end{aligned}\right\} \tag{1.75}$$

If we take the first terms in each series, we obtain the following expressions for the derivatives:

$$\left.\begin{aligned}
Dy_0 &= \frac{1}{h}(y_0 - y_{-1}) + 0(h) \\[2mm]
D^2 y_0 &= \frac{1}{h^2}(y_0 - 2y_{-1} - y_{-2}) + 0(h) \\[2mm]
D^3 y_0 &= \frac{1}{h^3}(y_0 - 3y_{-1} + 3y_{-2} - y_{-3}) + 0(h)
\end{aligned}\right\} \tag{1.76}$$

The symbol $0(h)$ stands for terms of order h and smaller. The expressions obtained are therefore backward difference approximations having an error of order h.

Similarly, formulas of order h^2 can be constructed by retaining the first two terms in the series:

$$\left.\begin{aligned}
D &= \frac{\nabla}{h} + \frac{hD^2}{2} + 0(h^2) \\[2mm]
D^2 &= \frac{\nabla^2}{h^2} + hD^3 + 0(h^2) \\[2mm]
D^3 &= \frac{\nabla^3}{h^3} + \frac{3hD^4}{2} + 0(h^2)
\end{aligned}\right\} \tag{1.77}$$

Eliminating D^2 from the first two equations gives

$$\begin{aligned}
D &= \frac{\nabla}{h} + \frac{h}{2}\left(\frac{\nabla^2}{h^2} + hD^3\right) + 0(h^2) \\[2mm]
&= \frac{\nabla}{h} + \frac{\nabla^2}{2h} + 0(h^2)
\end{aligned} \tag{1.78}$$

and, similarly,

$$D^2 = \frac{1}{h^2}(\nabla^2 + \nabla^3) + 0(h^2) \tag{1.79}$$

from which we get

$$Dy_0 = \frac{1}{2h}(3y_0 - 4y_{-1} + y_{-2}) + 0(h^2)$$

$$D^2y_0 = \frac{1}{h^2}(2y_0 - 5y_{-1} + 4y_{-2} - y_{-3}) + 0(h^2)$$

(1.80)

For formulas of order h^3, a similar procedure is followed by retaining terms up to the order h^3. We finally obtain

$$Dy_0 = \frac{1}{6h}(11y_0 - 18y_{-1} + 9y_{-2} - 2y_{-3})$$

$$D^2y_0 = \frac{1}{h^2}(2y_0 - 5y_{-1} + 4y_{-2} - y_{-3})$$

(1.81)

Houbolt employs these latter approximations of order h^3. Levy and Kroll[1] show that Houbolt's method will always converge; however, for good accuracy, it is still necessary to make Δt less than $0\cdot31/\omega$, where ω is the highest natural frequency of importance. For the equation governing a single mass system

$$m\ddot{y} + c\dot{y} + ky = F$$

(1.82)

this approximation gives, upon substitution of eqs. (1.81):

$$\left(\frac{2m}{(\Delta t)^2} + \frac{11c}{6\Delta t} + k\right)y_0 = \left(\frac{5m}{(\Delta t)^2} + \frac{3c}{\Delta t}\right)y_{-1} - \left(\frac{4m}{(\Delta t)^2} + \frac{3c}{2\Delta t}\right)y_{-2}$$

$$+ \left(\frac{m}{(\Delta t)^2} + \frac{c}{3\Delta t}\right)y_{-3} + F_0 \qquad (1.83)$$

From this equation, y_0 can be obtained once the past displacements y_{-1}, y_{-2}, y_{-3} are known.

Newmark's method

A somewhat different approach to obtaining the numerical solution to the governing differential equation (1.4) has been developed by Newmark.[5,7] Assume that the displacement, velocity, and acceleration of the system, denoted by y_0, $\dot{y}_0$, $\ddot{y}_0$, are known at time t_0. Let time $t_1 = t_0 + \Delta t$, where Δt is the small integration interval. Assume for the moment that $\ddot{y}_1$ is given a numerical value $\ddot{y}_1^*$. If we assume further that the acceleration is constant within the interval Δt and equal to the average of its initial and final values, the velocity $\dot{y}_1$ is

$$\dot{y}_1 = \dot{y}_0 + \tfrac{1}{2}(\ddot{y}_0 + \ddot{y}_1^*)(\Delta t)$$

(1.84)

Similarly, the displacement y_1 can be found by integration:

$$y_1 = y_0 + \dot{y}_0(\Delta t) + \tfrac{1}{4}(\ddot{y}_0 + \ddot{y}_1^*)(\Delta t)^2$$

(1.85)

The governing equation (1.4) at t_1 becomes

$$m\ddot{y}_1 + c\dot{y}_1 + ky_1 = F_1 \tag{1.86}$$

so that, solving for $\ddot{y}_1$, we have

$$\ddot{y}_1 = \frac{(F_1 - c\dot{y}_1 - ky_1)}{m} \tag{1.87}$$

where y_1 and $\dot{y}_1$ are obtained from eqs. (1.84) and (1.85). This process can now be repeated until convergence is achieved, starting again by using this new estimate for $\ddot{y}_1$ and repeating the same steps as before. The process is then continued sequentially for the next time step.

Alternatively, assume that the acceleration is allowed to vary linearly between t_0 and t_1. Then

$$\ddot{y} = \ddot{y}_0 + a(t - t_0)$$

so that, solving for a,

$$a = \frac{(\ddot{y} - \ddot{y}_0)}{\Delta t}$$

Integration now gives

$$\left.\begin{aligned} \dot{y}_1 &= \dot{y}_0 + \tfrac{1}{2}(\ddot{y}_1 + \ddot{y}_0)(\Delta t) \\ y_1 &= y_0 + \dot{y}_0 + \tfrac{1}{6}(2\ddot{y}_0 + \ddot{y}_1)(\Delta t)^2 \end{aligned}\right\} \tag{1.88}$$

These expressions can now also be substituted into the governing equation as before:

$$m\ddot{y}_1 + c\dot{y}_1 + ky_1 = F_1$$

and a solution obtained for $\ddot{y}_1$. The process can be repeated cyclically with this new estimate of $\ddot{y}_1$ until convergence is obtained.

Newmark has chosen something between the assumption in eq. (1.85) and that in eq. (1.88) by introducing the parameter β in the expressions

$$\left.\begin{aligned} \dot{y}_1 &= \dot{y}_0 + \tfrac{1}{2}(\ddot{y}_0 + \ddot{y}_1)(\Delta t) \\ y_1 &= y_0 + \dot{y}_1(\Delta t) + (\tfrac{1}{2} - \beta)\ddot{y}_0(\Delta t)^2 + \beta\ddot{y}_1(\Delta t)^2 \end{aligned}\right\} \tag{1.89}$$

These expressions can now be used to solve for $\ddot{y}_1$ in the governing equation

$$m\ddot{y}_1 + c\dot{y}_1 + ky_1 = F_1 \tag{1.90}$$

The process is repeated until convergence is achieved. The parameter β allows other variations in the acceleration or velocity within the time interval Δt. With $\beta = 1/4$, the average acceleration case is obtained. With $\beta = 1/6$, the acceleration varies linearly in the interval.

It can be shown[8] that Newmark's method is unconditionally stable if $\beta \geq 1/4$. It is conditionally stable when $\beta \leq 1/4$, the condition being that the time interval then has to be chosen small enough for the solution to remain stable. Houbolt's method is unconditionally stable.[9] The central difference method is conditionally stable,[1,10] in the sense that the time interval should be small enough compared with the highest natural frequency of the system, as shown in section 1.7. In all the methods a small time step is essential to accuracy up to the highest frequency present.

In application to complex structures, and, in particular, in applications in conjunction with finite element methods of structural analysis, it has been felt that the time interval necessary for stability at element frequencies is often too fine to allow such structures to be analyzed economically, and that inclusion of these high frequency components is not significant for accuracy. As a result, Wilson[6] has developed a method that extends certain features of Newmark's method and allows a scheme that is unconditionally stable. This situation has the advantage that the choice of time interval is no longer limiting with respect to stability. However, for the larger time intervals that are possible using Wilson's method, accuracy may still be a limiting feature, since, as in other methods, the accuracy depends on the fineness of the time interval.

Wilson's method

Wilson[6,11] introduces a variation of the Newmark formula with $\beta = 1/6$, which assumes that the acceleration varies linearly over an interval of time. He defines this interval to be of length $\tau = \theta \Delta t$ (with $\theta \geq 1$), where Δt is the desired integration interval. Then, assuming as before that, at t_0, y_0, $\dot{y}_0$, $\ddot{y}_0$ are known, and that, at $t_1 = t_0 + \tau$, $\ddot{y}_1$ is known, we have

$$
\left.
\begin{aligned}
\dot{y}_1 &= \dot{y}_0 + \frac{\tau}{2}(\ddot{y}_1 + \ddot{y}_0) \\[2mm]
y_1 &= y_0 + \tau\dot{y}_0 + \frac{\tau^2}{6}(\ddot{y}_1 + 2\ddot{y}_0)
\end{aligned}
\right\}
\qquad (1.91)
$$

Solving for $\dot{y}_1$ and $\ddot{y}_1$, we have

$$
\left.
\begin{aligned}
\ddot{y}_1 &= \frac{6}{\tau^2}(y_1 - y_0) - \frac{6}{\tau}\dot{y}_0 - 2\ddot{y}_0 \\[2mm]
\dot{y}_1 &= \frac{3}{\tau}(y_1 - y_0) - 2\dot{y}_0 - \frac{\tau}{2}\ddot{y}_0
\end{aligned}
\right\}
\qquad (1.92)
$$

Substitution into the governing equation:

$$
m\ddot{y}_1 + c\dot{y}_1 + ky_1 = F_1
\qquad (1.93)
$$

gives an expression for y_1:

$$
\left[\frac{6m}{\tau^2} + \frac{3c}{\tau} + k\right]y_1 = F_1 + m\left[\frac{6}{\tau^2}y_0 + \frac{6}{\tau}\dot{y}_0 + 2\ddot{y}_0\right] + c\left[\frac{3}{\tau}y_0 + 2\dot{y}_0 + \frac{\tau}{2}\ddot{y}_0\right]
\qquad (1.94)
$$

Thus, y can be found at the time $t = t + \theta \Delta t$. With y_1 known, $\ddot{y}_1$ and $\dot{y}_1$ are found from eq. (1.92). A linear interpolation gives the value of $\ddot{y}_1^*$ at $t = t + \Delta t$:

$$\ddot{y}_1^* = \ddot{y}_0\left(1 - \frac{1}{\theta}\right) + \frac{1}{\theta}\ddot{y}_1 \qquad (1.95)$$

The corresponding displacement y_1^* and velocity $\dot{y}_1^*$ are found from eq. (1.91):

$$\left.\begin{aligned}
\dot{y}_1^* &= \dot{y}_0 + \frac{(\Delta t)}{2}(\ddot{y}_1^* + \ddot{y}_0) \\[2mm]
y_1^* &= y_0 + (\Delta t)\dot{y}_0 + \frac{(\Delta t)^2}{6}(\ddot{y}_1^* + 2\ddot{y}_0)
\end{aligned}\right\} \qquad (1.96)$$

Wilson has shown that this method is unconditionally stable if $\theta \geq 1\cdot37$.

Further remarks

The computer programs in this book are written using the simple central difference approximations described in section 1.5. These approximations are stable if $\Delta t \leq 2/\omega$, where ω is the highest natural frequency of the system. In general, it will be found that, for the systems described in this book, it is desirable to find the dynamic response detail at time intervals that are considerably finer than the maximum allowable interval $2/\omega$. Thus, it is found convenient to use intervals of time considerably smaller than $2/\omega$ to retain the desired detail of solution. Where an initial model of a system includes a high frequency term of little importance, its mass can sometimes be lumped with nearby larger masses. In other cases, including only the low frequency modes is an effective approach.

A second factor enters into the choice of time interval. This is the factor of accuracy. Small time intervals are generally desirable and are conducive to accurate solutions. In time-stepping methods that are unconditionally stable, it will be possible at times to choose a very large time interval and have a solution that exists, but that is physically meaningless. In a conditionally stable situation, a time interval that is too large will produce a diverging solution, which can be detected as being incorrect. Detailed comparisons between the use of the central difference operator and the methods of Houbolt, Newmark, and Wilson have been made by McNamara[12] for a number of nonlinear problems. For a number of reasons, he concludes that the central difference operator can give better accuracy. Further comparisons and discussions are given by Belytschko et al.[13] and by Park[14].

We note, finally, that in the case of nonlinear problems the spring force depends nonlinearly on the displacement y. Thus, instead of the force ky appearing in the governing equations, we have the nonlinear force $R(y)$. With the central difference operator, the nonlinear function causes no difficulty since it is applied to the known displacement y_0 and not the unknown displacement y_1. In addition, as used in the component element method, it is never necessary to assemble the matrix for R since it is left in element form at all times. This observation will be important in systems with many degrees of freedom. In the case of Houbolt's method (eq. 1.83),

Newmark's method (eq. 1.87), and Wilson's method (eq. 1.94), the new displacement must be found from a resulting nonlinear equation. This calculation can be done by a local linearization of the force–deflection curve or by an iterative process. Regardless of which integration method is used, however, it has been found that it is desirable to model the nonlinear physical situations by using force–deflection relationships which are continuous (although their derivative may be discontinuous). This idea will play an important role when we come to develop the stop and friction elements in section 1.10.

1.9   TREATMENT OF INITIAL CONDITIONS

In general, initial conditions must be specified in terms of the initial displacement y_0 and the initial velocity $\dot{y}_0$. When this is done in the case of the central difference approximations (1.32) and (1.33), the velocity and accelerations become

$$\dot{y}_0 = \frac{(y_1 - y_{-1})}{2\Delta t} \tag{1.97}$$

$$\ddot{y}_0 = \frac{(y_1 - 2y_0 + y_{-1})}{(\Delta t)^2} \tag{1.98}$$

These expressions contain two unknown displacements y_1 and y_{-1}. The displacement y_1 may be solved for in eqs. (1.97) and (1.98) by eliminating y_{-1}, so that

$$y_1 = y_0 + \dot{y}_0\Delta t + (1/2)\ddot{y}_0(\Delta t)^2 \tag{1.99}$$

Since the governing equation at the initial time is

$$m\ddot{y}_0 + c\dot{y}_0 + ky_0 = F_0 \tag{1.100}$$

we have

$$\ddot{y}_0 = \frac{1}{m}(F_0 - c\dot{y}_0 - ky_0) \tag{1.101}$$

Substitution into eq. (1.99) gives a suitable expression for y_1 in terms of the initial conditions:

$$y_1 = y_0 + \dot{y}_0\Delta t + \frac{(\Delta t)^2}{2m}(F_0 - c\dot{y}_0 - ky_0) \tag{1.102}$$

Now knowing y_1 and y_0, the usual process can be used for the sequential calculation of $y_2, y_3, \ldots$. This process is used in the computer programs described in this book.

In the case of Houbolt's method, the situation is similar. For the initial time, we are given y_0 and $\dot{y}_0$. But the rearward displacements y_{-1}, y_{-2}, and y_{-3} are not known. Houbolt starts with a modified backward difference:

$$\dot{y}_0 = \frac{1}{6(\Delta t)}(2y_1 + 3y_0 - 6y_{-1} + y_{-2})$$

$$\ddot{y}_0 = \frac{1}{(\Delta t)^2}(y_1 - 2y_0 + y_{-1})$$

(1.103)

It is now desired to find y_{-1} and y_{-2} in terms of the initial conditions and in terms of y_1. From the governing equation written at the zero time, the initial acceleration is known to be

$$\ddot{y}_0 = \frac{1}{m}(F_0 - c\dot{y}_0 - ky_0)$$

(1.104)

Now, solving for y_{-1} and y_{-2} from eqs. (1.103), we get

$$y_{-1} - \ddot{y}_0(\Delta t)^2 + 2y_0 - y_1$$
$$y_{-2} = 6\ddot{y}_0(\Delta t)^2 + 6\dot{y}_0(\Delta t) + 9y_0 - 8y_1$$

(1.105)

Houbolt's approximation to the governing differential equation at time t_1 is, from eq. (1.83),

$$\left(\frac{2m}{(\Delta t)^2} + \frac{11c}{6\Delta t} + k\right)y_1 = \left(\frac{5m}{(\Delta t)^2} + \frac{3c}{\Delta t}\right)y_0 - \left(\frac{4m}{(\Delta t)^2} + \frac{3c}{2\Delta t}\right)y_{-1}$$

$$+ \left(\frac{m}{(\Delta t)^2} + \frac{c}{3\Delta t}\right)y_{-2} + F_1 \qquad (1.106)$$

Substitution of the displacements y_{-1} and y_{-2} from eq. (1.105) gives

$$\left(\frac{6m}{(\Delta t)^2} + \frac{3c}{\Delta t} + k\right)y_1 = \left(2m + \frac{c\Delta t}{2}\right)\ddot{y}_0 + 6\left(\frac{m}{\Delta t} + \frac{c}{3}\right)\dot{y}_0 + 3\left(\frac{2m}{(\Delta t)^2} + \frac{c}{\Delta t}\right)y_0 + F_1$$

(1.107)

Now y_1 can be found and the usual process repeated to find $y_2, y_3, \ldots$, by using eq. (1.83).

For the Newmark and Wilson methods, the starting values can be found by using the first backward differences described in eq. (1.103).

1.10 NONLINEAR ELEMENTS—SPRINGS, STOPS, FRICTION, AND OTHERS

In the foregoing sections we have discussed linear restoring forces associated with springs and dampers, where the restoring force is directly proportional to the displacement or velocity. A linear spring made of coil steel would give the force–displacement line of Fig. 1.8(a). In many instances, however, the restoring force is not at all linear in this fashion. Springs may harden or soften with displacement; the mass may strike a sudden stop after it moves to a certain point; and friction will introduce a force opposed to the direction of motion. Some of these effects

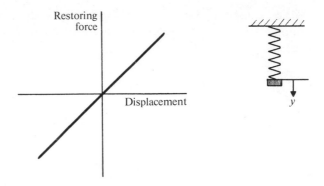

(a) Restoring force for coil steel spring

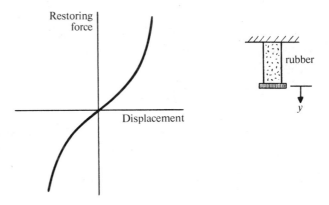

(b) Restoring force for rubber

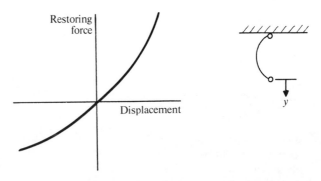

(c) Restoring force for C-spring

FIGURE 1.8
Types of springs

can be modeled by continuous algebraic expressions; others are described by discontinuous conditions that depend on the current displacement or velocity of the element.

Springs

Most springs become harder (or softer) as they stretch, particularly if the stretch is at all large. An example of a spring which becomes harder as it is extended is a rubber structure whose force–displacement curve is illustrated in Fig. 1.8(b).

A C-spring is an example of a spring which becomes harder as it is extended, and softer as it is compressed. Its force–displacement curve is illustrated in Fig. 1.8(c).

Algebraically, the linear relationship for the restoring force R is given by

$$R = ky \tag{1.108}$$

In the case of the nonlinear systems, the restoring force can be described by a series in y of the sort

$$R = k_1 y + k_2 y^2 + k_3 y^3 + \cdots \tag{1.109}$$

For example, the hardening rubber spring can be described by

$$R = k_1 y + k_3 y^3 \tag{1.110}$$

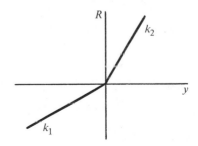

(a) Break at $y = 0$

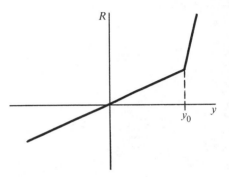

FIGURE 1.9
Bilinear spring representation   (b) Break at $y = y_0$

and the hard–soft C-spring by

$$R = k_1 y + k_2 y^2 \qquad (1.111)$$

Other spring systems can also be represented. For example, bilinear springs with varying positional discontinuities shown in Fig. 1.9 can equally well be represented. In Fig. 1.9(a), discontinuous expressions of the type

$$\left.\begin{array}{ll} R = k_1 y & (y < 0) \\ R = k_2 y & (y > 0) \end{array}\right\} \qquad (1.112)$$

represent the force. In Fig. 1.9(b), the force is represented by discontinuous expressions of the type

$$\left.\begin{array}{ll} R = k_1 y & (y < y_0) \\ R = k_1 y_0 + k_2 (y - y_0) & (y > y_0) \end{array}\right\}$$

Stops

Many pieces of equipment and packages have stops to limit the distance traveled by the spring. Figure 1.10 shows a typical stop. Near its original position, the mass

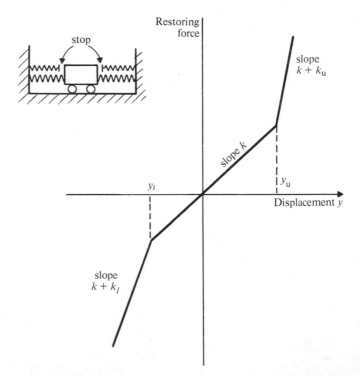

FIGURE 1.10
Stiffness increases by k_u at upper stop and by k_l at lower stop

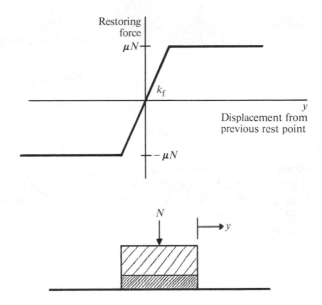

FIGURE 1.11
Dry friction causes a restraining force opposed to the direction of motion

is acted on by a linear spring of constant k. When the displacement y_l or y_u is reached, another spring comes into force, and the effective spring constant is increased. If the stop springs at the upper and lower excursion points are different, different restoring forces result, as shown in Fig. 1.10. Algebraically, these restoring forces can be expressed by the three conditions:

$$R = ky \qquad\qquad (y_l < y < y_u)$$
$$R = ky + k_u(y - y_u) \qquad (y > y_u) \qquad\qquad (1.113)$$
$$R = ky + k_l(y - y_l) \qquad (y < y_l)$$

where the subscripts u and l denote the upper and lower excursion points. A solid stop is obtained by making k_l or k_u very large; however, one should keep in mind that in physical systems no stop is infinitely rigid.

Friction

Dry friction, otherwise known as Coulomb friction, has the characteristic that the restraining force is always opposed to the direction of motion. For very small relative displacements, no actual sliding takes place. The relative displacement is accommodated by local flexibility at the point of contact. When the relative displacement becomes sufficiently large, however, the local force exceeds the frictional force, and sliding occurs. The relationship between the restoring, or frictional, force is illustrated in Fig. 1.11. During sliding, the magnitude of the friction force is governed by the coefficient of friction, μ, and by the total normal force, N, applied to the mass. Algebraically, if y denotes the displacement of the mass from its previous rest point, then

$$\left. \begin{array}{ll} R = \mu N & (k_f y > \mu N) \\ R = k_f y & (-\mu N < k_f y < \mu N) \\ R = -\mu N & (k_f y < -\mu N) \end{array} \right\} \qquad (1.114)$$

Elastic–plastic elements

If a spring or a beam support is allowed to yield plastically, then the force it exerts is likely to be of the form shown in Fig. 1.12(a). There is a period of linear elastic loading, whereupon, after further deformation, plastic yielding occurs. If the spring is unloaded, the behavior is again elastic until further compressive plastic yielding

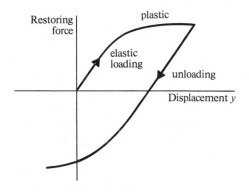

(a) General plastic behavior

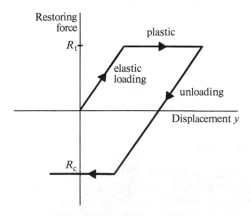

(b) Elastic–perfectly plastic behavior

FIGURE 1.12
Various elasto–plastic spring models

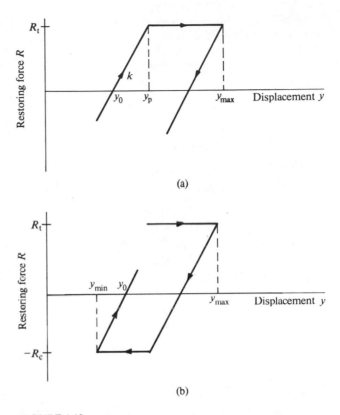

FIGURE 1.13
Elastic–perfectly plastic model

takes place. Cyclic loading and unloading can occur in this manner. Energy is dissipated during each cycle in an amount proportional to the area enclosed within the curve. The behavior can often be simplified by assuming a definite yield point. The behavior upon additional loading is such that there is no further load borne by the spring. Such behavior is known as elastic–perfectly plastic behavior and the corresponding force–deflection curve is shown in Fig. 1.12(b).

For the elastic–perfectly plastic case, expressions for the force are easily written down. They depend on whether the loading is such that the displacement is increasing ($\dot{y} > 0$) or decreasing ($\dot{y} < 0$). Referring to Fig. 1.13, we assume that the intercept y_0 is known (it is zero for a spring initially deflected from the origin). Then we compute the deflection y_p at which plastic deformation will take place:

$$y_p = y_0 + R_t/k \qquad (1.115)$$

Here R_t is the yield force in tension, while k is the elastic spring constant. We define R_c as the yield force in compression. R_c is positive when the force is compressive. The force R (positive when tensile, and negative when compressive) is as follows:

(a) If $\dot{y} > 0$, elastic loading or tensile plastic deformation occurs. Then,

$$R = R_t - (y_p - y)k \qquad (y < y_p) \atop = R_t \qquad\qquad\quad (y > y_p) \Bigg\} \qquad (1.116)$$

(b) If $\dot{y} < 0$, elastic unloading occurs. Define as y_{max} the point at which $\dot{y} = 0$. Then,

$$R = R_t - (y_{max} - y)k \qquad (y_{max} - (R_c + R_t)/k < y < y_{max}) \atop R = -R_c \qquad\qquad\qquad (y < y_{max} - (R_c + R_t)/k) \Bigg\} \qquad (1.117)$$

(c) If $\dot{y} > 0$, elastic loading again occurs. Define as y_{min} the point at which $\dot{y} = 0$. Then,

$$R = -R_c + (y - y_{min})k \qquad (y < y_{min} + R_c/k) \qquad (1.118)$$

When $y \geq y_{min} + R_c/k$, the cycle begins again with a new value of $y_0 = y_{min} + R_c/k$.

It now becomes clear that any general curve can be followed in the same manner. For example, if the curves in Fig. 1.12(a) were defined experimentally, they could be used in a tabular form by adapting the same logic as used above in the case of perfect plasticity. Linear work hardening could easily be accounted for in the above scheme by introducing a work hardening spring constant k_w that would come into play during the times of plastic flow.

1.11 SUMMARY

Some basic building blocks have been established which can now be used to study an extensive range of problems which involve the behavior of a single mass or of multiple masses which are controlled by springs and other elements of a linear or nonlinear nature. In the following chapter, we shall study the dynamics of single mass systems and see how a number of interesting phenomena arise when combinations of nonlinear elements and external forces come into play.

1.12 REFERENCES

1. S. Levy and W. D. Kroll, 'Errors introduced by finite space and time increments in dynamic response computation,' *Proc., First US Natl. Congress of Applied Mechanics*, pp. 1–8, 1951.
2. M. V. Salvadori and M. L. Baron, *Numerical Methods in Engineering*, Ch. 2, Prentice-Hall, Inc., Englewood Cliffs, New Jersey, 1961.
3. J. B. Scarborough, *Numerical Analysis*, 4th edn., The Johns Hopkins Press, Baltimore, 1958.
4. J. C. Houbolt, 'A recurrence matrix solution for the dynamic response of elastic aircraft,' *J. Aeronautical Sciences*, **17**, 540–550, 594, 1950.
5. N. M. Newmark, 'A method of computation for structural dynamics,' *Proc. ASCE, Journal of the Engineering Mechanics Div.*, **85**, 67–94, 1959; also *Trans. ASME*, **127**, Part 1, 1406–1435, 1962.
6. E. L. Wilson, I. Farhoomand, and K. J. Bathe, 'Nonlinear dynamic analysis of complex structures,' *Earthquake Engineering and Structural Dynamics*, **1**, 241–252, 1973.
7. N. M. Newmark and E. Rosenblueth, *Fundamentals of Earthquake Engineering*, Prentice-Hall, Inc., Englewood Cliffs, New Jersey, p. 15, 1971.
8. R. E. Nickell, 'On the stability of approximation operators in problems of structural dynamics,' *Int. J. Solids and Structures*, **7**, 301–319, 1971.
9. D. E. Johnson, 'A proof of the stability of the Houbolt method,' *AIAA J.*, **4**, 1450–1451, 1966.

10. J. W. Leech, P.-T. Hsu, and E. W. Mack, 'Stability of a finite-difference method for solving matrix equations,' *AIAA J., 3,* 2172–2173, 1966.
11. R. W. Clough and K. J. Bathe, 'Finite element analysis of dynamic response,' in J. T. Oden, R. W. Clough, and Y. Yamamoto (Eds.), *Advances in Computational Methods in Structural Mechanics and Design,* pp. 153–179, The University of Alabama in Huntsville Press, Huntsville, Alabama, 1972.
12. J. F. McNamara, 'Solution schemes for problems of nonlinear structural dynamics,' *J. Pressure Vessel Technology, Trans. ASME,* **96,** Ser J, 96–102, 1974.
13. T. Belytschko, N. Holmes, and R. Mullen, 'Explicit integration—stability, solution properties, cost,' *Finite Element Analysis of Transient Nonlinear Structural Behavior,* AMD-14, 1–21, ASME, New York, 1975.
14. K. C. Park, 'Evaluating time integration methods for nonlinear dynamic analysis,' *Finite Element Analysis of Transient Nonlinear Structural Behavior,* AMD-14, 35–58, ASME, New York, 1975.

2

DYNAMICS OF SINGLE MASS SYSTEMS

2.1 INTRODUCTION

A surprising variety of structures can be modeled as single mass systems. Water towers, equipment in boxes, and components or machines on foundations or in buildings are examples of things that are often represented by a single mass. Because of the relative simplicity of the single mass system as compared with systems with many masses, it is also useful as an aid to the understanding of the general modes of behavior of more complex structures. In this chapter, we study the behavior of a single mass that is restrained by various springs, both linear and nonlinear. The effects of stops and friction are also illustrated. The basic means of computation is the component element method. A computer program written in FORTRAN language is an integral part of this chapter. The numerical results discussed in the chapter are obtainable from this program. The program, together with two others concerning multi-mass systems that will be introduced later, is an important part of the process of the numerical calculations. It enables us to perform what would otherwise be rather tedious and highly repetitive calculations. Its application later in the chapter allows the inspection of such phenomena as subharmonic resonance, the appearance of instabilities in systems with nonlinear springs, and a number of effects concerning various excitation forces.

2.2   CHARACTERIZATION OF A NONLINEAR SINGLE MASS SYSTEM

Consider a mass as shown in Fig. 2.1. It is attached to a support by a number of force elements which constrain and modify its motion when it is acted on by an external force or by the motion of its support. The force elements are of four kinds:

(a) A nonlinear spring whose stiffness is

$$k_1(y - x) + k_2(y - x)^2 + k_3(y - x)^3$$

Here k_1, k_2, and k_3 are the spring stiffness coefficients, while $(y - x)$ is the stretch of the spring made up of the difference between the displacement y of the mass and the displacement x of the support.

(b) A viscous damper c, whose restraining force is $c(\dot{y} - \dot{x})$.

(c) A friction element with friction force F and of local stiffness k_f.

(d) A stop, with upper and lower clearances y_u and y_l, respectively.

This selection of elements is suitable for the purpose of demonstrating the dynamics of typical nonlinear single mass systems.

In order to speak of the dynamics further, a stimulus must be applied to the system—either as a force input on the mass, or as an acceleration of the support. Generally, time histories of the force or acceleration are given. However, in certain instances, such as in electrodynamic shakers, the input stimulus may be given in terms of a timewise sweep in frequency and amplitude. In the discursive examples which follow, we shall discuss system responses to both types of stimulus. The inputs will, in general, be given as functions of time, as illustrated in Fig. 2.2. Finally, both the mass and support are given initial conditions. It is necessary to know the initial displacement y_0 and the initial velocity $\dot{y}_0$ of the mass, and likewise the initial displacement x_0 and initial velocity $\dot{x}_0$ of the support.

FIGURE 2.1
Nonlinear single degree-of-freedom system

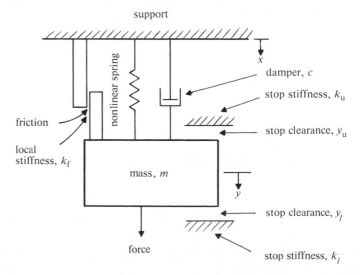

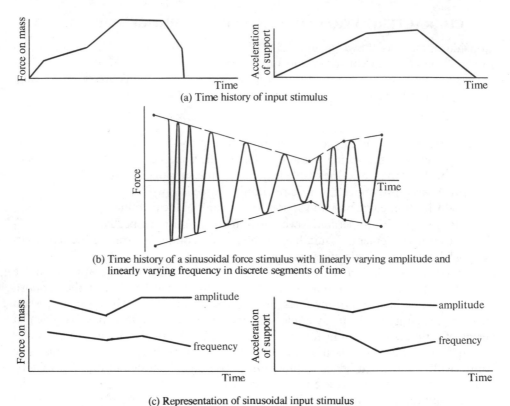

(a) Time history of input stimulus

(b) Time history of a sinusoidal force stimulus with linearly varying amplitude and linearly varying frequency in discrete segments of time

(c) Representation of sinusoidal input stimulus

FIGURE 2.2
Input stimuli for mass and support

The foregoing information is generally all that is required in order to calculate the motion of the mass as a function of time. The actual computation is done by the component element method with the step-by-step solution of the equations of motion. The detailed means by which this calculation can be made is the subject of the following section.

2.3   BASIC COMPUTING METHOD FOR NONLINEAR SINGLE MASS SYSTEMS

The basic differential equation governing the single degree-of-freedom system shown in Fig. 2.1 is

$$m\ddot{y} + c(\dot{y} - \dot{x}) + R[y - x] = f \tag{2.1}$$

Here, y is the displacement of the mass, while x is the support displacement. R is the restoring force imposed by the nonlinear spring, or the friction, or the stops. Generally, it is a nonlinear function of the relative displacement $(y - x)$ between the support and the mass. Its form has already been discussed for each of the element types in section 1.10.

To solve this equation numerically, we introduce the finite difference approximations of section 1.5. We wish to solve the equation under the excitation of an external force $f(t)$ and of the acceleration of the support $a(t)$. First, we note that the acceleration of the support is related to the support displacement x at various discrete instants of time. Thus, the support acceleration a_0 at time t_0 is given by

$$a_0 = \frac{x_1 - 2x_0 + x_{-1}}{(\Delta t)^2 g} \tag{2.2}$$

As a result, if the support displacement is known at the present time t_0, and at a previous instant t_{-1}, then the displacement x_1 at a future instant t_1 can be predicted:

$$x_1 = a_0 g (\Delta t)^2 + 2x_0 - x_{-1} \tag{2.3}$$

Now let us express the governing equation in terms of finite differences. We have

$$m\left(\frac{y_1 - 2y_0 + y_{-1}}{(\Delta t)^2}\right) + \frac{c}{2\Delta t}[(y_1 - y_{-1}) - (x_1 - x_{-1})] + R[y_0 - x_0] = f_0 \tag{2.4}$$

Rearranging this equation and solving for the displacement of the mass at the time t_1, we get

$$y_1 = \left(\frac{1}{1 + \dfrac{c\Delta t}{2m}}\right)\left\{2y_0 - y_{-1} + \frac{(\Delta t)^2}{m}(f_0 - R[y_0 - x_0])\right.$$

$$\left. + \frac{c\Delta t}{2m}(y_{-1} + x_1 - x_{-1})\right\} \tag{2.5}$$

As initial conditions, it is possible to specify both the initial displacements and velocities of the support and mass. For initial displacements, y_0 and x_0 are used directly. If the initial velocities are $\dot{y}_0$ and $\dot{x}_0$, then, according to the difference formulas,

$$\dot{y}_0 = \frac{(y_1 - y_{-1})}{2\Delta t} \tag{2.6}$$

so that we express y_{-1} and x_{-1} for the initial time step as

$$\left.\begin{array}{l} y_{-1} = y_1 - \dot{y}_0(2\Delta t) \\ x_{-1} = x_1 - \dot{x}_0(2\Delta t) \end{array}\right\} \tag{2.7}$$

Substitution into the governing equation (2.5) now allows the direct computation of y_1. Repetition of this process gives successive values of y.

The following section describes a simple computer program based on the equation given above. With this program, a number of examples are solved and discussed in subsequent sections. Those examples are intended to illustrate each feature of the computer program concerning the elements (springs, friction, and stops) and the excitation (force and acceleration stimuli). In addition, several examples will be given where detailed comparisons are made between the numerical results of the computer program and known analytical results published in the literature.

2.4   A COMPUTER PROGRAM FOR THE DYNAMIC RESPONSE OF SINGLE MASS SYSTEMS

The computer program described in this section calculates the response of a mass attached to a support by means of a nonlinear spring with stops, a viscous damper, and an element which provides a frictional resistance. The mass is excited by a time-dependent external force, which may be either a force defined by points on a force–time curve or a sinusoidal force. The support may be excited by an acceleration defined by points on an acceleration–time curve or by a sinusoidal acceleration. The response consists of the acceleration of the mass and the force experienced by the mass. It is printed out at controlled intervals during the calculation.

   The program is written in the FORTRAN computer language. This language is well known and its essentials are available to the reader.[1,2] The program is suitable

Table 2.1   DESCRIPTION OF INPUT VARIABLES

Variable	Symbol in text	Description
ITYPE		A control indicating type of excitation;
		0 = sinusoidal, nonzero indicates time history
TIMONE		Time at start of an interval of excitation
TIMTWO		Time at end of an interval of excitation
ACCEL1		Acceleration at TIMONE
ACCEL2		Acceleration at TIMTWO
FREQ1		Frequency at TIMONE
FREQ2		Frequency at TIMTWO
K	k_1	Linear spring constant
M	m	Mass
C	c	Damper
FORCE1		Force at TIMONE
FORCE2		Force at TIMTWO
FRIC	μN	Friction force
X	x_0	Initial support displacement
VELX	$\dot{x}_0$	Initial support velocity
KU	k_u	Stiffness of upper stop
YU	y_u	Clearance to upper stop
Y	y_0	Initial mass displacement
VELY	$\dot{y}_0$	Initial mass velocity
KL	k_l	Stiffness of lower stop
YL	y_l	Clearance to lower stop
DEL	Δt	Time step interval
TSTP		Time step for printing output
TTL		Total time
KSQ	k_2	Quadratic spring stiffness
KCUBE	k_3	Cubic spring stiffness
FRK	k_f	Local frictional stiffness
IOPT		1 for a new set of data; 0 for end of data

Table 2.2   INPUT DATA SYMBOLS AND FORMAT

Format	Variables					
(I1)	ITYPE					
(6F10·4)	TIMONE	TIMTWO	ACCEL1	ACCEL2	FREQ1	FREQ2
(6F10·4)	K	M	C	FORCE1	FORCE2	FRIC
(4F10·4)	X	VELX	KU	YU		
(4F10·4)	Y	VELY	KL	YL		
(6F10·4)	DEL	TSTP	TTL	KSQ	KCUBE	FRK
(4F10·4)	TIMTWO	ACCEL2	FORCE2	FREQ2	(as many similar cards as are needed)	
(I1)	IOPT					

for use on most computing machines. At this point, the reader is encouraged to load the program into a machine and to actually use it on the examples discussed later in this chapter.

A short list of the principal input variables and symbols used in the program is given in Table 2.1. The corresponding algebraic symbols that were used in the equations are also given. Input data cards are given in Table 2.2.

A flow diagram of the program is shown in Fig. 2.3. The program is an extremely simple one, containing one basic loop within which the response is computed at each discrete time interval. The computer program itself is given on the following pages. Comments are interspersed throughout the program to make it easy to understand what is being done at each step. The input required for the program is given in Table 2.2 as a list of FORTRAN symbols whose physical meanings are described in Table 2.1. Each line of input in Table 2.2 represents an input card with the given format. A detailed example of the use of the program is given in the next section.

It should be noted that the system of units used in this program is fixed. It is in pound, second, inch units. Other systems of units, e.g., the International System, can be accommodated easily by changing the numerical value of g (386·0 in./sec^2) to the value appropriate to the system. Thus, for the newton, second, metre system, the value of g should be 9·807 m/s^2.

Problems

2.1   If the support velocity $v(t)$ is given instead of the support acceleration a, how should the equations of motion be modified? What changes should be made to the computer program to account for support velocity excitation?

2.2   Similarly, what changes are required when the support is excited by a given displacement $d(t)$, as, for example, in the case of a cam-driven mass?

2.5   EXAMPLE—SYSTEM WITH SPRING, DAMPER, AND STOP

Consider a motor mounted on a vibrational isolator that consists of a linear spring and a damper as shown in Fig. 2.4. To restrain the motion, a stop is also added whose clearance is varied from 0·1 to 0·5 in. Our objective is to study the effect

Computer program listing

```
C         NONLINEAR SINGLE DEGREE OF FREEDOM SYSTEM
C
      COMMON  TIMONE,TIMTWO,ACCEL1,ACCEL2,FREQ1,FREQ2,FORCE1,FORCE2,
     1 K,KSQ,KCUBE,M,C,X,Y,XPREV,YPREV,DEL,TSTP,TTL,IOPT,ITYPE
      REAL  K,KSQ,KCUBE,KU,KL,M
C
 1000 FORMAT (6F10.4)
 1001 FORMAT (I1)
 1002 FORMAT(1H1)
  701 CONTINUE
      WRITE(6,1002)
      READ(5,1001) ITYPE
C
C   WHEN ITYPE IS 0, EXCITATION IS SINUSOIDAL, OTHERWISE
C   EXCITATION IS TIME-HISTORY.
C
      READ(5,1000) TIMONE,TIMTWO,ACCEL1,ACCEL2,FREQ1,FREQ2
C
C   WHEN EXCITATION IS SINUSOIDAL, (ITYPE=0) THE SUPPORT
C   ACCELERATION AMPLITUDE GOES FROM ACCEL1 TO ACCEL2
C   AND THE APPLIED FORCE AMPLITUDE GOES FROM FORCE1 TO FORCE2
C   WHILE TIME GOES FROM TIMONE TO TIMTWO AND FREQUENCY GOES FROM
C   FREQ1 TO FREQ2.
C   WHEN EXCITATION IS TIME-HISTORY (ITYPE NOT 0), THE ACCELERATION
C   GOES FROM ACCEL1 TO ACCEL2 AND THE APPLIED FORCE GOES FROM
C   FORCE 1 TO FORCE 2 WHILE TIME GOES FROM TIMONE TO TIMTWO
C   (VALUES OF FREQ1 AND FREQ2 ARE NOT NEEDED THEN.)
C
      READ(5,1000) K,M,C,FORCE1,FORCE2,FRIC
C
C   THE SPRING RATE IS K, THE MASS M AND THE DAMPING TO
C   GROUND C.
C   THE FRICTIONAL FORCE FRIC ACTS ON THE MASS IN A DIRECTION OPPOSITE
C   TO THE VELOCITY.
C
      ZFA=(ACCEL2-ACCEL1)/(TIMTWO-TIMONE)
      ZFF=(FREQ2-FREQ1)/(TIMTWO-TIMONE)
      ZFRC=(FORCE2-FORCE1)/(TIMTWO-TIMONE)
      IF (ITYPE) 1202,1203,1202
 1202 WRITE(6,1200)
C
C   INTERPOLATE TO FIND CURRENT FORCE AND ACCELERATION EXCITATION
C
      A=ACCEL1-TIMONE*ZFA
      FORCE=FORCE1-TIMONE*ZFRC
 1200 FORMAT(1H0,26HEXCITATION IS TIME-HISTORY)
      GO TO 1204
 1203 WRITE(6,1201)
      A=0.0
      FORCE=0.0
      FREQ=FREQ1-TIMONE*ZFF
 1201 FORMAT(1H0,24HEXCITATION IS SINUSOIDAL)
 1204 CONTINUE
      READ(5,1000) X,VELX,KU,YU
      READ(5,1000) Y,VELY,KL,YL
C
C   THE INITIAL DISPLACEMENT OF THE MASS IS Y AND THE INITIAL
C   VELOCITY OF THE MASS IS VELY. THE INITIAL DISPLACEMENT OF THE
C   SUPPORT IS X AND ITS INITIAL VELOCITY IS VELX.
C   YU IS THE CLEARANCE TO THE UPPER STOP AND KU IS ITS
C   STIFFNESS. YL IS THE CLEARANCE TO THE LOWER STOP
C   AND KL IS ITS STIFFNESS.
C
      READ(5,1000) DEL,TSTP,TTL,KSQ,KCUBE,FRK
C
```

```
C   THE SPRING FORCE IS K(Y-X)+KSQ(Y-X)SQUARE+KCURE(Y-X)CUBE.
C   THE DESIRED TOTAL TIME OF ANALYSIS IS TTL. THE TIME
C   STEP BETWEEN PRINT-OUTS IS TSTP. THE TIME STEP FOR
C   INTEGRATION IS DEL. (THE PROGRAM MAY DECREASE DEL
C   FOR COMPUTATIONAL ACCURACY.)
C   FRK IS THE LOCAL SPRING RATE AT THE FRICTIONAL CONTACT.
C
 1100 FORMAT(1H0,4HTIME,F10.4,3X,10HACCEL AMPL,F10.4,3X,10HFORCE AMPL,
     1F10.4,3X,4HFREQ,F10.4)
      WRITE(6,1100) TIMONE,ACCEL1,FORCE1,FREQ1
      WRITE (6,1100)TIMTWO,ACCEL2,FORCE2,FREQ2
      WRITE(6,1101) K,M,C,KSQ,KCUBE
 1101 FORMAT(1H0,6HSPRING,F10.4,3X,4HMASS,F10.4,3X,7HDAMPING,
     1F10.4,3X,3HKSQ,F10.4,3X,5HKCURE,F10.4)
      WRITE(6,1102)Y,VELY
      WRITE(6,1103)X,VELX
 1102  FORMAT(1H0,4HMASS,4X,20HINITIAL DISPLACEMENT,F10.4,3X
     1  16HINITIAL VELOCITY,F10.4)
 1103 FORMAT(1H0,7HSUPPORT,21H INITIAL DISPLACEMENT,F10.4,3X,
     1  16HINITIAL VELOCITY,F10.4)
      WRITE(6,1104) KU,YU
 1104 FORMAT(1H0,10HUPPER STOP,3X,6HSPRING,F12.4,3X,9HCLEARANCE,F10.4)
      WRITE(6,1105) KL,YL
 1105 FORMAT(1H0,10HLOWER STOP,3X,6HSPRING,F12.4,3X,9HCLEARANCE,F10.4)
      WRITE(6,1107) FRIC,FRK
 1107 FORMAT(1H0,8HFRICTION,F10.4,3X,15HFRICTION SPRING,F10.4)
C
C   ASSIGN DEL A SMALLER VALUE IF INPUT VALUE IS TOO LARGE
C
      IF (K)  20,22,20
   20 DELL=0.31*SQRT(M/K)
      IF(DEL-DELL)  22,22,21
   21 DEL=DELL
   22 CONTINUE
      IF (C)  23,25,23
   23 DELL=0.25*M/C
      IF (DEL-DELL)  25,25,24
   24 DEL=DELL
   25 CONTINUE
      IF (FRK)  26,28,26
   26 DELL=0.31*SQRT(M/FRK)
      IF (DEL-DELL)  28,28,27
   27 DEL=DELL
   28 CONTINUE
      WRITE(6,1106) DEL,TSTP,TTL
 1106 FORMAT(1H0,3HDEL,F10.6,3X,4HTSTP,F10.6,3X,3HTTL,F10.4)
C
C   INITIALIZE
C
      XPREV=X-VELX*DEL
      YPREV=Y-VELY*DEL
      SET=YPREV-XPREV
      WRITE(6,140)
  140  FORMAT(1H0,2X,4HTIME,3X,    22H      ACCELERATION,MASS,11X,20HACCELE
     2RATION,SUPPORT,5X,5HFORCE,5X,9HFREQUENCY/13X,7HABS MAX,8X,7HPRESEN
     3T,7X,7HABS MAX,8X,7HPRESENT/13X,7HIN STEP,9X,5HVALUE,8X,7HIN STEP,
     49X,5HVALUE)
      WW=0.0
      DELC=DEL*C*0.5/M
      GRAV=386.0
C
C   GRAV IS THE ACCELERATION OF GRAVITY IN POUND-INCH-SECOND UNITS
C   IN THE INTERNATIONAL SYSTEM USE 9.807 FOR NEWTON-METRE-SECOND UNITS
C
      DELSQI=1.0/(DEL*DEL*GRAV)
      DELM=DEL*DEL/M
      TPRINT=-0.000001
```

```
         TM=0.0
         GLEVS=0.0
         GLEVM=0.0
C
C   START TIME-STEP INTEGRATION LOOP
C
   10    CONTINUE
         FORCED=FORCE
C
C   ADD SPRING FORCE TO EXTERNAL FORCE
C
         FORCE=FORCE-(Y-X)*(K+(Y-X)*(KSQ+KCUBE*(Y-X)))
C
C   ADD FORCE IN STOPS
C
         IF(Y-X-YU) 71,71,70
   70    FORCE=FORCE-KU*(Y-X-YU)
   71    IF(-YL-Y+X)   73,73,72
   72    FORCE=FORCE-KL*(Y-X+YL)
   73    CONTINUE
C
C   ADD FORCE IN FRICTION ELEMENTS
C
         FYK=FRK*(Y-X-SET)
         FYKK=ABS(FYK)
         IF(FYKK-FRIC)   76,76,77
   76    FORCE=FORCE-FYK
         GO TO 75
   77    FYK=FRIC*FYK/FYKK
         SET=Y-X-FYK/FRK
         GO TO 76
   75    CONTINUE
C
C   COMPUTE NEW GROUND COORDINATE
C
         XNEW=A/DELSQI+X+X-XPREV
C
C   COMPUTE NEW MASS COORDINATE
C
         YNEW=(Y+Y-YPREV+DELM*FORCE+DELC*(YPREV-XPREV+XNEW))/(1.0+DELC)
C
C   CORRECT FOR INITIAL CONDITIONS
C
         IF(TM.GT.0.0)GO TO 175
         CORR=0.5*(XPREV+XNEW)-X
         XNEW=XNEW-CORR
         XPREV=XPREV+CORR
         CORR=0.5*(YPREV+YNEW)-Y

         YNEW=YNEW-CORR
         YPREV=YPREV+CORR
  175    CONTINUE
C
C   COMPUTE PRESENT MASS ACCELERATION IN G'S
C
         F=(YNEW-Y-Y+YPREV)*DELSQI
C
C   STORE MAX. GROUND AND MASS ACCNS THAT OCCUR WITHIN PRINT INTERVAL
C
         ACM=ABS(F)
         ACS=ABS(A)
         IF(ACM-GLEVM)60,60,61
   61    GLEVM=ACM
   60    CONTINUE
         IF(ACS-GLEVS) 62,62,63
   63    GLEVS=ACS
   62    CONTINUE
C   PRINT OUTPUT
```

```
      IF(TM-TPRINT) 110,110,64
   64 CONTINUE
      TPRINT=TPRINT+TSTP
   14 FORMAT(F10.6,3X,F10.6,3X,F10.6,4X,F10.6,3X,F10.6,1X,F10.2,5X,
     1F10.3)
      WRITE(6,14) TM,GLEVM,F,GLEVS,A,FORCED,FREQ
      GLEVM=0.0
      GLEVS=0.0
   40 CONTINUE
C
C  INCREMENT TIME AND SUPPORT DISPLACEMENTS FOR NEXT TIME STEP
C
  110 TM=TM+DEL
      XPREV=X
      X=XNEW
C
C  OBTAIN NEW VALUES OF EXCITATION
C
      IF(TM-TIL-.00002) 121,121,30
  121 IF (TM-TIMTWO-.00003) 11,11,120
  120 CONTINUE
   12 TIMONE=TIMTWO
      ACCEL1=ACCEL2
      FREQ1=FREQ2
      FORCE1=FORCE2
      READ (5,1000) TIMTWO,ACCEL2,FORCE2,FREQ2
      WRITE(6,1100) TIMTWO,ACCEL2,FORCE2,FREQ2
C
C  LINEAR INTERPOLATION OF FORCE AND ACCN. EXCITATION
C
      ZFA=(ACCEL2-ACCEL1)/(TIMTWO-TIMONE)
      ZFF=(FREQ2-FREQ1)/(TIMTWO-TIMONE)
      ZFRC=(FORCE2-FORCE1)/(TIMTWO-TIMONE)
   11 CONTINUE
      A=ACCEL1+(TM-TIMONE)*ZFA
      FORCE=FORCE1+(TM-TIMONE)*ZFRC
      IF (ITYPE)  50,51,50
   51 CONTINUE
C
C  FREQUENCY INTERPOLATION
C
C
      FREQQ=FREQ
      FREQ=FREQ1+(TM-TIMONE)*ZFF
      W=3.1416*(FREQ+FREQQ)
      WW=WW+W*DEL
      IF(WW-6.2831853) 52,52,53
   53 WW=WW-6.2831853
   52 CONTINUE
C
C  COMPUTE CURRENT VALUE OF SINUSOIDAL FORCE OR ACCN. INPUTS
C
      A=A*SIN(WW)
      FORCE=FORCE*SIN(WW)
   50 CONTINUE
C
C  INCREMENT MASS DISPLACEMENTS
C
      YPREV=Y
      Y=YNEW
C
C  LOOP BACK TO STATEMENT 10 FOR NEW TIME STEP
C
      GO TO 10
   30 READ(5,1001)IOPT
      IF(IOPT-1) 15,701,15
   15 STOP
      END
```

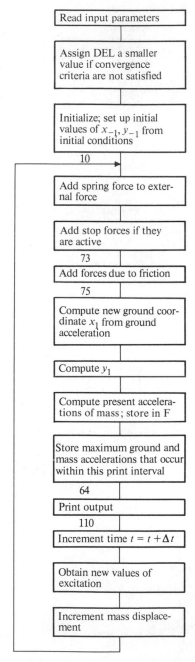

FIGURE 2.3
Flow chart for single degree-of-freedom
program

of the clearance on the acceleration of the motor. The motor is assumed to be initially at rest. It is then acted on by a sinusoidal imbalance force of 10 lb at a frequency of 25 rad/sec. For this case, the basic data required are given in Table 2.3.

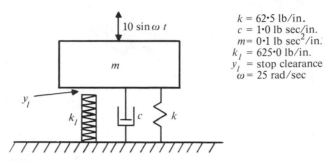

FIGURE 2.4
Motor mounted on vibration isolator and excited by unbalanced force of 10 lb with stop clearance y_l and stop stiffness k_l ($\omega - 2\pi f = 25$ rad/sec)

The time interval was chosen as 0·01 sec, which is somewhat less than the maximum interval recommended for accuracy,

$$\Delta t = 0\cdot31\sqrt{m/k} = 0\cdot0125 \text{ sec}$$

The appearance of the actual data required by the computer program is shown in Table 2.4 as an aid to the reader. We suggest that the interested reader use the computer program described in section 2.4 to actually solve this and future examples of this chapter. In this manner, the reader will perceive the many details of such calculations.

The computer results for the acceleration of the mass are shown in Fig. 2.5. We observe that for a clearance of 0·5 in. no contact is made with the stop. However, when the clearance is reduced to 0·3 in., contact is made by the third oscillation and a large acceleration occurs. With a smaller gap of 0·1 in., similar contact accelerations occur even sooner. The accelerations could be reduced by decreasing the stop stiffness k_l, but in that case, of course, the displacements would increase.

Table 2.3 INPUT DATA REQUIRED

C	1·0	TSTP	0·01
M	0·1	TTL	2·5
K	62·5	KU	0·0
FORCE1	10·0	YU	0·0
FORCE2	10·0	KL	625·0
FREQ1	3·98	YL	0·1†
FREQ2	3·98	ITYPE	0
TIMONE	0·0	KSQ	0·0
TIMTWO	2·5	KCUBE	0·0
FRIC	0·0	Y	0·0
ACCEL1	0·0	X	0·0
ACCEL2	0·0	VELY	0·0
FRK	0·0	VELX	0·0
DEL	0·01	IOPT	0‡

† Additional runs are made with YL = 0·3 and 0·5.
‡ To repeat a run with a new value of YL, one should set IOPT to 1, and input a new complete set of cards.

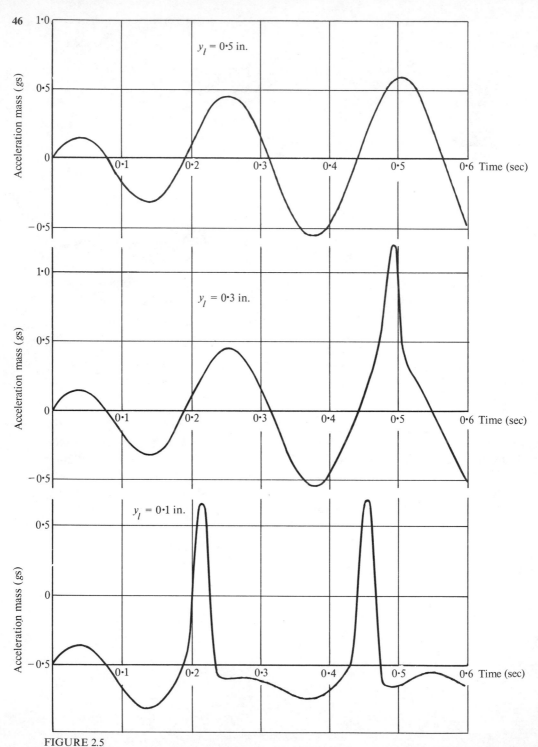

FIGURE 2.5
Effect of floor clearance y_l on accelerations of motor mounted on shock mount. 'Banging' occurs for $y_l = 0.3$ and 0.1 in.

Table 2.4 **INPUT DATA IN THE FORM REQUIRED BY THE COMPUTER PROGRAM FOR THE SYSTEM IN FIG. 2.4**

0					
0·0	2·5	0·0	0·0	3·98	3·98
62·5	0·1	1·0	10·0	10·0	0·0
0·0	0·0	0·0	0·0		
0·0	0·0	625·0	0·1		
0·01	0·01	2·5	0·0	0·0	0·0
1					
0					
0·0	2·5	0·0	0·0	3·98	3·98
62·5	0·1	1·0	10·0	10·0	0·0
0·0	0·0	0·0	0·0		
0·0	0·0	625·0	0·3		
0·01	0·01	2·5	0·0	0·0	0·0
1					
0					
0·0	2·5	0·0	0·0	3 98	3·98
62·5	0·1	1·0	10·0	10·0	0·0
0·0	0·0	0·0	0·0		
0·0	0·0	625·0	0·5		
0·01	0·01	2·5	0·0	0·0	0·0
0					

As a second example, we consider an equipment package having an internal transformer weighing 20 lb. The transformer is suspended from the outer case by a spring with a stiffness of 3000 lb/in., as shown in Fig. 2.6. The damping coefficient is 1·5 lb sec/in. The transformer is free to move through a 0·2 in. clearance before contacting the case. The case stiffness is 15 000 lb/in. It is estimated that the

FIGURE 2.6
Transformer mounted in case with 0·2 in. clearance. Excitation 10g frequency sweep from 20 Hz to 60 Hz

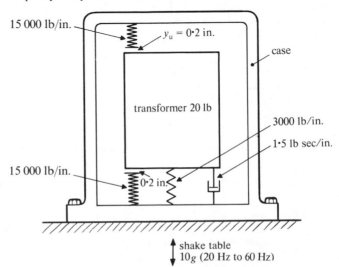

15 000 lb/in.

$y_u = 0·2$ in.

case

transformer 20 lb

3000 lb/in.

1·5 lb sec/in.

15 000 lb/in.

0·2 in.

shake table
10g (20 Hz to 60 Hz)

transformer is initially at rest, but that the case has an initial velocity of 30·7 in./sec, corresponding to the velocity at time zero for a sinusoidal acceleration at $10g$ and 20 Hz.

The excitation is that of a shake table driven at $10g$ over a linear sweep of frequency from 20 to 60 Hz in 1 sec. We want to observe the growth of the acceleration amplitude sustained by the transformer during the frequency sweep. The basic input data is much the same as before and is given in Table 2.5.

Table 2.5 INPUT DATA FOR TRANSFORMER

Quantity	Symbol	Magnitude
Mass	m	0·05181 lb sec^2/in.
Spring stiffness	k	3000 lb/in.
Damping	c	1·5 lb sec/in.
Stop stiffness	k_u	15 000 lb/in.
	k_l	15 000 lb/in.
Stop clearance	y_u	0·2 in.
	y_l	0·2 in.

The results are plotted in Fig. 2.7 (some of the 1 sec time period is omitted to save space). We note that the upper stop is contacted first at about 0·315 sec. In the next half-cycle, the lower stop is contacted. Continued periodic contact with the stops occurs as the frequency is increased to 60 Hz. The maximum acceleration reaches $110g$ during the contacts. The steady-state amplitude would have been only about $17g$ if there had been no stops, or if the stops had not been contacted in going to 60 Hz.

As a final example, the impact due to dropping a package 5 ft from a truck platform will be studied. It strikes the pavement with a velocity of 215 in./sec. The package contains a mass of 40 lb (0·1036 lb sec^2/in.). The mass is supported by packaging having an equivalent spring constant of 100 lb/in. The damping is estimated at 2 lb sec/in. The effective wall stiffness of the package is 1000 lb/in. The clearance to the outside of the package is to be investigated. Should it be 4 in. or 2 in.? The basis of choice is the level of acceleration experienced by the package. The mathematical model is shown in Fig. 2.8.

The result of the calculation is shown in Fig. 2.9. The initial deceleration of just over $10g$ is due to the damping of 2 lb sec/in. between package and case. In other words, a force of $2 \times 215 = 430$ lb develops, which is the equivalent of $430/40 = 10·75g$. It is evident that with a 2 in. clearance, the packaged article hits the case bottom at time 0·02 sec after initial contact and then hits the case top at 0·09 sec after initial contact. The peak accelerations reach $37g$s. With a 4 in. clearance, the peak acceleration is $18g$ and the top of the case is not impacted at all. With larger clearance, the peak acceleration is about $14g$, and no impact with the case occurs. Clearly, the 4 in. package clearance is advantageous.

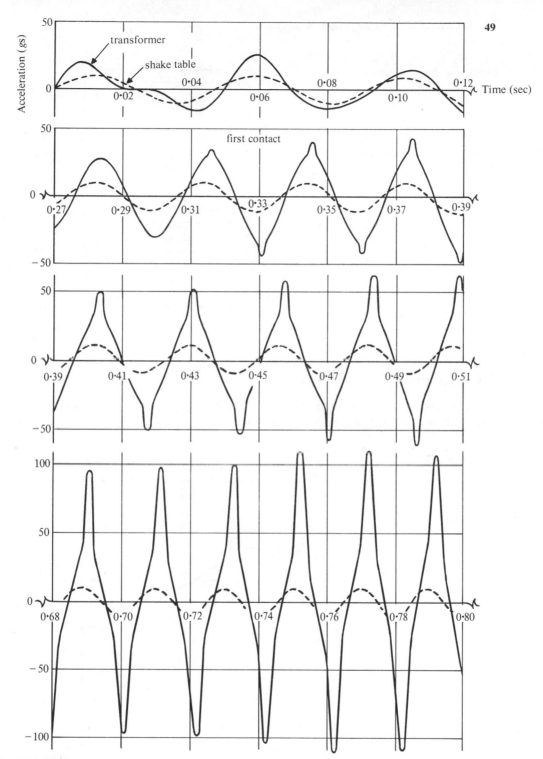

FIGURE 2.7
Response to frequency sweep of a system with stops

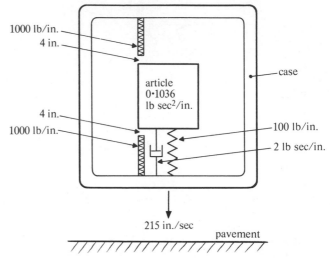

FIGURE 2.8
Packaged article strikes pavement after falling 5 ft from tailgate of truck

2.6   THE EFFECTS OF FRICTION

Frictional effects can be studied with the aid of the computer program of section 2.4. We shall first use the program to reproduce a number of solutions already found

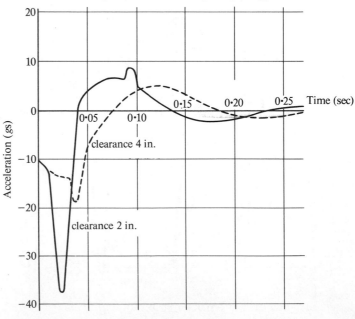

FIGURE 2.9
Effect of clearance on peak accelerations of package dropped from truck

in the literature. First, we consider the movements of a mass m attached to a spring of stiffness k and resting on a surface with friction force R. The mass is given an initial displacement y_0 and starts oscillating with zero initial velocity. During each successive swing, the frictional force will change direction. The mass will exhibit unsteady sinusoidal motions before coming to rest, as shown, for example, by Jacobsen and Ayre.[3] They show that the governing equation for the system is

$$m\ddot{y} \pm R + ky = 0 \tag{2.8}$$

Defining Δ as k/R, we have

$$m\ddot{y} + k(y \pm \Delta) = 0 \tag{2.9}$$

which, since $\ddot{\Delta} = 0$, is the same as writing

$$m(\ddot{y} \pm \ddot{\Delta}) + k(y \pm \Delta) = 0 \tag{2.10}$$

The sign of Δ depends on the direction of the velocity and always opposes the motion. The solution to this equation is

$$y \pm \Delta = A \cos \omega t + B \sin \omega t \tag{2.11}$$

where A and B are constants. For a start from position y_0, we have

$$y = (y_0 \pm \Delta) \cos \omega t \mp \Delta \tag{2.12}$$

The upper sign is used when the velocity is positive, and the lower sign is used when it is negative. The frequency $\omega = \sqrt{k/m}$ is the natural frequency of the system, which is not changed by the frictional effect. It will be observed from eq. (2.12) that the motion is a sinusoidal oscillation about the point $y = \pm\Delta$, depending on the velocity in that swing. The end of the first swing occurs at a time $t = \pi/\omega$, so that the displacement, from eq. (2.12), is

$$y = -(y_0 - \Delta) + \Delta = -(y_0 - 2\Delta) \tag{2.13}$$

During the second swing, the sinusoid is centered about $y = -\Delta$ and the displacement at the end of the swing is $y = (y_0 - 4\Delta)$. It is seen that the displacement at the end of each swing is reduced by an amount of 2Δ. The numerical results obtained by means of the computer program are shown in Fig. 2.10. Observe that the motion is indeed that predicted by the theory. In this example, $m = 1$, $k = 1$, and the initial displacement was 4 in. The local spring stiffness k_f was taken as 10 000 and the time step increment $\Delta t = 0{\cdot}0025$ sec. The motion is seen to cease at the end of the second swing, which is also predictable through eq. (2.12).

Now consider a mass m resting on a frictional surface which generates a frictional resisting force R. Let the mass be driven by an external sinusoidal force $F \sin \omega t$. Tou and Schultheiss[4] have given solutions for this situation. The interesting features of the motion are that if $R/F < 2/\pi$, the motion is roughly sinusoidal, but has discontinuities in acceleration. If $R/F > 2/\pi$, then the motion is sporadic, there being so-called dead bands within which no motion occurs. And when $R/F > 1$, no motion is possible, except for an initial transient. Cunningham[5] also gives a solution for this sort of motion. Briefly, he states that the velocity v is given by

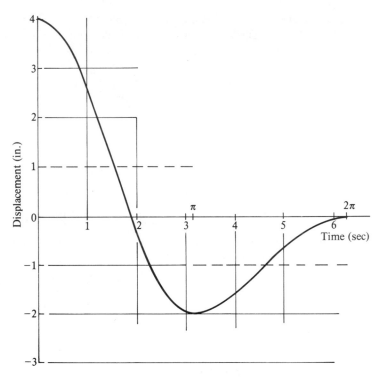

FIGURE 2.10
Oscillation of a mass with frictional restraint, under an initial displacement of
4 in.

$$v = \frac{F}{m\omega}\left[\sin \omega t - \frac{R}{F}\left(\omega t - \frac{\pi}{2} - \beta\right)\right] \tag{2.14a}$$

when $v \geq 0$ and $\beta \leq \omega t < (\pi + \beta)$, and

$$v = \frac{F}{m\omega}\left[\sin \omega t + \frac{R}{F}\left(\omega t - \frac{3\pi}{2} - \beta\right)\right] \tag{2.14b}$$

when $v \leq 0$ and $(\pi + \beta) \leq \omega t \leq (2\pi + \beta)$, where β is defined by the relationship

$$\sin \beta = -\pi R/2F$$

The computer program was used to obtain these solutions when $m = 1$, $k = 1$, and
for the special case when $R/F = 1/\pi$ (when $\beta = -30°$). An initial velocity of
$v_0 = 26\cdot8$ was applied to the mass. The resulting velocity is shown in Fig. 2.11(a).
There we observe the discontinuity as v goes through zero. These numerical results
are identical with those obtained through the evaluation of eqs. (2.14). When the
driving force is reduced so that $R/F = 2/\pi$, the motion has periodic dead bands as
predicted by Tou and Schultheiss. This phenomenon is illustrated by the numerical
results of Fig. 2.11(b).
    A remark is in order at this point concerning the success of the numerical

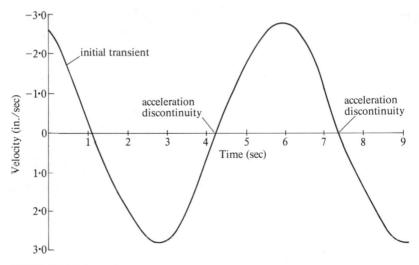

FIGURE 2.11(a)
Response of a mass with friction, $R/F = 1/\pi$

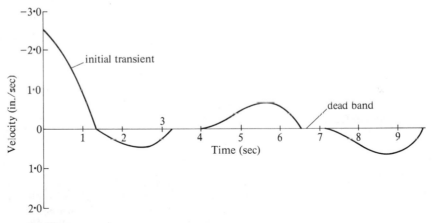

FIGURE 2.11(b)
Response of a mass with friction, $R/F = 2/\pi$

scheme in the computer program as it concerns frictional effects. The key to the frictional model is the use of the local stiffness k_f. This spring stiffness avoids a step discontinuity as the velocity changes sign and allows the numerical integration to proceed smoothly. The stiffness k_f represents the local stiffness of the surfaces in contact. It is thus usually substantially larger than the dominant system springs.

Problem

2.3   The computer program in section 2.4 can be modified to accept any desired resistive force $R(y)$. In this section we have studied the resistive force typical of dry friction, or Coulomb friction, where the force is constant and changes sign as the velocity does. This model

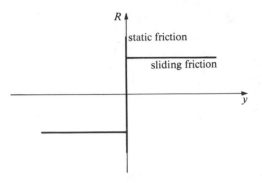

(a) Static and sliding friction

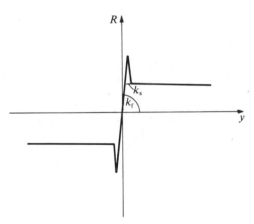

(b) Model of static and sliding friction

FIGURE 2.12
Static and sliding friction representations

is adequate where only sliding friction is considered. It is known, however, that at the instant before sliding takes place, the resistive force can be considerably higher than the force during sliding. Such an initial frictional resistance is known as static friction. Tou and Schultheiss[4] have developed analytical solutions to special situations involving static friction in servomechanisms. A schematic diagram of the forces involved is shown in Fig. 2.12(a). The phenomenon can be modeled with two local springs k_f and k_s, as shown in Fig. 2.12(b). Develop the necessary algebraic descriptions of the element, corresponding to eq. (1.114), in the case of simple friction. Insert these expressions into the computer program. What effect does static friction have on the velocity discontinuities discussed previously?

Now consider again a package dropped from the tailgate of a truck. Include, however, the effect of frictional restraint between the internal article and the case. Take the clearance to the walls of the case quite large so as not to obscure the effects of friction. Figure 2.13 shows the dynamic system. The frictional force F is taken at 100 lb in one example and 400 lb in another. The local spring constant k_f is assumed to be 300 lb/in. The internal mass again weighs 40 lb.

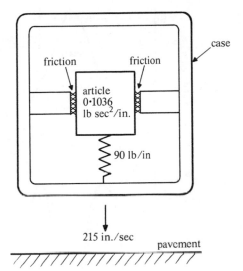

FIGURE 2.13
Packaged article strikes pavement after
falling 5 ft from tailgate of truck. Friction
values of 100 lb and 400 lb

The response is shown in Fig. 2.14. In the upper figure, the initial deceleration
is at 2·5gs, which is to be expected with a 40 lb mass (0·1036 lb sec²/in.) and a
100 lb frictional force. In the lower figure, the initial deceleration is 10gs, which
might be expected from a 40 lb mass with a 400 lb frictional force. The early
motion is a distorted sinusoid. After a time of about 0·35 sec in the upper figure
and 0·15 sec in the lower figure, no further sliding takes place and the motion is a
sinusoid. The amplitude, as might be expected, is approximately the frictional force
divided by the weight.

Incidentally, two values of the time step interval were used—0·01 sec and
0·005 sec—with negligible differences in results. This observation shows that even
discontinuous forces such as friction are amenable to the time-stepping procedure,
and that the convergence criteria discussed previously for linear systems can yield
satisfactory results also for discontinuous nonlinear systems.

Problems

2.4   What is the effect of adding damping, $c = 0·61$ lb sec/in., to the example problem with
$F = 100$ lb?

2.5   What is the effect of adding a nonlinear spring with $k_3 = 4·5$ lb/in.³ to the example
problem with $F = 100$ lb? (*Note:* The springing for a packaging material is frequently
nonlinear. With $k = 90$ lb/in. and $k_3 = 4·5$ lb/in.³, the spring force for a 4 in. deflection is
$90 \times 4 + 4·5 \times 4^3 = 360 + 284 = 644$ lb rather than the 360 lb from a linear spring of
90 lb/in. rate.)

2.6   What is the effect of adding a stop to the example problem with $F = 100$ lb, when the
stop stiffness is 1000 lb/in. and the initial clearance is 4 in.?

As another example of the effect of Coulomb friction, consider again the
package in Fig. 2.13, but in this case place it on a shake table, as in Fig. 2.15.

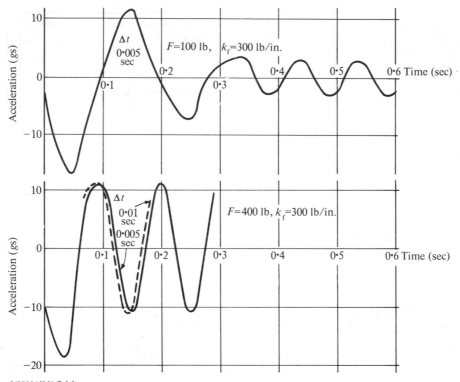

FIGURE 2.14
Response of system in Fig. 2.13 for different values of the frictional restraint

The excitation of the shake table is $5g$ at 5 Hz. Since the weight of the internal article is 40 lb and the inertia force could reach $5 \times 40 = 200$ lb, one should expect some shifting motion of the article in the case. The initial velocity of the table is computed directly from the given $5g$ acceleration as $-(5 \times 386)/(2\pi 5)$ $\cos (2\pi 5t)$ with $t = 0$, giving 61·56 in./sec.

The response is shown in Fig. 2.16. When there is a reversal in the direction of travel of the internal article with respect to the case, the frictional force reverses sign. Thus, the change in force acting on the internal article is from -100 lb to $+100$ lb, or a 200 lb change. Since the weight of the internal article is 40 lb, one might expect a jump in g-level of $200/40 = 5g$. At reversals (a–a) in Fig. 2.16, the jump is indeed $5g$.

The response in Fig. 2.16 is out of phase with the excitation. It tends to be maximum when the excitation is zero and is itself zero when the excitation is maximum. This behavior is typical of damped response near resonance. The un-damped natural frequency of the system in Fig. 2.13 is $(1/2\pi)\sqrt{k/m} =$ $(1/2\pi)\sqrt{90/0·1036} = 4·7$ Hz. This is quite close to the excitation frequency of 5 Hz. In the absence of friction, the steady-state response of the system in Fig. 2.15 would be $5·0/(1 - 5^2/4·7^2) = -5·0/0·1318 = -38g$. The minus sign indicates that at fre-

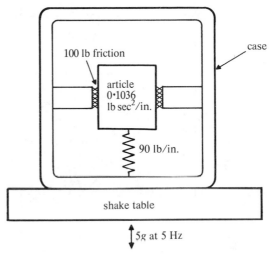

FIGURE 2.15
Packaged article with friction to case

quencies above the resonance frequency, the response is opposed to the excitation. The response at 0·6 sec, of the order of 20g, would have increased to near 38g if the record had been continued. In this case, the friction does not cause much damping of the motion.

Problems

2.7   Repeat the example problem with twice the excitation; i.e., 10g acceleration of the shake table at 5 Hz. Is the response doubled as it would be in a linear system? (*Note:* Take the values of $\dot{y}_0$ and $\dot{x}_0$ at $-123·12$ in./sec.)

2.8   Repeat the example problem with a nonlinear spring, changing k_3 from 0 to 4·5 lb/in.^{3}. What effect does this spring have on the acceleration levels?

2.9   Repeat the example problem with increased friction, changing F from 100 lb to 150 lb. What effect does this have on the acceleration levels? Is there now a significant damping effect? (*Hint:* Take the time record to 3·0 sec to allow time for a steady-state condition to develop.)

As a final example of the effect of Coulomb friction, consider forces applied to the same mass, as shown in Fig. 2.17. Again, take the friction at 100 lb. The excitation force of 200 lb is twice the frictional force. The frequency of excitation is 5 Hz, as compared with an undamped natural frequency of 4·7 Hz.

The response is shown in Fig. 2.18. The initial response through the first 1·4 sec is similar to that in Fig. 2.16; it is omitted to show greater detail near steady-state conditions. At reversals in velocity (*a–a*) in Fig. 2.18, we note a jump in g-level by 5g. It is similar to the jump found in Fig. 2.16. The reason is also the same; namely, that when the velocity changes sign, the reversal in the direction of the frictional force causes a 200 lb change in force on the mass. The weight of the mass is 40 lb. The g-level change is therefore $200/40 = 5g$.

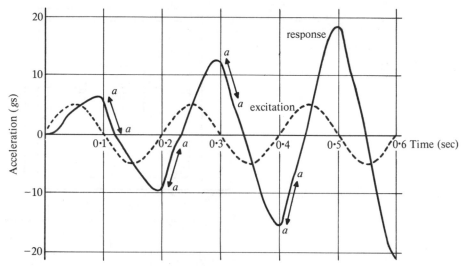

FIGURE 2.16
Response of internal article of package in Fig. 2.15. Effect of friction is to cause discontinuities in the acceleration curve (*a–a*) at the reversals in velocity

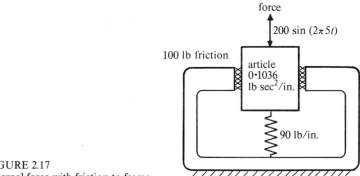

FIGURE 2.17
External force with friction to frame

Again the response is out of phase with the force and lagging. In the absence of friction, the steady-state response would be $(200/40)/(1 - 4\cdot7^2/5^2) = 5/0\cdot118 = 42g$. There is little improvement in this case, therefore, due to the presence of friction.

Problems

2.10 Repeat the example problem with the excitation frequency changed from 5 Hz to the undamped resonant frequency 4·7 Hz. What is the effect on the acceleration levels?

2.11 Repeat problem 2.10 with the frictional force F increased from 100 lb to 175 lb and determine the effect on accelerations.

2.12 Repeat the example problem with the frequency change to 10 Hz. Does the friction now have a significant effect on the g-level? (*Hint:* The steady-state g-level in the absence of friction would be $(200/40)/(1 - 4\cdot7^2/10^2) = 5/0\cdot779 = 6\cdot42g$.)

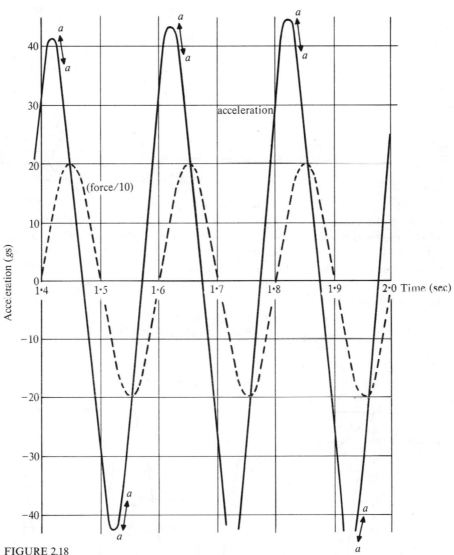

FIGURE 2.18
Effect of friction on response to sinusoidal force excitation. Velocity reverses sign at (a–a)

2.7   VARIABLE AMPLITUDE FORCE EXCITATION

Occasionally, a small article, which can be considered as a single degree-of-freedom system, is attached to a larger system. The motions of the larger system then determine the forces on the small article.

Consider the system in Fig. 2.19. The force applied to the mass decreases exponentially. The initial value is 200 lb. The 'decay' drop-off is 20 per cent each 0·05 sec.

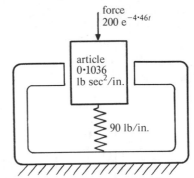

FIGURE 2.19
Exponentially decaying force applied to mass

The response is shown in Fig. 2.20. The initial acceleration of the mass is $5g$, as one might expect with an initial force of 200 lb and a weight of 40 lb. The subsequent motion has an almost sinusoidal variation with $5g$ amplitude. This motion is typical for an applied force which decays only a moderate amount during a cycle of natural oscillation (0·2 sec in this case). For a somewhat more rapid decay of applied force, the final response would be higher. For a much more rapid decay, the final response would be lower.

In some cases a large system is subjected to a disturbance, causing it to execute a decaying sinusoidal motion. A smaller system attached to the larger may then be

FIGURE 2.20
Response to exponentially decaying force of system in Fig. 2.19

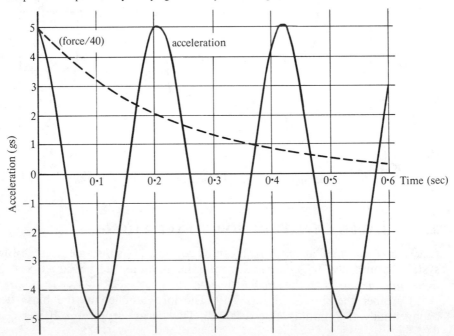

subjected to a sinusoidally varying force of decaying amplitude. Consider again the system in Fig. 2.19, but let the force be sinusoidal at 5 Hz and be given by

$$\text{force} = 200\,e^{-4.46t}\sin(2\pi 5t)$$

The response is plotted in Fig. 2.21. Since the natural frequency of the system is 4·7 Hz and the excitation is at 5 Hz, there is a substantial build-up of response before the force decays to negligible values. The ratio of maximum force to weight, 200/40, is only 5g. The maximum response is of the order of 15g.

Problems

2.13   Repeat the example problem in Fig. 2.19 with forces dropping by 40 per cent in each 0·05 sec time period. Does the final response increase in comparison with the example problem?

2.14   Repeat the example problem with a force that varies linearly from 200 lb initially to zero after 0·3 sec.

2.15   Repeat the example problem with a force that increases linearly from zero to 200 lb in 0·1 sec and then decreases linearly to zero at 0·3 sec.

2.16   Repeat the example described in Fig. 2.20 with the decaying sinusoidal force at 10 Hz rather than 5 Hz. What is the maximum response amplitude in this case in a 0·6 sec time period?

FIGURE 2.21
Response to decaying sinusoidal force (natural frequency 4·7 Hz)

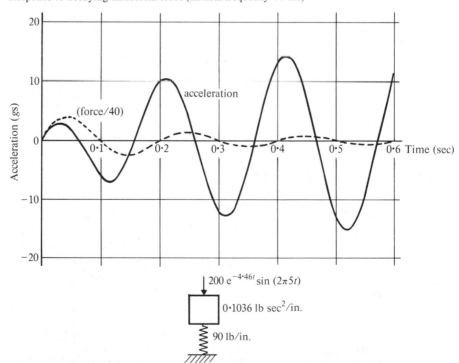

2.8 NONLINEAR SPRINGS

Jump conditions

A linear system under a sinusoidal excitation is governed by the equation of motion

$$m\ddot{y} + c\dot{y} + ky = F \sin \omega t \tag{2.15}$$

For such a system, a *resonance curve* can be constructed, as shown in section 1.4 and in Fig. 1.2. When the spring restoring force R is nonlinear, the resonance curves are considerably changed, and a phenomenon known as a jump condition occurs.

Consider the case where the spring force is nonlinear and is given by

$$R = k_1 y + k_3 y^3 \tag{2.16}$$

When $k_3 > 0$, the spring stiffness increases with y, while, with $k_3 < 0$, the stiffness decreases with y. The springs with $k_3 > 0$ and $k_3 < 0$ are known as hardening and softening springs, respectively. The force–deflection curves of each sort of spring are shown in Fig. 2.22.

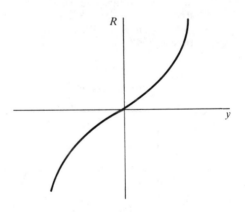

(a) Hardening spring, $k_3 > 0$

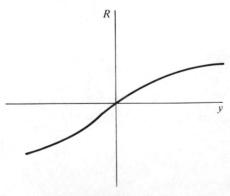

FIGURE 2.22
Hardening and softening restoring forces

(b) Softening spring, $k_3 < 0$

It has been shown in many texts[6,7] that a system with a hardening spring has a resonance curve whose central spine is tilted to the right, as shown in Fig. 2.23(b). The resonance curve shown there is obtained for a given magnitude of the excitation force F. In practice, this resonance curve shows areas of instability. If the frequency of the excitation is increased, the response follows the resonance curve to a certain point A shown in Fig. 2.23(c), whereupon the motion becomes unstable and abruptly jumps to a smaller amplitude B. Similarly, if the frequency is reduced through the resonance frequency, the response will pass along the lower curve, past point B, to the inflection point C, whereupon the response magnitude will abruptly increase to the point D. Further decrease in frequency causes the response to follow the former curve once more. If the spring has softening characteristics, then the spine of the response curve will be tilted towards the left, as shown in Fig. 2.23(d). In this case, the jumps occur in the reverse fashion.

The phenomenon is further complicated by the fact that the response curves are sensitive to the *rate* at which the frequency changes. Mitropolskii[7] has shown that in the case of a hardening system, the lines AB and CD are no longer vertical but are distorted in a way which depends upon the frequency sweep rate. These phenomena are easily studied by applying the computer program of section 2.4.

To study the resonance curves numerically, we consider a particular spring mass system which has a unit mass, a spring restoring force given by $R = y + y^3$, and damping $c = 0.2$. The magnitude of the exciting force is unity. Since the rate at which the frequency changes with time is an important parameter, we express the excitation force mathematically as $\sin(\omega_0 + \beta t)t$. Here, ω_0 is the initial or starting frequency, while β is the parameter that controls the change in frequency with time.

Three cases are studied, with $\beta = 0.0001$, $\beta = 0.001$, and $\beta = 0.0025$. Roughly, the three frequency sweep rates are such that in passing through the resonant condition the system undergoes 600, 60, and 25 cycles, respectively. When $\beta = 0.0001$, the response curve of Fig. 2.24(a) is obtained. In this curve, we note that as the frequency increases, so does the resonant amplitude. This amplitude follows closely the steady-state response curve ($\beta = 0$) until roughly $\omega = 1.2$, whereupon there is a fairly abrupt decrease in amplitude. During this decrease, there are some secondary maxima and minima in the response curve. Upon decreasing the frequency, the lower branch of the steady-state response curve is more or less followed until there is an abrupt increase in amplitude, followed by several minor minima and maxima.

For larger values of β, when the sweep is faster, the response curves diverge further still from the steady-state response curves. Such responses with $\beta = 0.001$ and $\beta = 0.0025$ are shown in Fig. 2.24(b) and Fig. 2.24(c). At some points, the calculated curves diverge somewhat from the results obtained analytically by Mitropolskii (shown by dotted lines), seeming, however, to oscillate about his results. There is a reason for this difference in predictions. In Mitropolskii's work, it is assumed that the start of the frequency sweep is taken distant enough that any starting transient has died out, and steady state is reached; i.e., the initial conditions and starting transients are not present. In our computer approach, it is inherent that these starting transients are part of the solution. Thus, returning to Fig. 2.24(c),

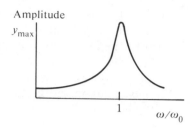

(a) Linear system $R = ky$

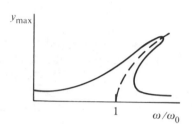

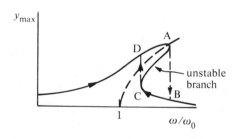

(b) Nonlinear system (hardening spring)

(c) Jump conditions (hardening spring)

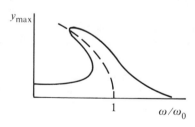

(d) Nonlinear system (softening spring)

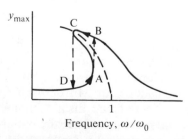

(e) Jump conditions (softening spring)

FIGURE 2.23
Illustrations of idealized resonance curves for nonlinear systems

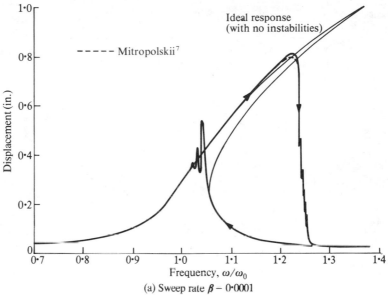

(a) Sweep rate $\beta = 0.0001$

FIGURE 2.24
Response of a mass with a cubic-hardening spring

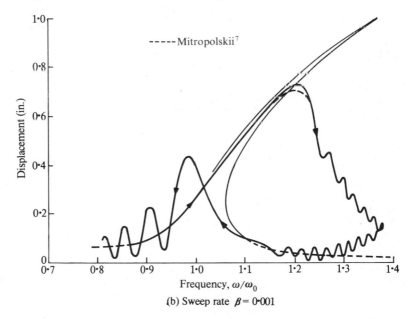

(b) Sweep rate $\beta = 0.001$

we observe that at the beginning of the frequency sweep, near $\omega = 0.8$, there is some divergence in results. At the end of the sweep, when the frequency again decreases, there is again considerable oscillation. If the end of the frequency sweep is extended to $\omega = 2.0$ (the results of which are shown in Fig. 2.25), then the numerical results coincide well with Mitropolskii's.

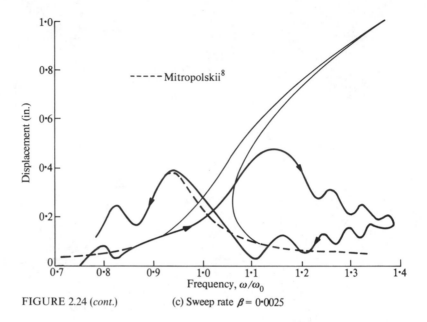

FIGURE 2.24 (*cont.*)   (c) Sweep rate $\beta = 0.0025$

Not only can the frequency vary with time, so also can the magnitude of the external force. If a system with a cubic spring is subjected to a rising and a lowering force, its response will be different during the two periods in a manner analogous with the case of rising and lowering frequency that we have just discussed. This situation occurs also because of inherent instabilities in the response curves.

FIGURE 2.25
Response of a mass with a cubic-hardening spring, and sweep rate $\beta = 0.0025$. The frequency sweep was extended to $\omega/\omega_0 = 2$ in order to eliminate the initial transients

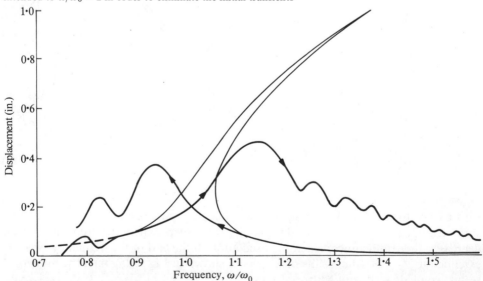

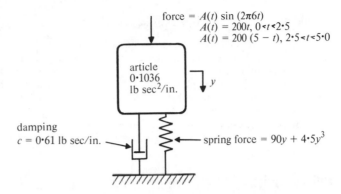

FIGURE 2.26(a)
Nonlinear system with increasing exciting force at 6 Hz (amplitude at
which secant spring rate would cause resonance at 6 Hz is 3·6 in.)

To illustrate the situation we consider the system in Fig. 2.26(a). There, the
spring is symmetrically hardening with amplitude. The force excitation on the mass
is sinusoidal with a 6 Hz frequency, while its amplitude varies with time, rising
linearly to 500 lb in 2·5 sec and then decreasing linearly again to zero.

Figure 2.26(b) shows a plot of the acceleration amplitude of the mass as a
function of the applied force amplitude. Note that the curves for the increasing
and decreasing force regimes are quite different. The curve for decreasing exciting
force is above that for increasing force at force levels below about 200 lb, the value
where a rapid rise in response occurs. The perturbations in the rising force curve
are a result of having a relatively rapid rise in force. The tendency of the response
to remain at relatively high levels as the force decreases to levels below about
200 lb is characteristic of this type of nonlinearity.

FIGURE 2.26(b)
Results for damped case, $c = 0·61$ lb sec/in. of system in Fig. 2.26(a). The
response amplitude is different for increasing and for decreasing force
amplitudes

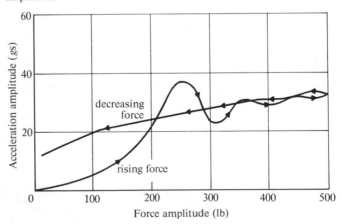

Problems

2.17   Repeat the example in Fig. 2.26(a) with the rate of rise and the rate of decrease half as great. Do these rates change the plot in Fig. 2.26(b) significantly?

2.18   Repeat the example in Fig. 2.26(a) with reduced damping by taking $c = 0.305$. Does this damping change the plot significantly? Repeat with increased damping $c = 1.22$.

2.19   Repeat the example with $k_3 = 2.25$. What effect does this spring have on the maximum g-levels?

2.20   Repeat the example with the frequency of excitation increased to 7 Hz. Note the increase in force level at which a marked rise of response occurs and the change in maximum response.

Bilinear springs

In order to illustrate a further application of the program of section 2.4 to nonlinear spring systems, we shall discuss here the special case of a bilinear spring where an exact solution is available for comparison with the results of the computer program.

An exact solution is obtainable for some special situations that involve the bilinear stop spring described by eq. (1.113). Consider the case when the system consists of a mass m connected to a rigid support by such a spring. If the mass is given an initial displacement y_i and then released, the subsequent oscillations can be traced very simply by a closed form solution that is made up of linear solutions to two piecewise linear problems. The solution is developed as follows,[8] with reference to Fig. 2.27(a). In the initial time interval during which $y_u < y < y_i$, the mass moves with simple harmonic motion with a center about the point A, so that

$$y = a + (y_i - a) \cos \omega_2 t \tag{2.17}$$

FIGURE 2.27(a)
Geometric descriptions necessary in the analytical solution

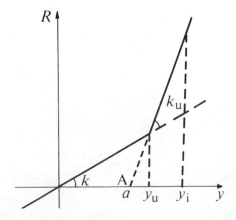

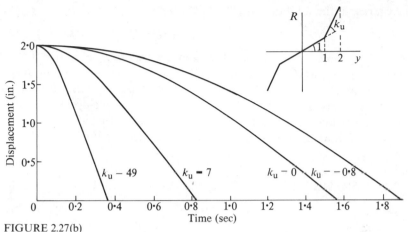

FIGURE 2.27(b)
Initial response of a bilinear spring system

where

$$a = y_u \left(1 - \frac{k}{k \mid k_u} \right)$$

$$\omega_2 = \sqrt{\frac{(k + k_u)}{m}}$$

The time t_2 taken to reach y_u from y_i is then obtained from eq. (2.17) by substituting y_u for y. We obtain, as a result,

$$t_2 = \frac{1}{\omega_2} \cos^{-1} \left(\frac{y_u - a}{y_i - a} \right) \tag{2.18}$$

Now, when $0 < y < y_u$, we require that both displacement and velocity are matched at time t_2, so that

$$\begin{aligned}
(y)_{t=t_2} &= y_u \\
(\dot{y})_{t=t_2} &= -(y_i - a)\omega_2 \sin \omega_2 t_2
\end{aligned} \tag{2.19}$$

The motion is now simple harmonic about a center $y = 0$, so that the solution in this interval is

$$y = y_u \cos \omega_1 (t - t_2) - \frac{(y_i - a)\omega_2 \sin \omega_2 t_2}{\omega_1} \sin \omega_1 (t - t_2) \tag{2.20}$$

where

$$\omega_1 = \sqrt{\frac{k}{m}}$$

To reach the origin from y_u takes a time

$$t_1 = \frac{1}{\omega_1} \tan^{-1} \left[\frac{y_u \omega_1}{(y_i - a)\omega_2 \sin \omega_2 t_2} \right]$$

As a result, the total period of oscillation is

$$\tau = 4(t_1 + t_2)$$

The computer program of section 2.4 was used to study the oscillations of a system with a unit mass, an initial clearance $y_u = 1$ in., and an initial displacement $y_i = 2$ in. The spring stiffness k was unity, while k_u was varied from 49 to -0.8. The results are shown in Fig. 2.27(b) for the first quarter-cycle. Subsequent oscillations will repeat this basic cycle. The numerical results coincide with those obtained from eqs. (2.17) and (2.20), within the accuracy of the plotting.

2.9   VARIABLE FREQUENCY FORCE EXCITATION

Where rotating machinery operates at variable speed, the exciting forces may change frequency with time. We consider a system subjected to a constant force amplitude of 200 lb at frequencies increasing linearly from 0 to 5 Hz in 0·6 sec. This system is shown in Fig. 2.28.

A small item is noted on Fig. 2.28 that might otherwise be done incorrectly. When one says that a sine function is at a frequency f, one means that the argument θ of the sine function changes at the rate

$$\frac{d\theta}{dt} = 2\pi f$$

For f constant, this gives on integration

$$\theta = 2\pi f t + (\text{constant of integration})$$

When f itself varies linearly with time, however, from zero at time 0 to F at time T,

$$\frac{d\theta}{dt} = 2\pi \left(\frac{F}{T}\right) t$$

FIGURE 2.28
System excited by variable frequency force (undamped natural frequency 4·7 Hz)

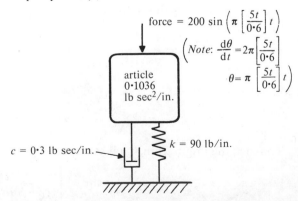

$$\text{force} = 200 \sin\left(\pi \left[\frac{5t}{0 \cdot 6}\right] t\right)$$

$$\left(\textit{Note: } \frac{d\theta}{dt} = 2\pi \left[\frac{5t}{0 \cdot 6}\right]\right.$$

$$\left.\theta = \pi \left[\frac{5t}{0 \cdot 6}\right] t\right)$$

article
0·1036
lb sec²/in.

$k = 90$ lb/in.

$c = 0·3$ lb sec/in.

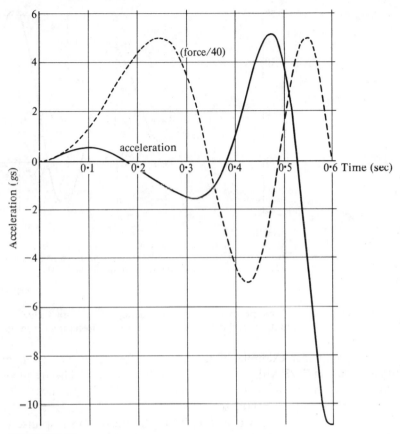

FIGURE 2.29
Response of system in Fig. 2.28 to variable frequency excitation

and

$$\theta = \pi\left(\frac{Ft}{T}\right)t + \text{(constant of integration)}$$

The results of the computation are shown in Fig. 2.29. Large amplitudes occur as the frequency passes the resonant value at about the end of the computation time.

Problems

2.21   Repeat the computation in the example problem with a rise in frequency in 6·0 sec rather than 0·6 sec. What effect does this have on the maximum response? Does the maximum acceleration now approach the steady-state resonance value at 4·7 Hz of $(200/0·3)$ $(2\pi \, 4·7/386) = 51g$?

2.22   Repeat the example problem with a rise in frequency to 10 Hz in 10 sec. How does the maximum g-level compare with that in the preceding problem?

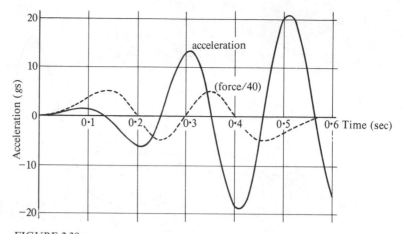

FIGURE 2.30
Response of system shown schematically in Fig. 2.28. Excitation frequency increases
from 0 to 5 Hz in 0·2 sec, holds at 5 Hz until 0·6 sec, and then decreases back to
0 Hz at 0·6 sec (undamped natural frequency 4·7 Hz)

2.23   Repeat the example problem with the frequency going from 4 Hz to 6 Hz in 5 sec. How
does the maximum g-level compare with that in the preceding examples?

As a further example of the effect of frequency variation, we consider the
system in Fig. 2.28 with a changed excitation history. The frequency is considered
to rise to 5 Hz in the first 0·2 sec, to hold at 5 Hz for the next 0·2 sec, and to drop
back to 0 Hz in the final 0·2 sec.

The results are shown in Fig. 2.30. The maximum response is now over $20g$.
This response is more than was found for the frequency history in Fig. 2.29. The
principal reason is that more time is spent at frequencies near the 4·7 Hz undamped
resonant frequency.

Problems

2.24   Repeat the computation in the example with a rise in frequency in 1·0 sec, a hold at
5 Hz for the four following seconds, and then a drop back to 0 Hz in the final second.
Does the maximum response exceed the steady-state $51g$ value for excitation by a 200 lb
force at 4·7 Hz?

2.25   In some disturbances the initial frequency of excitation is quite high and drops to lower
values as time progresses. Repeat the computation in the example with an initial frequency
of 20 Hz, dropping to 0 Hz at 0·6 sec. (*Hint:* Take $\Delta t = (1/20)$ of the initial excitation
period or $(1/20)(0·05) = 0·0025$ sec to adequately follow the excitation.)

2.26   Repeat problem 2.25 with a frequency drop from 20 Hz to 5 Hz in the first 0·45 sec
and a drop from 5 Hz to 0 Hz at 1·0 sec. Does this cause a higher g-level than
observed in problem 2.25?

As a final example of the effect of a variable frequency of excitation, consider
the nonlinear system in Fig. 2.31. The excitation frequency rises to 6 Hz in the first

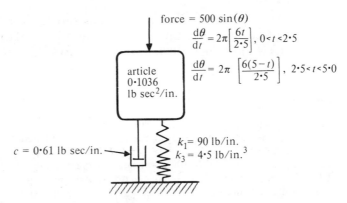

$$\text{force} = 500 \sin(\theta)$$

$$\frac{d\theta}{dt} = 2\pi\left[\frac{6t}{2\cdot5}\right], \quad 0 < t < 2\cdot5$$

article
0·1036
lb sec²/in.

$$\frac{d\theta}{dt} = 2\pi\left[\frac{6(5-t)}{2\cdot5}\right], \quad 2\cdot5 < t < 5\cdot0$$

$c = 0\cdot61$ lb sec/in.

$k_1 = 90$ lb/in.
$k_3 = 4\cdot5$ lb/in.³

FIGURE 2.31
Nonlinear system with variable frequency force (undamped natural frequency at low amplitude 4·7 Hz)

2·5 sec and drops back to 0 Hz in the next 2·5 sec. The undamped, low amplitude, natural frequency is 4·7 Hz. The damping is 0·61 lb sec/in. in comparison with the low amplitude critical damping $2\sqrt{km} = 2\sqrt{0\cdot1036 \times 90} = 6\cdot1$ lb sec/in.

The results are plotted in Fig. 2.32. This plot is not the usual amplitude versus time type. Instead, it gives, as a function of frequency, the maximum amplitude that occurs within every 0·1 sec interval. It is evident that at frequencies above about 4 Hz, there is some resonant build-up in response. It has not fully built up, however,

FIGURE 2.32
Maximum acceleration amplitude in 0·1 sec time step for nonlinear system in Fig. 2.31 subjected to variable frequency excitation. Frequency goes from 0 to 5 Hz in 2·5 sec and returns to 0 Hz at 5·0 sec (weight of mass 40 lb)

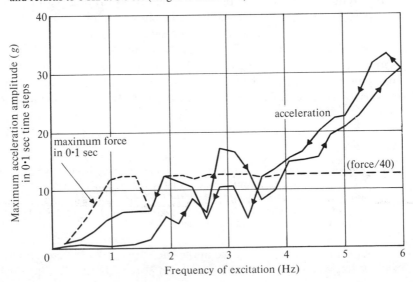

at 6 Hz. The spring nonlinearity has substantially increased the effective natural frequency from its low amplitude value of 4·7 Hz. The response for decreasing frequencies is a little higher than that for increasing frequencies in the resonant range above 4 Hz.

Problems

2.27 Repeat the example problem with an increase in frequency range to 10 Hz. What is the effect on the maximum amplitude and on the difference between the curves for rising and falling frequency?

2.28 Repeat problem 2.27 with half the damping. Does this double the maximum g-level?

2.29 Repeat problem 2.27 with half the rate of frequency change. Does this change the nature of the response?

2.10 EXCITATION BY SUPPORT MOTION

In many cases, the excitation of a dynamic system is the result of support motion. Examples abound in the transportation field. The bouncing of cars going over a pothole or railroad tracks causes the dynamic response of passengers and equipment riding in the car. The disturbance of an aircraft by wind gusts similarly excites the dynamic response of passengers and equipment in the airplane. Ship motion in a sea-state is another source of dynamic disturbance. And, finally, a very important example of support motion is found in the study of the response of structures to seismic shock and earthquake tremors. It was to indicate the treatment for this broad class of problems that provision was made for excitation by support motion in the computer program in section 2.4.

As an illustration of some relatively simple problems of support motion, we wish to repeat the three examples of the previous section on variable frequency force excitation (section 2.9), but where the support rather than the mass experiences the stimulus. First consider the example shown in Fig. 2.28, but let the excitation be a $5g$ motion of the support over the same frequency range, rather than a 200 lb external force. Note that $5g$ on the 40 lb mass is also 200 lb. The results are shown in Fig. 2.33. As might have been predicted, the support motion minus the mass motion in Fig. 2.33 is the same as the mass motion in Fig. 2.29. As the frequency passes 4·7 Hz, the undamped natural frequency, the response increases to its largest value.

Problems

2.30 Repeat the example problem with the frequency going from 0 to 5 Hz in 6 sec rather than 0·6 sec. Does the maximum response now approach the steady-state resonance value of about $51g$?

2.31 Repeat the example with the frequency going from 4 Hz to 6 Hz in 5 sec. Compare the maximum g-level with that in problem 2.30.

As the second example, we allow the support to have a $5g$ acceleration level, with a frequency content that rises from 0 to 5 Hz in the first 0·2 sec, holds at

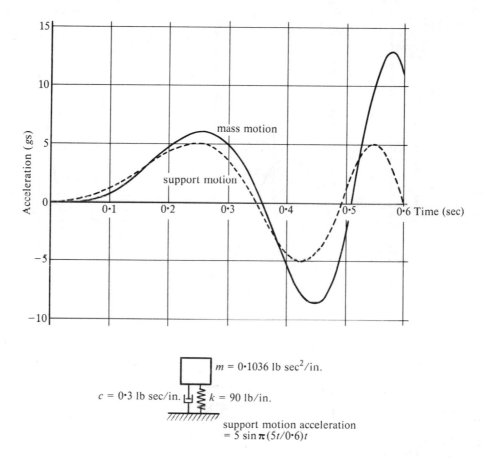

FIGURE 2.33
Response of system in Fig. 2.28 to excitation by support motion

5 Hz for the next 0·2 sec, and drops back linearly to 0 Hz during the final 0·2 sec. This example is identical to that of section 2.9, except that once again the support supplies the stimulus.

The response is shown in Fig. 2.34. Again, we observe a close relationship between Fig. 2.34 for support motion excitation at $5g$ and Fig. 2.30 for forced excitation at 200 lb. The support motion minus the mass motion in Fig. 2.34 equals the mass motion in Fig. 2.30.

Problems

2.32 Repeat the computation in the example with the frequency going from 0 to 5 Hz in the first second, holding at 5 Hz until the time is 5 sec, and returning to 0 Hz at 6·0 sec. Does the maximum response exceed the steady-state $51g$?

2.33 Repeat the example computation with an initial frequency of 20 Hz dropping to 0 Hz at 0·6 sec. (*Hint*: Use $\Delta t = 0·0025$ sec to obtain close following of the 20 Hz excitation.)

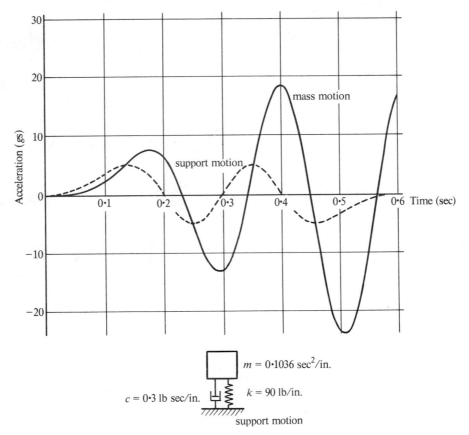

FIGURE 2.34
Response to excitation at 5g (undamped natural frequency 4·7 Hz)

As the third example, we reconsider the system of Fig. 2.31, which has a non-linear hardening spring. The frequency of the 12·5g support motion rises to 6 Hz in the first 2·5 sec, and drops back down to 0 Hz in the next 2·5 sec. Again, 12·5g is chosen to correspond with the 500 lb force that the mass experienced in the previous case (12·5 × 40 lb = 500 lb).

A plot of the results is given in Fig. 2.35. The plot is similar to that in Fig. 2.32, when account is taken of the 12·5g input motion. An exact correspondence at the lower frequencies is masked by the fact that the plots give maximum values in a time step, and these may not occur simultaneously. Again we observe that above 4 Hz there is a resonant build-up. This build-up would be greater at still higher frequencies because of the cubic term in the spring stiffness.

Problems

2.34   Repeat the example problem with an increase in frequency range to 10 Hz.
2.35   Repeat problem 2.34 with half the value of damping. Is there a tendency for the system

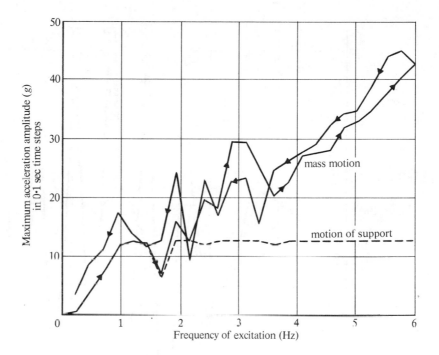

Frequency (Hz)	0·0	6·0	0·0
Time (sec)	0·0	2·5	5·0

FIGURE 2.35
Maximum acceleration in a 0·1 sec time step for excitation by support motion at variable frequency (system in Fig. 2.31)

to fall out of resonance for frequencies well above the resonant value with rising frequency? Is there a tendency for the system to fall out of resonance for frequencies well below the resonant value with falling frequency?

2.36   Repeat problem 2.35 with zero damping.

2.11   TIME HISTORY EXCITATION

Occasionally, a dynamic system is exposed to a time history of support motion. Examples occur in the case of displacements controlled by a cam follower, when a vehicle traverses a bumpy road, with automatic machine tools, and in assembly operations.

As an example of such an excitation, we consider a system which has motions of the support such that the support accelerations are 0, $+1$, -1, $+1$, -1, $+1$, 0, and $0g$ at times 0, 0·05, 0·08, 0·09, 0·10, 0·13, 0·18, and 0·20 sec, respectively, shown as dotted lines in Fig. 2.36.

The results are also plotted in Fig. 2.36. During the relatively gradually changing acceleration of the support up to 0·05 sec, the mass and support move almost

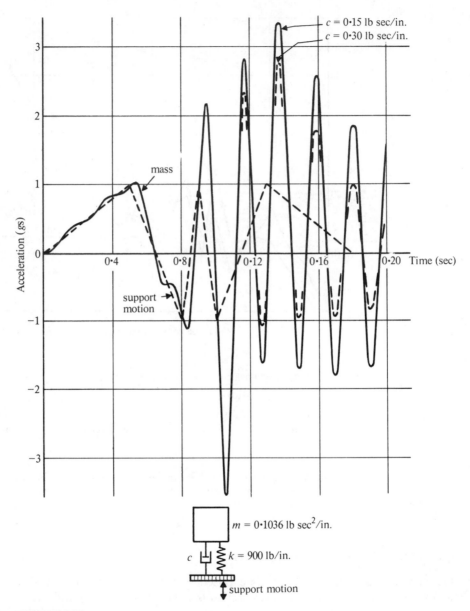

FIGURE 2.36
Response to erratic motion of support (time history disturbance)

together. During the quicker (in comparison with the system period of $2\pi\sqrt{m/k} = 0.0213$ sec) reversals in the support acceleration between 0.05 sec and 0.12 sec, there is build-up of dynamic response. This build-up reaches about three times the support acceleration at $t = 0.106$ sec and at $t = 0.138$ sec. After the support acceleration stops at $t = 0.18$ sec, the system continues to execute a damped oscillation.

Problems

2.37 In the example problem let the time history of support acceleration be 0, 1, -1, 1, -1, 1, -1, 1, -1, 1, 0, and $0g$ at times 0, 0·05, 0·06, 0·07, 0·08, 0·09, 0·10, 0·11, 0·12, 0·13, 0·18, and 0·20 sec, respectively. Does this cause the g-level to become larger? What is the peak value?

2.38 In the example problem let the time history of support acceleration be 0, 1, 0, 0, -1, 0, 0, -1, 0, 0, 1, 0, and $0g$ at times 0, 0·01, 0·02, 0·04, 0·05, 0·06, 0·08, 0·09, 0·10, 0·12, 0·13, 0·14, and 0·20 sec, respectively. What is the peak value of mass acceleration? (This motion would be typical of a cam follower which is first caused to move to the right and stop, and then is caused to move to the left and stop.)

2.39 Repeat problem 2.38 without damping and compare the motion with that in problem 2.38. (*Note:* For transient motions of this type, the effect of damping is less significant than for steady-state sinusoidal motion.)

2.12 SEISMIC EXCITATION

The characteristics of earthquake shock and other large-scale disturbances can be readily duplicated with the time-stepping scheme. A seismic disturbance may, for example, initially involve high frequency components of low level, gradually changing to moderate frequencies and high levels, and finally returning to low frequencies at low levels. To illustrate such a seismic event, we consider the following characteristics, which correspond to the input stimuli of Fig. 2.2(c):

time (sec)	0·0	0·2	0·4	0·6
frequency (Hz)	20·0	5·0	5·0	0·0
acceleration (g)	0·0	0·6	0·6	0·0

We will study the effect of this disturbance on a dynamic system of natural frequency 4·7 Hz, shown in the inset of Fig. 2.37.

Figure 2.37 shows the response of the system to this stimulus. The maximum response is about four times the ground motion and occurs at about 0·5 sec after the worst of the tremor is over. Damping reduces the peak response only a small amount in going from $c = 0·15$ to 0·30 lb sec/in. The 0·30 damping is about 5 per cent of the critical damping which is given by $2\sqrt{mk} = 2\sqrt{0·1036 \times 90} = 6·1$ lb sec/in. In an actual earthquake there are many more cycles than occur for this simple example. In such cases damping has a greater influence on the peak responses.

Problems

2.40 In the example problem change the frequency history to 20·0, 10·0, 10·0, and 0·0 Hz at times 0·0, 0·2, 0·4, and 0·6 sec. What is the effect on peak response?

2.41 Repeat the example with a history of 20·0, 3·0, 3·0, 0·0 Hz at times 0·0, 0·2, 0·4, 0·6 sec. What is the effect on peak response?

2.42 Repeat the example problem with the times changed to 0·0, 0·2, 0·7, 1·0 sec. Note the increased response due to the increased number of cycles at 5 Hz.

2.43 A certain building installation can be simulated as a 40 000 lb weight on a 90 000 lb/in. spring. Consider the damping to be 6000 lb sec/in. and determine the response to the seismic event in the example problem.

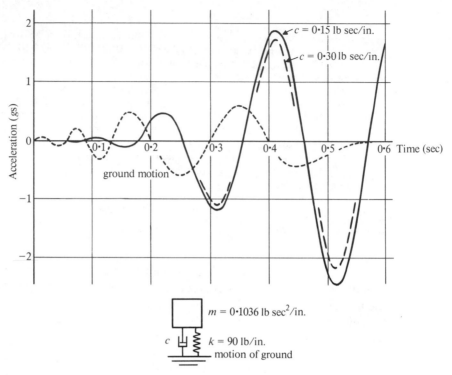

FIGURE 2.37
Response of 4·7 Hz system to seismic event

2.44 Repeat problem 2.43 for the seismic event in problem 2.42. What is the maximum g-level experienced?

2.13 FREQUENCY SWEEP EXCITATION

Specifications for equipment frequently call for a test on a shake table. In such a test the frequency is slowly swept through a frequency range. The g-level is ordinarily low at the two ends of the frequency range and higher at intermediate values of frequency. The reduced values at the low frequency end are usually the result of shake table amplitude limitations to about 0·5 in. travel. The 'roll-off' at high frequencies is due to the drop in excitation at high frequencies for most environments.

As an example of how equipment might respond to a sweep test on a shake table, we consider a piece of equipment weighing 4 lb situated on a support structure with 900 lb/in. stiffness. The damping constant is 0·15 lb sec/in. in one case and 0·30 lb sec/in. in another. We allow the excitation to be an acceleration which varies linearly from $1g$ to $5g$ and the frequency from 5 Hz to 30 Hz in the first 0·5 sec. Up to time 2·0 sec, the g-level holds steady at $5g$ while the frequency rises linearly to 60 Hz. Between 2·0 and 3·0 sec, the g-level drops linearly to $1g$ while the frequency goes linearly to 100 Hz.

The response is plotted in Fig. 2.38. A maximum occurs just below 47 Hz, the undamped resonant frequency. As shown in section 1.4, the magnification factor for constant frequency excitation is $\sqrt{mk}/c$ or $\sqrt{0.01036 \times 900}/0.15 = 20.4$ for the case with $c = 0.15$ and 10.2 for the case with $c = 0.30$. Figure 2.38 shows that during the sweep of frequency the magnifications were nearly these values. They are smaller because the frequency sweep rate is so high. This effect is more evident for the low damping case. Above about 70 Hz, the response is less than the shake table motion. Vibration theory states that this crossover for lightly damped systems should occur at $\sqrt{2}$ times the natural frequency. In this case, that would be 67 Hz.

Problems

2.45 Repeat the example problem with a one octave per second sweep rate by going from 7·5 Hz to 15 Hz in the first second, 15 Hz to 30 Hz in the second, 30 Hz to 60 Hz

FIGURE 2.38
Response to shake-table excitation

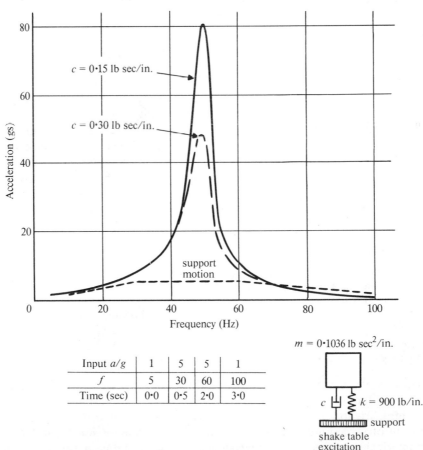

Input a/g	1	5	5	1
f	5	30	60	100
Time (sec)	0·0	0·5	2·0	3·0

$m = 0.1036$ lb sec^2/in.

in the third, and 60 Hz to 120 Hz in the fourth. Let the g-level be $0.31g$ at $t = 0$, $1.25g$ at $t = 1.0$, $5.0g$ at $t = 2.0$, $5.0g$ at $t = 3.0$, and $0.0g$ at $t = 4.0$. This description corresponds to constant displacement amplitude excitation below 30 Hz, and a roll-off to $0g$ between 60 Hz and 120 Hz.

2.46   In some cases at resonance, the forces are severe enough to cause a bolt to lift off its seat. In such a case, the effective spring stiffness drops markedly at lift-off. Rerun the example problem with $c = 0.15$ lb sec/in. and add $k_u = -800$ lb/in., $y_u = 0.222$ in., corresponding to lift-off at $50g$ ($900 \times 0.222/4 = 50g$). The total stiffness after lift-off is $900 - 800 = 100$ lb/in. How does the peak amplitude compare with the value in Fig. 2.38?

2.47   Repeat problem 2.46 for decreasing frequency. (*Hint:* Let frequency be 100, 60, 30, 5 and g-level be 1, 5, 5, 1 for times 0.0, 1.0, 2.5, and 3.0, respectively.) Note the difference in response for increasing and decreasing frequency for this nonlinear system.

2.48   Repeat problems 2.46 and 2.47 with $c = 0.20$. Does the increased damping cause a marked decrease in amplitude for this nonlinear problem?

2.14   SUBHARMONIC RESONANCE

Subharmonic resonance can occur in mechanical systems if the spring force is non-linear. As an example, we will consider a system for which

$$\text{spring force} = 90(y - x) + 10(y - x)^2$$

We take the mass as 0.1036 lb $\text{sec}^2/\text{in.}$, so that the low amplitude natural frequency is 4.7 Hz. The system is excited by a support movement at 9.4 Hz.

The response is shown in Fig. 2.39. The response is quite nonlinear with both $c = 0.0$ and $c = 0.3$; however, it is evident that the 4.7 Hz component is much larger than the 9.4 Hz component. An interesting feature of the response curves is their tendency to treat the $+5g$ level as a barrier. In the spring force sketch, we note that the maximum compressive force developed by the nonlinear spring is 202.5 lb at 4.5 in. compression. Here m represents a weight of 40 lb, so the spring can at most apply $202.5/40 = 5.06g$ of upward acceleration. The damper might modify this value slightly. For the spring elongation, however, the restoring force becomes quite high and results in downward accelerations as high as $15g$.

Problems

2.49   Repeat the example problem with the support g-level increased from $10g$ to $15g$.
2.50   Repeat the example problem with the exciting frequency at three times the low amplitude natural frequency.
2.51   Repeat problems 2.49 and 2.50 with $k_2 = 5$ lb/in. instead of 10 lb/in.

2.15   SUMMARY

This chapter has concerned itself with the single mass system. In general, restoring forces of the system were nonlinear, consisting of nonlinearly hardening or softening springs, friction, or stops. Numerous examples of such systems were studied by means of a computer program given in FORTRAN in section 2.4 of this chapter. The numerical examples were designed to illustrate each feature of the computer program:

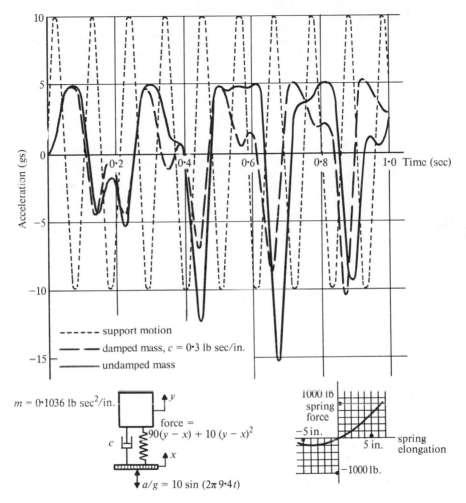

FIGURE 2.39
Response of nonlinear system with spring force = 90 (displacement) + 10 (displacement)^2 to excitation at 9·4 Hz. Low amplitude natural frequency $\omega_0 = 4·7$ Hz. Response shows subharmonic resonance effect

namely, each element (springs of various types, stops, friction), and each type of excitation (forces on the mass, acceleration of the support, and time-varying excitations of one kind or another). In addition, we made comparisons between the numerical results of the computer program and some of the known solutions existing in the literature for nonlinear systems.

We have reached a point where we can inquire into the behavior of a system having more than one mass, each mass interacting with others by means of springs, dampers, friction, and stops. Such multi-mass systems will be discussed in the next chapter, where computer programs suitable for their study will also be given.

2.16 REFERENCES

1. M. V. Farina, *FORTRAN IV Self-taught,* Prentice-Hall, Inc., Englewood Cliffs, New Jersey, 1966.
2. J. T. Golden, *FORTRAN IV Programming and Computing,* Prentice-Hall, Inc., Englewood Cliffs, New Jersey, 1965.
3. L. S. Jacobsen and R. S. Ayre, *Engineering Vibrations,* McGraw-Hill Book Company, Inc., New York, pp. 209–210, 1958.
4. J. Tou and P. M. Schultheiss, 'Static and sliding friction in feedback systems,' *J. Appl. Phys.,* **24,** 1210–1217, 1953.
5. W. J. Cunningham, *Introduction to Nonlinear Analysis,* McGraw-Hill Book Company, Inc., New York, pp. 208–209, 1958.
6. N. W. McLachlan, *Theory of Vibrations,* Dover Publications, Inc., New York, pp. 46–61, 1951.
7. Y. A. Mitropolskii, *Problems of the Asymptotic Theory of Nonstationary Vibrations,* Israel Program for Scientific Translations, Jerusalem, pp. 89–102, 1965.
8. S. Timoshenko and D. H. Young, *Vibration Problems in Engineering,* 3rd edn., D. VanNostrand Company, Inc., Princeton, New Jersey, pp. 134–135, 1955.

THE COMPONENT ELEMENT METHOD— DYNAMICS OF MANY DEGREE-OF-FREEDOM SYSTEMS

3.1 INTRODUCTION

Systems with many masses can be studied in a manner similar to the component element procedure that was applied to the single mass system of the previous chapter. Because of the greater variety that is possible in multi-mass systems, however, it is necessary to introduce a number of concepts which allow a systematic development in a form general enough to be applicable to most systems likely to be encountered. In this chapter, the system dynamics are described in terms of component elements.

   Component elements allow a convenient description of the dynamic system. The system is described by three qualities: its mass or inertial properties; its internal force elements such as springs; and the generalized coordinates that describe the positions of each inertial property, such as displacements, angular motions, or orthogonal modal amplitudes. The inertial property may consist of a point mass, an inertia, or a modal mass. Any one of these inertial properties is given the broader name of generalized mass. Generalized masses are connected to one another by elements, such as springs, dampers, stops, frictional elements, and others which will be encountered, like beam springs. These elements are so chosen that their descriptions allow the setting up of equations of motion for each generalized mass that are uncoupled from one another. The connections between each generalized mass and the component elements that affect it are made through a coupling ratio, the construction of which will be described in this chapter. The manner in which the component elements are chosen and the equations of motion are set up through the use of coupling ratios is given the generic name of the component element method, after which this book is named. The application of the component element method allows the study of the dynamic response of such diverse things as systems with point masses, systems formed of rigid bodies having inertial properties, and systems such as vibrating beams, which consist of distributed masses, and are described through the properties of their modes of free vibration.

Once again, at the heart of the discussion of multi-mass systems are two computer programs. They are described in general terms within the chapter, and are given in sections 3.13 and 3.16 of this chapter, in FORTRAN. One program is specifically designed to calculate the response of systems having two or three mass component elements. The other is more general and can be applied to systems having up to 65 mass component elements. In the latter program, specific attention is paid to how certain quantities are stored and retrieved during the computations, making the program quite efficient from the computational viewpoint. In both programs, provision is made for a substantial number of force component elements (springs, dampers, stops, etc.).

A number of examples are given wherein the responses are calculated with the aid of these programs. Several examples concern the vibrations of beams that are parts of nonlinear systems, illustrating the use of normal modes as coordinates. More complex situations are discussed in chapter 4, which concerns vehicle dynamics.

3.2   EQUATIONS OF MOTION FOR LINEAR MULTI-MASS SYSTEMS—NATURAL FREQUENCIES, NATURAL MODES

Consider the three mass system shown in Fig. 3.1(a). The system consists of the masses m_1, m_2, m_3, connected by springs of stiffness k_1, k_2, k_3. Each mass is acted on by a force F_1, F_2, F_3. By applying Newton's second law to the system, the equations of motion which govern its behavior are given by

$$\left.\begin{array}{l} m_1\ddot{y}_1 + k_1 y_1 - k_2(y_2 - y_1) = F_1 \\ m_2\ddot{y}_2 + k_2(y_2 - y_1) - k_3(y_3 - y_2) = F_2 \\ m_3\ddot{y}_3 + k_3(y_3 - y_2) = F_3 \end{array}\right\} \tag{3.1}$$

Otherwise arranged, this system of equations can be written in matrix form as

$$\begin{bmatrix} m_1 & 0 & 0 \\ 0 & m_2 & 0 \\ 0 & 0 & m_3 \end{bmatrix} \begin{Bmatrix} \ddot{y}_1 \\ \ddot{y}_2 \\ \ddot{y}_3 \end{Bmatrix} + \begin{bmatrix} k_1 + k_2 & -k_2 & 0 \\ -k_2 & k_2 + k_3 & -k_3 \\ 0 & -k_3 & k_3 \end{bmatrix} \begin{Bmatrix} y_1 \\ y_2 \\ y_3 \end{Bmatrix} = \begin{Bmatrix} F_1 \\ F_2 \\ F_3 \end{Bmatrix} \tag{3.2}$$

These equations form a *coupled* system in y. Each displacement y depends on the others. We note that the matrix containing the mass terms is a diagonal matrix,

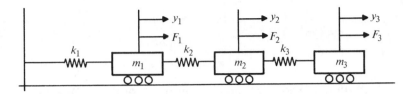

(a) A three mass system

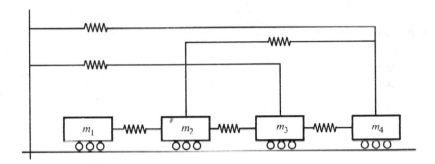

(b) A more generally connected system

FIGURE 3.1
Multi-mass system

and that the matrix containing the spring stiffness terms is symmetric. The symmetry of the stiffness matrix is a result of reciprocity in linear systems.

Springs can extend from any of the masses to any others, as illustrated in Fig. 3.1(b), so that the stiffness matrix is usually full. In general, when a system has n masses, the equations of motion can be written as

$$\left. \begin{array}{l} m_1\ddot{y}_1 + k_{11}y_1 + k_{12}y_2 + \cdots + k_{1n}y_n = F_1 \\ m_2\ddot{y}_2 + k_{21}y_1 + k_{22}y_2 + \cdots + k_{2n}y_n = F_2 \\ \quad\vdots \\ \quad\vdots \\ m_n\ddot{y}_n + k_{n1}y_1 + k_{n2}y_2 + \cdots + k_{nn}y_n = F_n \end{array} \right\} \tag{3.3}$$

When the system of equations is written in matrix notation, the mass matrix will be diagonal, while the stiffness matrix will again be symmetric. Again, these equations are coupled in the displacements y.

To actually solve this system of equations for arbitrary forces is not our present purpose; that will be possible with the computer tools that will be arrayed later in this chapter. We wish at this point simply to inspect the steady-state behavior of this system under a force excitation at a single harmonic frequency ω, which means that

$$\left.\begin{array}{l} F_i = f_i \sin \omega t \\ y_i = a_i \sin \omega t \\ \ddot{y}_i = -\omega^2 a_i \sin \omega t \end{array}\right\} \tag{3.4}$$

Substitution into eq. (3.3) and rearrangement of the terms gives, in matrix notation,

$$\begin{bmatrix} (k_{11} - m_1\omega^2) & k_{12} & \cdots & k_{1n} \\ k_{21} & (k_{22} - m_2\omega^2) & \cdots & k_{2n} \\ \vdots & \vdots & & \\ k_{n1} & k_{n2} & \cdots & (k_{nn} - m_n\omega^2) \end{bmatrix} \begin{bmatrix} a_1 \\ a_2 \\ \vdots \\ a_n \end{bmatrix} = \begin{bmatrix} f_1 \\ f_2 \\ \vdots \\ f_n \end{bmatrix} \tag{3.5}$$

The solution to this system is

$$\begin{Bmatrix} a_1 \\ a_2 \\ \vdots \\ a_n \end{Bmatrix} = [D]^{-1} \begin{Bmatrix} f_1 \\ f_2 \\ \vdots \\ f_n \end{Bmatrix} \tag{3.6}$$

where

$$[D] = \begin{bmatrix} (k_{11} - m_1\omega^2) & k_{12} & \cdots & k_{1n} \\ k_{21} & (k_{22} - m_2\omega^2) & \cdots & k_{2n} \\ \vdots & \vdots & & \\ k_{n1} & k_{n2} & \cdots & (k_{nn} - m_n\omega^2) \end{bmatrix} \tag{3.7}$$

The matrix $[D]$ is known as the frequency matrix. The meaning of this statement becomes clear when it is observed that if the determinant $|D|$ becomes zero, then the amplitudes will become infinite. This situation is reached when the system is excited at a frequency ω which is equal to one of its natural frequencies of free vibration. Indeed, the natural frequencies of the system can be calculated by finding those frequencies at which

$$|D| = 0 \tag{3.8}$$

This equation is known as the frequency equation. For simple systems, it is feasible to evaluate the determinant algebraically. This operation leads to a polynomial in ω^2 which has n real roots. Thus, n natural frequencies are obtained. These roots are known as *characteristic values* or *eigenvalues* of the system of equations. As the number of masses increases, the determination of the roots by a method like a polynomial expansion becomes impractical. Many procedures have been developed to obtain eigenvalues of a large system. Some investigators have used various iterative methods (see, for example, Crandall[1]). We favor a method known as the Jacobi method, which is suitable for use with digital computers, and which will be described subsequently (section 6.5).

The concept of natural frequencies, at which the system has very large amplitudes regardless of the applied forces, leads us to inquire into the behavior of the system when all forces are zero. In that case, the equations of motion become

$$[D] \begin{Bmatrix} a_1 \\ a_2 \\ \vdots \\ a_n \end{Bmatrix} = \{0\} \tag{3.9}$$

These equations have a solution only when the determinant $|D|$ is zero. The solution yields values of the coefficients a_{ij}. These values, each set of which corresponds to one eigenvalue ω_j, are known as the *characteristic vectors*, or *eigenfunctions*, of the system. Each eigenfunction can be multiplied by a common factor and still continue to satisfy eq. (3.9). Multiplication by such a factor is called normalization. Sometimes, the factor is chosen to make the largest value unity. The Jacobi method automatically gives both the eigenvalues and eigenfunctions.

3.3 MATRIX NOTATION

It will be convenient from time to time to refer to the equations of motion of a large system in a shorthand form. To this end, we introduce the matrix notation as follows.

The equations of motion are written as

$$[m]_D\{\ddot{y}\} + [K]\{y\} = \{F(t)\} \tag{3.10}$$

Here $[m]_D$ is a diagonal mass matrix, while $[K]$ is a square symmetric stiffness matrix. If the system is vibrating freely in one of its natural modes, then

$$[m]_D\{\ddot{y}\} + [K]\{y\} = \{0\} \tag{3.11}$$

Letting $\{y\} = \{a_n\} \sin \omega_n t$ represent the nth natural mode, we get

$$[[K] - \omega_n^2[m]_D]\{a_n\} = \{0\} \tag{3.12}$$

A solution exists when the determinant is equal to zero; thus,

$$|[K] - \omega_n^2[m]_D| = 0 \tag{3.13}$$

At the characteristic frequencies ω_n, where this equation is satisfied, there exist also characteristic vectors $\{a_n\}$ which represent the shape of the mode of vibration.

The modes of free vibration of the undamped system are also known as *normal modes*. The reason for this nomenclature is that the modes possess a very important characteristic—they are *orthogonal* to one another. This characteristic is of great consequence in the study of the dynamic response of the system to external forces.

3.4 ORTHOGONALITY

To show what is meant by orthogonality, we proceed as follows. Take any two modes at frequencies ω_i and ω_j. Then, according to the equations of motion,

$$\omega_i^2 [m]_D \{a_i\} = [K]\{a_i\} \tag{3.14a}$$

and

$$\omega_j^2 [m]_D \{a_j\} = [K]\{a_j\} \tag{3.14b}$$

Here, the symbol $[\ ]_D$ stands for a diagonal matrix. Take the transpose (the transpose of a matrix product $[A][B]$ is $[B]^T[A]^T$) of the first equation, and postmultiply by $\{a_j\}$:

$$\omega_i^2 \{a_i\}^T [m]_D^T \{a_j\} = \{a_i\}^T [K]^T \{a_j\} \tag{3.15}$$

Premultiply the second equation by $\{a_i\}^T$:

$$\omega_j^2 \{a_i\}^T [m]_D \{a_j\} = \{a_i\}^T [K]\{a_j\} \tag{3.16}$$

Now since both $[m]_D$ and $[K]$ are symmetric,

$$[m]_D^T = [m]_D$$
$$[K]^T = [K] \tag{3.17}$$

so that, upon subtracting eq. (3.16) from eq. (3.15), we get

$$(\omega_i^2 - \omega_j^2)\{a_i\}^T [m]_D \{a_j\} = 0 \tag{3.18}$$

But since $\omega_i \neq \omega_j$

$$\{a_i\}^T [m]_D \{a_j\} = 0 \quad \text{and} \quad \{a_i\}^T [K]\{a_j\} = 0 \tag{3.19}$$

This is the condition of orthogonality. Its importance will become apparent as we proceed to look at how the modes of vibration are related to one another and to the dynamic response as a whole.

Equation (3.18) is satisfied identically when $i = j$. In that case,

$$\{a_i\}^T [m]_D \{a_i\} \neq 0 \tag{3.20}$$

Upon multiplying out this number, we get

$$\{a_i\}^T [m]_D \{a_i\} = a_{1i}^2 m_1 + a_{2i}^2 m_2 + \cdots + a_{ni}^2 m_n$$

$$= \sum_{r=1}^{n} m_r a_{ri}^2 \tag{3.21a}$$

$$= M_i$$

By premultiplying eq. (3.14a) by $\{a_i\}^T$ and using eq. (3.21a), we get

$$\{a_i\}^T [K]\{a_i\} = \omega_i^2 M_i \tag{3.21b}$$

The quantity in eq. (3.21a), as will soon become clear, is known as the modal mass of the ith mode. It is formed as the sum of the products of each mass and the square of the magnitude of the eigenfunction at that mass point. The quantity in eq. (3.21b) is known as the modal spring stiffness of the ith mode.

The magnitude of the modal amplitudes is ordinarily adjusted so that the modal mass is either unity or, in some cases, the actual mass of the system. This adjustment is known as normalizing the mode.

3.5   DYNAMIC RESPONSE—THE MODAL EQUATIONS

The equation of motion when external forces act on the system is

$$[m]_D\{\ddot{y}\} + [K]\{y\} = \{F(t)\} \tag{3.22}$$

The response y can be found in terms of a linear combination of the eigenfunctions, or characteristic vectors. This fact arises from the property of orthogonality of the eigenfunctions. We write

$$\{y\} = \{a_1\}A_1(t) + \{a_2\}A_2(t) + \cdots + \{a_n\}A_n(t) \tag{3.23}$$

where $a_1, a_2, \ldots, a_n$ represent the characteristic vectors of the $1, 2, \ldots, n$th mode of free vibration. $A_1, A_2, \ldots, A_n$ are the modal amplitudes, the time-varying coefficients which determine how the normal modes must be added to determine the total response of the system. Thus,

$$\{y\} = [\{a_1\}\{a_2\}\ldots\{a_n\}]\begin{Bmatrix} A_1 \\ A_2 \\ \vdots \\ A_n \end{Bmatrix}$$

$$= [a]\{A\} \tag{3.24}$$

Substituting into the equations of motion, we get

$$[m]_D[a]\{\ddot{A}\} + [K][a]\{A\} = \{F(t)\} \tag{3.25}$$

Premultiplying by the transpose of $[a]$,

$$[a]^T[m]_D[a]\{\ddot{A}\} + [a]^T[K][a]\{A\} = [a]^T\{F(t)\} \tag{3.26}$$

However, by using the orthogonality condition

$$\begin{aligned} \{a_i\}^T[m]_D\{a_j\} &= 0 &&(i \neq j) \\ &= M_i &&(i = j) \end{aligned} \Bigg\} \tag{3.27}$$

it can be shown that

$$[a]^T[m]_D[a] = [M]_D \tag{3.28}$$

And since the equations of free vibration state that

$$[K]\{a_i\} = \omega_i^2[m]_D\{a_i\} \tag{3.29}$$

then, also,

$$[K][a] = [m]_D[a][\omega^2]_D \tag{3.30}$$

Premultiplying by $[a]^T$:

$$[a]^T[K][a] = [a]^T[m]_D[a][\omega^2]_D$$
$$= [M]_D[\omega^2]_D$$
$$= [M\omega^2]_D \tag{3.31}$$

Thus, upon substitution of eqs. (3.28) and (3.31) into the modified equation, we finally obtain

$$[M]_D\{\ddot{A}\} + [M\omega^2]_D\{A\} = [a]^T\{F(t)\} \tag{3.32}$$

Writing this equation out in a fuller notation for the ith mode, we find that

$$M_i\ddot{A}_i + M_i\omega_i^2 A_i = \{a_i\}^T\{F\} \tag{3.33}$$

This equation is an *uncoupled* equation in A_i. This fact is very important in structural dynamics, since it allows one to solve for each modal amplitude independently of the others. This situation comes about solely because the modes of vibration of a linear structure are orthogonal. The equation is a second-order differential equation in time. Its form is identical to the differential equation for a basic single mass system. It was for this reason that, by analogy with the single mass system, we called M_i the modal mass of the ith mode. Again, by analogy, we call $M_i\omega_i^2$ the modal stiffness, or modal spring. And we denote $\{a_i\}^T\{F\}$ as the modal force. In full,

$$\{a_i\}^T\{F\} = a_{1i}F_1 + a_{2i}F_2 + a_{3i}F_3 + \cdots + a_{ni}F_n \tag{3.34}$$

Thus, the modal force in the ith mode is the sum of the products of the force at each mass and the magnitude of the eigenfunction or characteristic vector at that mass.

3.6   SYSTEM COORDINATES

Thus far, we have inspected systems that have been described by rectangular coordinates. Consider further a number of particles. To describe their positions at any time, it is necessary to give three coordinates x, y, z in a rectangular coordinate system. Thus, for n particles, there are $3n$ quantities which must be given to fully describe the position of the particles. We say that the system has $3n$ *degrees of freedom*. If, now, the n particles are restrained in some manner, let us say, for example, that they are constrained to move only in the x–y plane, then fewer coordinates are required to describe their position—only $2n$ in this case, because each particle has one constraint which precludes motion in the z-direction. In general, if there are k constraints in a system (let us imagine, for example, that k of the particles are rigidly connected), then we only require $3n - k$ coordinates to fully describe the system's position, so that the system has $3n - k$ degrees of freedom.

Consider a double pendulum, shown in Fig. 3.2. There are two masses, so in rectangular coordinates its position can be described by the four coordinates $x_1, x_2,$

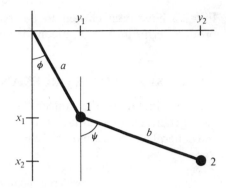

FIGURE 3.2
Double pendulum system showing generalized coordinates ϕ and ψ

y_1, and y_2. However, because the masses are joined by rigid links of length a and b, these provide constraints in the form

$$\left.\begin{array}{l} a^2 = x_1^2 + y_1^2 \\ b^2 = (x_2 - x_1)^2 + (y_2 - y_1)^2 \end{array}\right\} \tag{3.35}$$

Thus, in reality, there are only two degrees of freedom for this system.

The angular displacements ϕ and ψ can be used to describe the system. These coordinates are related geometrically to the rectangular coordinates. Coordinates such as these, which are independent of one another, are known as *generalized coordinates*. Generalized coordinates are not necessarily dimensions of length. They may be angular coordinates, as they are here, or they may have other descriptive properties. They play an important role in the development of structural dynamics.

We have spoken of the normal modes of a system. Upon reflection, it will be seen that the amplitudes in these normal modes can themselves be considered as descriptive of the system position. Indeed, the amplitudes in normal modes can be used as coordinates and, by doing so, many interesting structural problems can be solved in a relatively simple manner. Normal modes satisfy the important additional condition that they are neither spring- nor mass-coupled in the dynamic equations, because this is precisely what the orthogonality condition means. The choice of coordinates is frequently based on avoiding either mass or spring coupling. For example, when one chooses the displacements at the center of gravity of a mass and the rotations about the center of gravity as coordinates, the purpose is to avoid mass coupling.

Examples of systems of two and three degrees of freedom are shown in Fig. 3.3. First, in Fig. 3.3(a) and 3.3(d), the degrees of freedom are taken as displacements of the masses. In Fig. 3.3(b) and 3.3(e), the degrees of freedom are taken as displacements of masses or rotations of masses about the center of gravity. Finally, in Fig. 3.3(c) and 3.3(f), one degree of freedom is a displacement; in Fig. 3.3(f), one is a rotation; and in Fig. 3.3(c) and 3.3(f), one is a mode of vibration, which in this case is the lowest mode of free vibration of the beam. All the coordinates in

Fig. 3.3 have been chosen to be free of mass coupling, i.e., there are no cross-product terms in the expression for kinetic energy.

3.7   GENERALIZED FORCES AND COUPLING RATIOS

We now introduce the concept of a generalized force, which is associated with each generalized coordinate. If the external forces on a body are known, then, once the generalized coordinate z is chosen, the generalized force Q is defined as follows.

FIGURE 3.3
Systems with two and three degrees of freedom (these generalized coordinates have no mass coupling with each other)

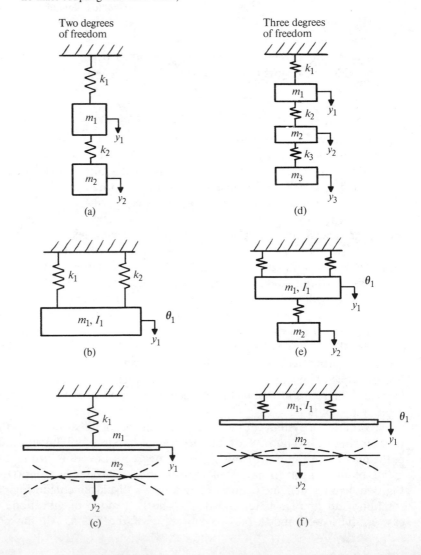

Upon making a small increment δz in the generalized coordinate, while keeping all other coordinates constant, the work done by the external forces is $Q\,\delta z$. Thus, for example, the systems shown in Fig. 3.4 have the following generalized forces.

(a) For a generalized coordinate y_1 representing a linear displacement, Fig. 3.4(a), the generalized force Q_1 is the sum of the components of the actual forces in the direction of the displacement.

FIGURE 3.4
Examples of generalized forces

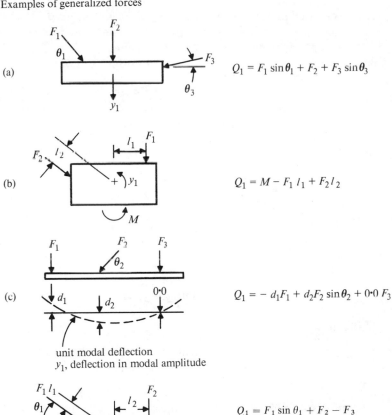

(a) $Q_1 = F_1 \sin\theta_1 + F_2 + F_3 \sin\theta_3$

(b) $Q_1 = M - F_1\,l_1 + F_2\,l_2$

(c) $Q_1 = -\,d_1 F_1 + d_2 F_2 \sin\theta_2 + 0{\cdot}0\,F_3$

unit modal deflection
y_1, deflection in modal amplitude

(d) $Q_1 = F_1 \sin\theta_1 + F_2 - F_3$
       $Q_2 = F_1 l_1 - F_2 l_2 - F_3 l_3$

y — generalized displacements
Q — generalized forces
F — actual forces
M — actual moments
d — modal deflection at station
l — coupling ratio

(b) For a generalized coordinate y_1 which is a rotation, Fig. 3.4(b), the generalized force Q_1 is the moment.

(c) For a generalized coordinate y_1 which is the amplitude in a mode of vibration, Fig. 3.4(c), the generalized force Q_1 is the sum of the components of the actual forces in the direction of the modal displacement multiplied by the modal displacement.

(d) An actual body may displace linearly y_1 and rotate y_2, Fig. 3.4(d). The actual forces will contribute to each of the corresponding generalized forces Q_1 and Q_2.

An important characteristic of relationships like those in Fig. 3.4 is that the coefficients used in going from actual forces to generalized forces are identical with those used in going from generalized coordinates to actual displacements. The coefficients are referred to in this book as *coupling ratios*. This situation is a consequence of the principle of virtual work. Thus, in Fig. 3.4(a):

$$\text{displacement of } F_1 = y_1 \sin \theta_1$$
$$\text{displacement of } F_2 = y_1$$
$$\text{displacement of } F_3 = y_1 \sin \theta_3$$

Similarly, in Fig. 3.4(b):

$$\text{displacement of } F_1 = -l_1 y_1$$
$$\text{displacement of } F_2 = l_2 y_1$$
$$\text{rotation of } M \quad = y_1$$

In Fig. 3.4(c):

$$\text{displacement of } F_1 = -d_1 y_1$$
$$\text{displacement of } F_2 = d_2 y_1 \sin \theta_2$$
$$\text{displacement of } F_3 = 0 \cdot 0 \; y_1$$

Finally, in Fig. 3.4(d):

$$\text{displacement of } F_1 = y_1 \sin \theta_1 + y_2 l_1$$
$$\text{displacement of } F_2 = y_1 - l_2 y_2$$
$$\text{displacement of } F_3 = -y_1 - y_2 l_3$$

Note the correspondence between these equations and the coefficients shown in Fig. 3.4.

Coupling ratio rule

The coupling ratio between a force and a generalized coordinate is the displacement of the force due to unit displacement in the generalized coordinate.

For efficient computation, it is desirable to treat masses, moments of inertia, and modal response intermixed. We will, therefore, find it convenient to have a program which is written in terms of generalized coordinates on the inside, but may look like ordinary coordinates on the outside.

3.8   GENERALIZED MASS

Another useful concept is the generalized mass. It is determined by the kinetic energy of that part of the system governed by each generalized coordinate. Thus, the generalized mass times the generalized velocity squared is equal to the actual mass times its velocity squared.

For example, for a generalized coordinate which is the linear displacement of the center of gravity of a body, the generalized mass is identical with the actual mass. For a generalized coordinate which is a rotation about a principal axis through the center of gravity, the generalized mass is the corresponding moment of inertia.

In the case when modal motion of a body is used as a generalized coordinate, the generalized mass is the integral of the distributed mass times the square of the modal deflection. Thus, if the nth mode shape is described by y_n, the generalized mass corresponding to that mode shape is

$$m_n = \int_{\text{vol}} m y_n^2 \, dv \tag{3.36}$$

where m is the mass per unit volume. (This quantity will be recognized as the modal mass described previously.) The concept of a generalized mass in the case of a modal motion is so useful that we will find it valuable to enlarge on it somewhat in the case of beams. By this means, it will be shown how beam vibrations can be included in the study of the motion of systems with two or three degrees of freedom.

3.9   VIBRATIONS OF BEAMS

A beam of elastic modulus E and sectional inertia I is shown in Fig. 3.5. The moment M in the beam is given by the well-known differential equation

$$EI \frac{d^2 y}{dx^2} = -M \tag{3.37}$$

By successively differentiating with respect to x, we find

$$EI \frac{d^3 y}{dx^3} = -\frac{dM}{dx} = -Q \tag{3.38a}$$

$$EI \frac{d^4 y}{dx^4} = -\frac{dQ}{dx} = f(x) \tag{3.38b}$$

Here, Q is the shear force, and f is the applied distributed external load per unit length. If the beam is in motion, then f is equated to the inertia force, so that

$$f(x) = -\frac{A}{g} \rho \frac{\partial^2 y}{\partial t^2} \tag{3.39}$$

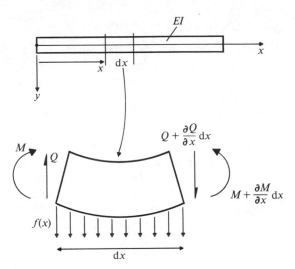

FIGURE 3.5
Forces acting in a beam

where A is the cross-sectional area and ρ is the weight per unit volume. Thus, the differential equation of motion is

$$\frac{d^2 y}{dt^2} + b^2 \frac{d^4 y}{dx^4} = 0 \qquad (3.40)$$

where $b^2 = EIg/A\rho$.

It can be shown by direct substitution that, in the case of harmonic motion, the general solution to this fourth-order differential equation is

$$y = (A_1 \cos \kappa x + B_1 \sin \kappa x + C_1 \operatorname{ch} \kappa x + D_1 \operatorname{sh} \kappa x)\, e^{i\omega t} \qquad (3.41)$$

where $\kappa^4 = \omega^2/b^2 = \omega^2 \rho A/EIg$, and ω is the frequency of vibration in rad/sec.

Another possibly more convenient form of this solution is

$$y = [A_2(\cos \kappa x + \operatorname{ch} \kappa x) + B_2(\cos \kappa x - \operatorname{ch} \kappa x) + C_2(\sin \kappa x + \operatorname{sh} \kappa x)$$

$$+ D_2(\sin \kappa x - \operatorname{sh} \kappa x)]\, e^{i\omega t} \qquad (3.42)$$

This form is useful because two terms usually vanish for most of the common conditions of end fixity encountered in beams. The natural frequencies of the beam are found by determining the value ω for which the constants A_2, B_2, C_2, D_2 can differ from zero and satisfy the beam end conditions.

3.10 SIMPLY SUPPORTED BEAM

In the case of a simply supported beam, the displacements and bending moments at each end are zero. The end conditions are thus

$$y = 0$$
$$\left.\frac{d^2y}{dx^2} = 0\right\} \text{at } x = 0 \text{ and } x = l \qquad (3.43)$$

Substituting the general solution of eq. (3.42) into these conditions, we find that at $x = 0$, $A_2 = 0$ and $B_2 = 0$. Similarly, at $x = l$,

$$\left. \begin{array}{l} C_2(\sin \kappa l + \text{sh } \kappa l) + D_2(\sin \kappa l - \text{sh } \kappa l) = 0 \\ C_2(-\sin \kappa l + \text{sh } \kappa l) + D_2(-\sin \kappa l - \text{sh } \kappa l) = 0 \end{array} \right\} \qquad (3.44)$$

from which

$$C_2 = D_2$$

and

$$2C_2 \sin \kappa l = 0$$

However, $C_2 \neq 0$, so that for the end conditions to be satisfied, we demand that

$$\sin \kappa l = 0 \qquad (3.45)$$

This condition is satisfied only at certain values of ω—the natural frequencies of the system. The condition is, in fact, the frequency equation of the system. The natural frequencies are given by

$$\omega_n = \frac{n^2 \pi^2}{l^2} b \quad (n = 1, 2, 3, \ldots) \qquad (3.46)$$

Since C_2 is arbitrary in free oscillation, we choose to normalize the modal deflection so that the generalized mass is equal to the actual mass of the beam (as will be shown in a moment) The modal deflections are given by

$$y_n - \sqrt{2} \sin \kappa x = \sqrt{2} \sin \frac{n\pi x}{l} \qquad (3.47)$$

The first three of these mode shapes are sketched in Fig. 3.6.
      For the lowest mode, the generalized mass is, using eq. (3.47),

$$m_1 = \frac{A\rho}{g} \int_0^l y_1^2 \, dx$$

$$= 2 \frac{A}{g} \rho \int_0^l \left(\sin \frac{\pi x}{l}\right)^2 dx$$

$$= \frac{A\rho l}{g} = \text{total mass of beam} = m \qquad (3.48)$$

Similarly, the generalized mass can be shown to be m for the other modes.

3.11   SOLUTIONS FOR OTHER CONDITIONS OF END FIXITY

By following the steps described for the simply supported beam, other beams with various conditions of end fixity can also be studied. In Table 3.1, we summarize

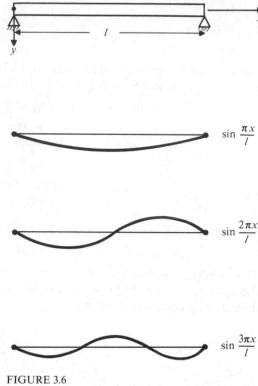

FIGURE 3.6
First three natural mode shapes of a simply supported beam

the results obtainable for a number of conditions, namely, for free beams, clamped beams, and those with combinations of such support conditions.[2,3] The table gives the imposed boundary conditions, the resulting frequency equation together with the two lowest eigenvalues $\kappa_n l$. These results are useful in studying the dynamics of systems involving beams, as will be illustrated shortly.

Concerning the generalized mass, it may be shown that for all the modal deflections shown in Table 3.1, it is equal to the mass m of the beam. This relationship follows directly from integration for the given modes, since

$$\int_0^l y^2 \, dx = l \tag{3.49}$$

3.12   THE COMPONENT ELEMENT METHOD— A COMPUTER APPROACH

We propose in this section to gather together the concepts we have discussed previously and unify them so that they can be used in a logical fashion in a computing program.

Table 3.1   MODAL CHARACTERISTICS OF BEAMS

Supports	Boundary conditions	Frequency equation	κl $n = 1$	κl $n = 2$	Modal deflection in lowest mode
Free–free	at $x = 0$ and $x = l$ $y'' = 0,\ y''' = 0$	$\cos \kappa l\ \mathrm{ch}\ \kappa l = 1$	0	4·730	$n = 1$: rigid body motion for $n = 2$: $\cos \kappa x + \mathrm{ch}\ \kappa x$ $-0·9825\ (\sin \kappa x + \mathrm{sh}\ \kappa x)$
Clamped–free	$x = 0 \begin{cases} y = 0 \\ y'' = 0 \end{cases}$ $x = l \begin{cases} y'' = 0 \\ y''' = 0 \end{cases}$	$\cos \kappa l\ \mathrm{ch}\ \kappa l = -1$	1·875	4·694	$\cos \kappa x - \mathrm{ch}\ \kappa x$ $-0·7341\ (\sin \kappa x - \mathrm{sh}\ \kappa x)$
Simply supported	at $x = 0$ and $x = l$ $y = 0,\ y'' = 0$	$\sin \kappa l = 0$	3·142	6·283	$1·414 \sin \kappa x$
Clamped–supported	$x = 0 \begin{cases} y = 0 \\ y' = 0 \end{cases}$ $x = l \begin{cases} y = 0 \\ y'' = 0 \end{cases}$	$\tan \kappa l - \mathrm{th}\ \kappa l = 0$	3·927	7·069	$\cos \kappa x - \mathrm{ch}\ \kappa x$ $-1·0008\ (\sin \kappa x - \mathrm{sh}\ \kappa x)$
Clamped–clamped	at $x = 0$ and $x = l$ $y = 0,\ y' = 0$	$\cos \kappa l\ \mathrm{ch}\ \kappa l = 1$	4·730	7·853	$\cos \kappa x - \mathrm{ch}\ \kappa x$ $-0·9825\ (\sin \kappa x - \mathrm{sh}\ \kappa x)$

System representation and assembly

We seek to describe a system of three uncoupled generalized masses. The generalized masses can, as we saw, be either point masses in the case of linear displacements of a center of gravity, or a mass moment of inertia in the case of a rotational motion, or a modal mass in the case of a vibratory motion. It is with these three categories that we shall concern ourselves in the computer program.

The equations of motion for a system consisting of such generalized masses can be written as

$$m_i \ddot{z}_i = Q_i \quad (i = 1, \ldots, N) \tag{3.50}$$

where there are N masses m_i, whose generalized coordinates are z_i. The generalized forces Q_i act on these masses. The generalized forces include not only external forces but also those due to friction, springs, stops, etc. The generalized forces are given by the summation

$$Q_i = \sum_r F_r B_{i,r} \tag{3.51}$$

Here, F_r are the forces that act on the generalized mass. The coefficients $B_{i,r}$ are the coupling ratios. They are the deflections at r in the direction of F_r for a unit value of the ith generalized coordinate. Examples of these coupling ratios were given in section 3.7.

To clarify these concepts, let us inspect the structure of several three mass systems and define for each one the coupling ratios $B_{i,r}$. First consider a system of three point masses, shown in Fig. 3.7. The masses are connected by a number of force elements denoted by s_j in the figure. The force elements can be connected to a ground support, as shown in the case of s_1 and s_2, or between masses, as shown in the case of s_3, s_5. The force elements may be of several types: spring–damper elements, stop elements, or friction elements. These equivalents are shown in Fig. 3.8. The forces in each force element are denoted by F_r, and it is assumed for the moment that they are known at any time. Writing the equation of motion for mass m_1, for example, we have

$$m_1\ddot{z}_1 = -F_2 - F_3 - F_4 - F_6 \tag{3.52}$$

The convention used here is that the forces F_r are positive if the force in the element is compressive. Thus, for a spring, $F = -kx$ where k is the spring constant and x is the spring extension. Similar equations can be written for the other masses:

$$\left.\begin{array}{l} m_2\ddot{z}_2 = -F_1 + F_3 + F_4 - F_5 \\ m_3\ddot{z}_3 = +F_5 + F_6 \end{array}\right\} \tag{3.53}$$

As a result, we observe that the coupling ratios can be written in matrix form as

$$[B] = \begin{bmatrix} 0 & -1 & -1 & -1 & 0 & -1 \\ -1 & 0 & 1 & 1 & -1 & 0 \\ 0 & 0 & 0 & 0 & 1 & 1 \end{bmatrix} \tag{3.54}$$

Note that the coupling ratio is $(+)$ when the displacement of the mass lengthens the element, and it is $(-)$ when the displacement of the mass shortens the element.

Now consider the three mass system shown in Fig. 3.9. It consists of a rigid beam of mass m_1 whose center of gravity translates vertically, and a mass moment of inertia m_2 which rotates about the beam center of gravity. A third mass m_3 rides atop the beam and is connected to it by a force element at a distance l_3 away from the center of gravity of the beam. Three force elements connect the beam to the

FIGURE 3.7
Three mass system with connecting force elements

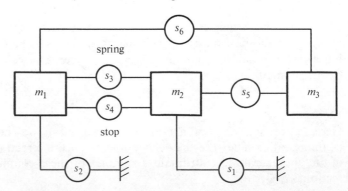

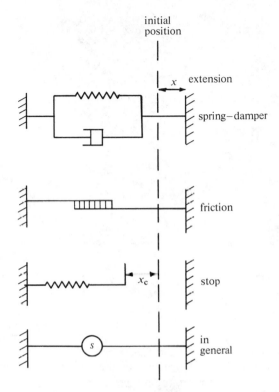

FIGURE 3.8
Force element extension for spring-damper, friction
element, stop, or general element

ground. Their lines of action are located at distances l_1, l_2 and l_4 from the center
of gravity of the beam.

The equations of motion of the system can be written as

$$m_i \ddot{z}_i = Q_i = \sum_r F_r B_{i,r} \tag{3.55}$$

Here, the matrix of coupling ratios is given by

$$[B] = \begin{bmatrix} 1 & 1 & -1 & 1 \\ l_1 & -l_2 & -l_3 & -l_4 \\ 0 & 0 & 1 & 0 \end{bmatrix} \tag{3.56}$$

If the external force F_{ext} were to act on the system at a location l_f, as shown,
then

$$[B] = \begin{bmatrix} 1 & 1 & -1 & 1 & 1 \\ l_1 & -l_2 & -l_3 & -l_4 & -l_f \\ 0 & 0 & 1 & 0 & 0 \end{bmatrix} \tag{3.57}$$

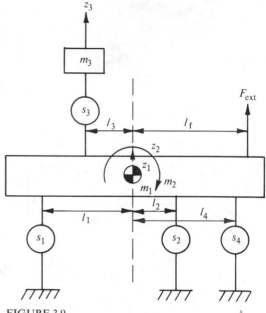

FIGURE 3.9
System with translational and rotational coordinates

The last column indicates the action of the force F_{ext}. In the computer program of the next section, the matrix elements are read directly for the force elements, but for simplicity they are not read for the applied forces. Instead, the actual external force acting on each generalized coordinate is read as input. Thus, a force F_{ext} acts on the beam mass translation, while a generalized force (actually a moment) of $-l_f F_{ext}$ acts on the beam rotational coordinate.

When a flexible beam is part of the system, it too can be modeled by its modal displacements by using them as generalized coordinates. Thus, in the system shown in Fig. 3.10(a), we have a flexible beam held by two springs to the ground. One additional spring element must be included to represent the modal spring, $m\omega^2$. Its coupling ratio is $1\cdot0$ to the modal generalized coordinates and zero to all others. The three generalized masses are then the beam mass m_1 in translation, the mass moment of inertia m_2 in rotation, and the modal mass m_3 in the first vibrational mode. In this case, the coupling ratios for the first two masses are as given in eq. (3.56) for s_1 and s_2. For the modal forces the equation of motion is

$$m_3\ddot{z}_3 = F_{ext}d_f - F_1d_1 - F_2d_2 + F_3 \qquad (3.58)$$

where F_3 is the internal modal force corresponding to s_3, the modal spring is $m\omega^2$, and d_1 and d_2 represent the shortening of s_1 and s_2 due to unit modal displacement. Thus,

$$[B] = \begin{bmatrix} 1 & 1 & 0 & 1 \\ l_1 & -l_2 & 0 & -l_f \\ -d_1 & -d_2 & 1 & d_f \end{bmatrix} \qquad (3.59)$$

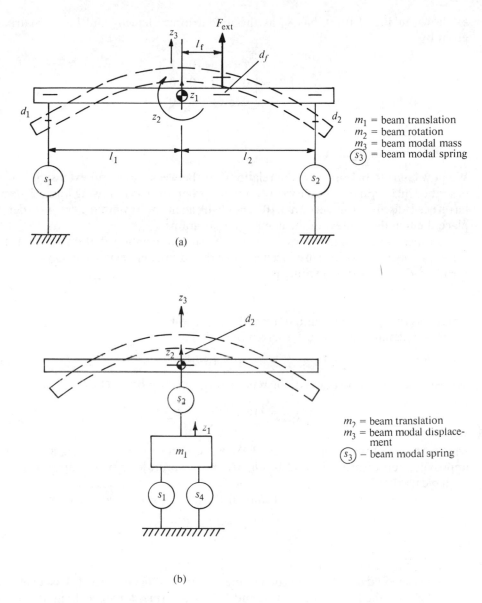

FIGURE 3.10
Vibrating beam systems

The last column again indicates the effect of the external force F_{ext}. In the computer program to be described in the next section, the force input is direct. Thus, F_{ext} acts on the beam mass translation, a moment $-l_f F_{\text{ext}}$ acts on the beam rotation, and a generalized force $d_f F_{\text{ext}}$ acts on the first vibratory mode of the beam.

In a similar vein, a system with a mass and vibrating beam mounted centrally, as shown in Fig. 3.10(b), has s_3 as the modal force element and has a matrix B given by

$$[B] = \begin{bmatrix} 1 & -1 & 0 & 1 \\ 0 & 1 & 0 & 0 \\ 0 & d_2 & 1 & 0 \end{bmatrix} \tag{3.60}$$

Force element description

We now come to discuss how the relationships between force and extension can be described in a manner suitable for a computer program. The following force elements have been discussed in section 1.10. They will again be reviewed here in order to place them in the context of the computing program.

In the case of a linear spring, we use the sign convention that positive force F is compressive and positive displacement x of the spring is an extension. On this basis, the force F generated in a spring is

$$F = -kx \tag{3.61}$$

where k is the spring constant and x is the extension.

For a damper, the force F is given by

$$F = -c\dot{x} \tag{3.62}$$

We replace $\dot{x}$ by the backward difference approximation of eq. (1.80):

$$\dot{x}_0 = \frac{(3x_0 - 4x_{-1} + x_{-2})}{2\Delta t} \tag{3.63}$$

where Δt is the time-stepping interval. As we saw in section 1.8, this is a difference approximation of order h^2 and is slightly more accurate than the approximation we have used previously.

Thus, we arrive at the relationship for the force F generated in a damper:

$$F = \frac{-c(3x_0 - 4x_{-1} + x_{-2})}{2\Delta t} \tag{3.64}$$

where c is the damping coefficient.

The frictional force F is opposed to the relative sliding velocity. It has a value of μN (where μ is the friction coefficient and N is the normal force), as long as there is a substantial sliding displacement. The frictional force is controlled by a local stiffness k_f when the sliding displacement is small. Let the position of the element when the last sliding took place be d_0. When the displacement d_n is enough to make $k_f(d_n - d_0) > \mu N$, then the frictional force is

$$F = -\mu N, \text{ if } k_f(d_n - d_0) > \mu N \tag{3.65}$$

and the new value of d_0 is $d_1 = d_n - \mu N/k_f$. When $0 < k_f(d_n - d_0) < \mu N$, however, the frictional force is

$$F = -k_f(d_n - d_0), \text{ if } 0 < k_f(d_n - d_0) < \mu N \qquad (3.66)$$

and d_0 remains unchanged. When $d_n < d_0$, the sequence is reversed.

The characteristic of a stop is to allow a free motion within the clearance, and beyond that point, to build up a restraining force given by

$$\left.\begin{array}{ll} F = -k_s(x + x_c), & x < (-x_c) \\ F = 0, & x > (-x_c) \end{array}\right\} \qquad (3.67)$$

Here k_s is the spring constant for closure in excess of the clearance x_c.

In the case of modal deformation, the equation of motion is

$$m_n\ddot{z}_n + m_n\omega_n^2 z_n = Q_n \qquad (3.68)$$

Here, z_n is the modal amplitude, m_n is the modal mass, ω_n is the modal natural frequency, and Q_n is the generalized force. The term $m_n\omega_n^2$ is analogous to a spring constant, and the modal deformation force, which acts on the generalized mass m_n, is given by

$$F = -m_n\omega_n^2 z_n \qquad (3.69)$$

Time-stepping

The equations of motion of a system have now been arranged into a form

$$m_i\ddot{z}_i = Q_i; \quad \ddot{z}_i = Q_i/m_i \quad (i = 1, \ldots, N) \qquad (3.70)$$

where Q_i is the generalized force for coordinate i. A finite difference equation can now be written as

$$\frac{(z_{-1} - 2z_0 + z_1)}{(\Delta t)^2} = \ddot{z}_0 \qquad (3.71)$$

A subscript i pertains to all these quantities, relating them to the ith generalized mass. Solving for the new displacement z_1 in terms of the previous displacement z_{-1} and present displacement z_0, we have

$$z_1 = 2z_0 - z_{-1} + \ddot{z}_0(\Delta t)^2 \qquad (3.72)$$

Now, we recall that for the force elements, Q_0 is a function only of the present and previous element extensions, so z_1 can be determined from Q_0, z_0, and z_{-1}.

The initial conditions are accounted for by setting, for the initial time interval,

$$\left.\begin{array}{l} z_1 = z_0 + \dot{z}_0\Delta t + \frac{1}{2}\ddot{z}_0(\Delta t)^2 \\ z_0 = z_0 \\ z_{-1} = z_0 - \dot{z}_0\Delta t + \frac{1}{2}\ddot{z}_0(\Delta t)^2 \\ z_{-2} = 2z_{-1} - z_0 + \ddot{z}_0(\Delta t)^2 \end{array}\right\} \qquad (3.73)$$

The latter condition arises from the assumption that the acceleration prior to time t_0 is $\ddot{z}_0$, so that

$$\frac{(z_0 - 2z_{-1} + z_{-2})}{(\Delta t)^2} = \ddot{z}_0 \qquad (3.74)$$

It is now a simple matter to step forward in time by sequentially calculating new values of z_1 at successive time steps by keeping track of the previous displacements z_0, z_{-1}, and z_{-2}. This calculation and storage is easily done for a system of many degrees of freedom. In the next section, a computer program for three degrees of freedom will be presented, while in section 3.16 one will be given that is capable of computing the response of a system having up to 65 generalized masses.

3.13 A COMPUTER PROGRAM FOR THE DYNAMIC RESPONSE OF TWO AND THREE DEGREE-OF-FREEDOM SYSTEMS

In this section, a computer program is described whose purpose is to calculate the response of two or three degree-of-freedom systems to external stimuli. These stimuli may take the form of excitation forces applied to either the masses or to the system supports. This program is written in the FORTRAN language.

The principal inputs required for the program concern the generalized masses, the specification of the restraining elements (spring–dampers, friction elements, and stops), and the description of the excitation stimuli. A list of the principal input variables used in the program is given in Table 3.2, together with their meanings in mathematical or physical terms. The order in which these quantities are read into the program and the format required is shown in Table 3.3. It should be noted that the system of units used in the program is fixed. It is in pound, second, and inch units, except for acceleration, which is given in gs. Other systems of units, for example the International System, can be accommodated easily by changing the numerical value of g (386 in./sec^2) used in the program to the appropriate value compatible with that system. Thus for the newton, second, metre system, the value of g should be 9·807 m/s^2.

Table 3.2 DESCRIPTION OF INPUT VARIABLES

Variable	Description
IDONE	0 = do another problem
	1 = stop
NEWLDS	0 = new problem has new force elements
	1 = change only force and initial conditions
NZ	Number of generalized coordinates
MSV	Number of points in the displacement input of SV, up to 8; if MSV < 2, SV is not applied
MFV	Number of points in the force input of FV, up to 8; if MSV < 2, FV is not applied
MSHAKE	A value other than 0 means excitation is by a shaker or shock machine
ITYPE	0 = shaker sinusoidal excitation with variable amplitude and frequency
	1 = shock machine pulse excitation, time history
ISINEF	If not 0, sinusoidal forces are input
ISINES	If not 0, sinusoidal support displacements are input
NRAND	If not 0, excitation of support is 'random'
DEL	Time step, Δt
TSTP	Time step for printing output
TTL	Total time of analysis
ZM	Generalized masses

Table 3.2—*continued*

Variable	Description
INDX	1 = spring damper elements follow 2 = friction elements follow 3 = stop elements follow 4 = no further elements follow
NG	Number of masses on which a particular force element acts. (For example, for a spring force element between two bodies, NG = 2, since the spring applies force to both. If one body can rotate, NG = 3, since the spring can apply force on its rotational mass.)
NC	Numbers identifying the masses acted on by a particular force element (zero for any values not used)
C1, C2	For a damper—C1 is damping, C2 is the spring rate For a friction element—C1 is the friction force, C2 is the local spring stiffness For a stop—C1 is the clearance, C2 is the spring rate
CC	Coupling ratios, or multipliers, to obtain generalized forces on the corresponding masses in NC; force in a force element is positive when it is compressive
Z	Initial displacements of masses
ZDOT	Initial velocities of masses
FV	Time history forces on masses, up to 8 points
SV	Time history displacements on the first three spring–damper supports, up to 8 points
FSIN	Amplitude and frequency of sinusoidal forces applied to masses
SSIN	Amplitude and frequency of sinusoidal displacements applied to the first three spring–damper supports
OMEGA	Frequencies used in 'random' support excitation
PSI	Damping constant used in 'random' excitation
AMP	Amplitudes of 'random' excitation components
TIMONE	Time t_1 at start of excitation step for shaker or shock machine excitation
TIMTWO	Time t_2 at end of excitation step for shaker or shock machine excitation
ACCEL1	Acceleration at t_1
ACCEL2	Acceleration at t_2
FREQ1	Frequency at t_1
FREQ2	Frequency at t_2
XX	Initial displacement of shaker or shock machine
VELX	Initial velocity of shaker or shock machine

This program is structured to accept input for up to three generalized masses, which are connected to one another and to supports, or ground, by up to 10 spring–dampers, up to 10 friction elements, and up to 10 stops. Not all of them have to be used or specified, of course.

The application of the excitation forces or displacements allows for a reasonably broad range of types of input. These inputs are described in a flow chart in Fig. 3.11. From this chart, it is seen that the type of excitation input is controlled by three indices: MSHAKE, NRAND, and ITYPE. When MSHAKE is zero, the excitations are by either, or both, forces acting on the masses, or by displacements which act on the supports of the first three springs specified in the input as force elements. These forces FV and displacements SV are time-dependent. For example, FV is dimensioned as FV $(8, 2, 3)$ allowing for up to eight force–time points for each of the three masses. Thus, for the first time point (t_1) on mass 1 in Fig. 3.12(a), FV would be specified as $FV(1, 1, 1) = f_1$, $FV(1, 2, 1) = t_1$. For the second, $FV(2, 1, 1) = f_2$

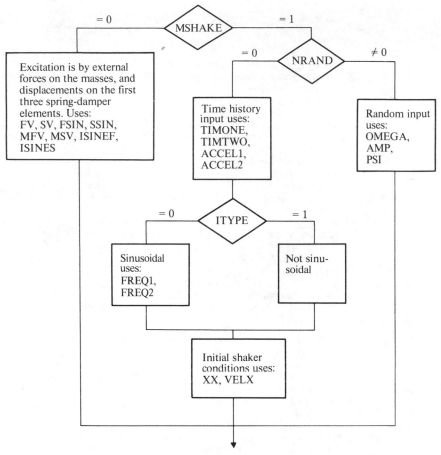

FIGURE 3.11
Flow diagram for excitation inputs

and $FV(2, 2, 1) = t_2$, etc. There are MFV points. For SV, displacement is similarly given. Note, therefore, that the arrays FV and SV contain both the excitation magnitude and time. In addition, the excitation can be sinusoidal. In that case, the input arrays FSIN and SSIN are used instead. For example, FSIN is dimensioned FSIN(3, 2), allowing for a force amplitude and corresponding frequency on each mass. The array SSIN is dimensioned SSIN(3, 2), allowing for a displacement amplitude and corresponding frequency to act at the support of the first three springs. Specifying ISINEF = 1 activates FSIN, and ISINES = 1 activates SSIN. Both the excitation amplitude and frequency of the excitation acting on each mass or spring support are specified in each array.

   When MSHAKE is not zero, the input excitation is obtained from a shaker or a shock machine, which induces a specified excitation at the support of the first spring element. The type of shaker excitation is determined by the indices

NRAND and ITYPE. If NRAND $\neq$ 0, the acceleration excitation is random. If NRAND = 0 and ITYPE = 0, the acceleration is sinusoidal, as in a shaker. When NRAND = 0 and ITYPE = 1, it is a time history, as in a shock machine. In either of the cases where NRAND = 0 the input is read through the variables TIMONE, TIMTWO, ACCEL1, ACCEL2, and in the case of shaker excitation, also through

FIGURE 3.12
Specification of force or displacement excitations

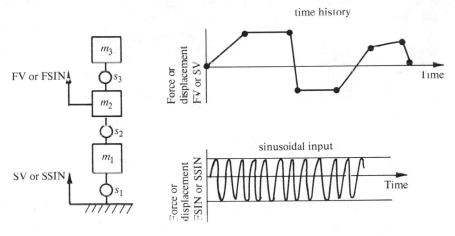

(a) Inputs for MSHAKE = 0: force excitation of masses, or displacement excitation of spring supports, either time history or with sinusoidal excitation

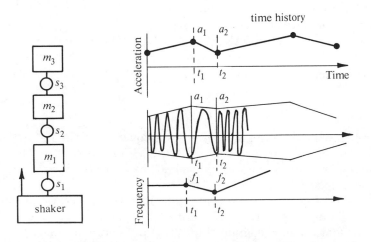

(b) Inputs for MSHAKE = 1, NRAND = 0: excitation by a shaker (ITYPE = 0), or by a shock machine (ITYPE = 1)

Table 3.3   INPUT DATA SYMBOLS AND FORMAT

Format	Variables
(15I2)	IDONE, NEWLDS, NZ, MSV, MFV, MSHAKE, ITYPE, ISINEF, ISINES, NRAND
(3F10.4)	DEL, TSTP, TTL
(8F10.4)	(ZM(J), J = 1, NZ)
(5I2, 5F10.4)	INDX, NG, (NC(J), J = 1,3), C1, C2, (CC(J), J = 1, NG)
	As many cards as are needed, one for each force element and one to indicate that no more force element cards will be read in
(6F10.4)	(Z(1, J), J = 1, 3), (ZDOT(J), J = 1, 3)
	Read the following four lines only if excitation is not by a shaker:
(8F10.4)	(((FV (J, K, L), J = 1, 8), K = 1, 2), L = 1, 3)
(8F10.4)	(((SV (J, K, L), J = 1, 8), K = 1, 2), L = 1, 3)
(8F10.4)	((FSIN (J, K), J = 1, 3), K = 1, 2)
(8F10.4)	((SSIN (J, K), J = 1, 3), K = 1, 2)
	Read the following two lines for 'random' support motion input by a shaker:
(8F10.4)	(OMEGA (J), J = 1, NRAND), PSI
(8F10.4)	(AMP (J), J = 1, NRAND)
	Read the following two lines for other than 'random' support input by a shaker or shock machine:
(6F10.4)	TIMONE, TIMTWO, ACCEL1, ACCEL2, FREQ1, FREQ2
(2F10.4)	XX, VELX
	Read additional values when time exceeds previous value of TIMTWO for input by a shaker or shock machine:
(3F10.4)	TIMTWO, ACCEL2, FREQ2

FREQ1 and FREQ2. The specification is illustrated in Fig. 3.12(b). There, the non-sinusoidal time history case would specify the times TIMONE, TIMTWO and the accelerations, ACCEL1, ACCEL2. The alternate case of a sinusoidal variation in acceleration is also illustrated. In this case, the frequency may also vary with time, and this is allowed for by specifying FREQ1 and FREQ2. In each case where MSHAKE = 1, the initial conditions of displacement and velocity of the testing machine are given by XX and VELX.

Finally, when MSHAKE = 1 and NRAND $\neq$ 0, a model of a 'random' excitation of the shaker base is possible. The acceleration of the base is modeled by a series of decaying sinusoids of up to 20 terms:

$$A = \sum_{n=1}^{NRAND} a_n \exp(-\psi 2\pi f_n t)[\cos(2\pi f_n t) - \psi \sin(2\pi f_n t)] \qquad (3.75)$$

The acceleration amplitudes a_n in gs, frequencies f_n in Hz, and decay constant ψ can be specified in the arrays AMP, OMEGA, and PSI. This sort of excitation input is sometimes used for specialized shaker tests where a deterministic model of a random shock input is specified.

The program listing is given in the following pages. Comments are interspersed throughout the program to describe the input variables and to illustrate the flow of the logic. A flow diagram of the program is given in Fig. 3.13.

A typical three mass system will be studied in the next section (3.14), where the details of the necessary inputs will be shown.

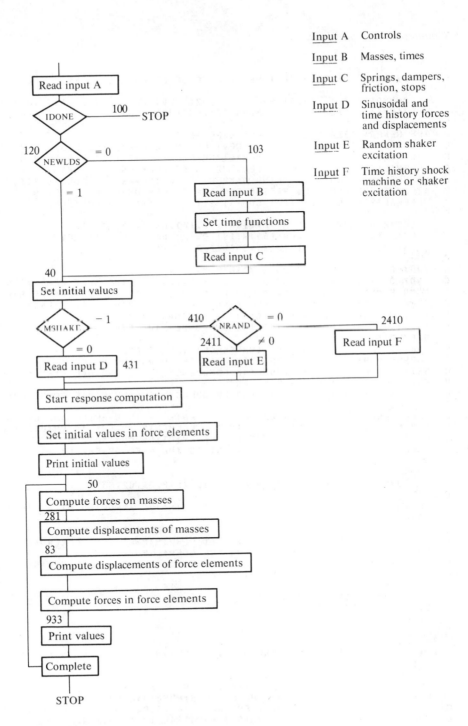

FIGURE 3.13
Flow chart for three degree-of-freedom computer program

Computer program listing

```
C     DYNAMIC ANALYSIS OF THREE DEGREE OF FREEDOM SYSTEMS
C
      COMMON ZM(3),Z(3,3),FRC(3),NZ,JZ(3,20),AZ(3,20),ZDOT(3),NZZ(3),
     1ZACC(3),ZMACC(3),PSI,OMEGA(20),AMP(20),NRAND
C
C     THE GENERALIZED MASSES ARE STORED IN ZM.
C     Z CONTAINS THE PRESENT AND PREVIOUS TWO GENERALIZED DISPLACEMENTS.
C     FRC PROVIDES FOR THE 3 GENERALIZED FORCES. NZ IS THE NUMBER OF
C     GENERALIZED COORDINATES USED. JZ IS A SELF-GENERATED MAP OF THE
C     MASSES ON WHICH THE SPRING DAMPER, FRICTION, STOP, FORCES ACT.
C     AZ CONTAINS COUPLING RATIOS FOR THE FORCES USED IN COMPUTING THE
C     GENERALIZED FORCES. ZDOT CONTAINS THE INITIAL VELOCITIES.
C
      COMMON INDX,MX,MY,MW,FX(10),X(3,10),CX(3,10),FY(10),
     CY(2,10),CY(2,10),FW(10),W(10),CW(2,10),FV(8,2,3),MFV,
     CSV(8,2,3),MSV,FSIN(3,2),SSIN(3,2)
C
C     INDX OF 1,2,3 CORRESPONDS WITH X,Y,W TYPE FORCES.
C     INDX OF 4 IS END OF FORCE DATA. MX,MY,MW ARE THE SELF GENERATED
C     NUMBER OF ELEMENTS OF EACH TYPE. FX,FY,FW ARE THE FORCES OF EACH
C     TYPE. CX,CY,CW ARE CORRESPONDING SELF GENERATED COEFFICIENTS,
C     FV GIVES THE EXTERNAL FORCE-TIME HISTORY AT THE MASSES.
C     SV GIVES THE SUPPORT DISPLACEMENT UNDER THE FIRST THREE SPRING-DAMPER
C     ELEMENTS.
C     FOR SV AND FV 8 POSITION (OR FORCE) VALUES ARE STORED WITH 8
C     CORRESPONDING TIME VALUES FOR 3 SPRING ENDS (OR MASSES).
C     SSIN STORES THE AMPLITUDE AND FREQUENCY OF SINUSOIDAL DISPLACEMENT
C     FOR THE FIRST THREE SPRING ENDS
C     FSIN STORES THE AMPLITUDE AND FREQUENCY OF SINUSOIDAL FORCE FOR 3
C     MASSES.
C     WHEN READING X (SPRING-DAMPER) ELEMENTS C1 IS DAMPING AND C2 IS
C     SPRING RATE. WHEN READING FRICTION ELEMENTS, Y, C1 IS FRICTION FORCE,
C     C2 IS THE LOCAL SPRING RATE AT THE FRICTION CONTACT.
C     WHEN READING STOP ELEMENTS, W, C1 IS THE FREE SPACE, C2 IS THE
C     SPRING RATE.
C
      COMMON NG,NC(3),CC(3),DEL,TM,TTL,TSTP,C1,C2,C3
C
C     NG IS THE NUMBER OF GENERALIZED COORDINATES AFFECTED BY A PARTICULAR
C     FORCE
C     NC IS THE COORDINATE NUMBER AND CC IS THE DISPLACEMENT AT THE FORCE
C     PER UNIT GENERALIZED COORDINATE (1 FOR TRANSLATION, LEVER ARM FOR
C     ROTATION, DISPLACEMENT IN MODE FOR NORMAL MODE, ETC.)
C     DEL IS THE TIME STEP, TM THE TIME, TTL THE TOTAL TIME OF INTEREST,
C     TSTP THE TIME BETWEEN OUTPUT PRINTS.
C
 1103 FORMAT(8F10.4)
 1101 FORMAT(15I2)
 1701 FORMAT(1H1)
C
  701 CONTINUE
  101 READ(5,1101) IDONE,NEWLDS,NZ,MSV,MFV,MSHAKE,ITYPE,ISINEF,ISINES
     1    ,NRAND
C
C     WHEN MSHAKE IS NOT ZERO,EXCITATION IS BY A SHAKER OR SHOCK MACHINE.
C     WHEN ISINEF IS NOT ZERO, THERE ARE SINUSOIDAL FORCES.
C     WHEN ISINES IS NOT ZERO, THERE ARE SINUSOIDAL SUPPORT DISPLACEMENTS.
C     WHEN ITYPE IS 0 EXCITATION IS SINUSOIDAL IN A SHAKER.
C     WHEN ITYPE IS 1 EXCITATION IS PULSE IN A SHOCK MACHINE.
C     WHEN NRAND IS NOT ZERO, SHAKER EXCITATION IS A SUMMATION OF TERMS.
```

```
C   NRAND IS THE NUMBER OF TERMS.
C   PSI IS THE PERCENTAGE OF CRITICAL DAMPING IN THE TERMS.
C   OMEGA IS THE FREQUENCY OF THE TERMS IN HERTZ.
C   AMP IS THE AMPLITUDE OF THE TERM IN G UNITS.
C
        IF(IDONE) 102,102,100
    102 CONTINUE
        WRITE(6,1701)
        IF(NEWLDS) 103,103,40
    103 CONTINUE
        READ(5,1103) DEL,TSTP,TTL
        READ(5,1103) (ZM(J),J=1,NZ)
   1130 FORMAT(1H0,6HMASSES,3F16.4)
        WRITE(6,1130)(ZM(J),J=1,NZ)
   1131 FORMAT(1H0,3HDEL,F10.6,3X,4HTSTP,F10.6,3X,3HTTL,F10.4)
        WRITE(6,1131) DEL,TSTP,TTL
C
C   SET INITIAL VALUES
C
   2003 FORMAT(5I2,5F10.4)
        DO 4 J=1,3
      4 NZZ(J)=0
        MX=0
        MY=0
        MW=0
        DELSQI=DEL**2
        DELSQ=1./DELSQI
        GRAV=386.0
C
C   GRAV IS THE ACCELERATION OF GRAVITY IN POUND-INCH-SECOND UNITS
C   IN THE INTERNATIONAL SYSTEM USE 9.807 FOR NEWTON-METRE-SECOND UNITS
C
        DELACC=DEL*DEL*GRAV
        DELI=0.5/DEL
C
C   READ FORCE ELEMENT DATA
C
      2 READ(5,2003) INDX,NG,(NC(J),J=1,3),C1,C2,(CC(J),J=1,NG)
        GO TO (10,12,14,16),INDX
C
C   GENERATE SPRING-DAMPER ELEMENT DATA
C
     10 MX=MX+1
        WRITE(6,1019) (MX,C1,C2,((NC(J),CC(J)),J=1,NG))
        CX(1,MX)=-DELI*3.*C1-C2
        CX(2,MX)=DELI*4.*C1
        CX(3,MX)=-DELI*C1
        DO 11 K=1,NG
        NCK=NC(K)
        NZZ(NCK)=NZZ(NCK)+1
        NZK=NZZ(NCK)
        JZ(NCK,NZK)=MX
        AZ(NCK,NZK)=CC(K)
     11 CONTINUE
        GO TO 2
C
C   GENERATE FRICTION ELEMENT DATA
C
     12 MY=MY+1
        WRITE(6,1020) (MY,C1,C2,((NC(J),CC(J)),J=1,NG))
```

```
            CY(1,MY)=C1
            CY(2,MY)=-C2
            DO 13 K=1,NG
            NCK=NC(K)
            NZZ(NCK)=NZZ(NCK)+1
            NZK=NZZ(NCK)
            JZ(NCK,NZK)=MY+1000
            AZ(NCK,NZK)=CC(K)
      13 CONTINUE
            GO TO 2
C
C   GENERATE STOP ELEMENT DATA
C
      14 MW=MW+1
            WRITE(6,1021) (MW,C1,C2,((NC(J),CC(J)),J=1,NG))
            CW(1,MW)=C1
            CW(2,MW)=C2
            DO 15 K=1,NG
            NCK=NC(K)
            NZZ(NCK)=NZZ(NCK)+1
            NZK=NZZ(NCK)
            JZ(NCK,NZK)=MW+2000
            AZ(NCK,NZK)=CC(K)
      15 CONTINUE
            GO TO 2
      16 CONTINUE
      40 CONTINUE
C
C   ZERO CERTAIN MATRICES
C
            CALL ZEROM(X,3,10)
            CALL ZEROM(ZACC,3,1)
            CALL ZEROM(ZMACC,3,1)
            CALL ZEROM(Y,2,10)
            CALL ZEROM(W,1,10)
            XNEW=0.0
C
C   READ INITIAL CONDITIONS
C
            READ(5,1103)(Z(1,J),J=1,3),(ZDOT(J),J=1,3)
C
C   COMPUTE Z(-1), Z(-2) FROM CONDITIONS
C
            DO 41 J=1,3
            Z(2,J)=Z(1,J)-DEL*ZDOT(J)
      41 Z(3,J)=2.0*Z(2,J)-Z(1,J)
            IF (MSHAKE)  430,431,430
C
C   WHEN MSHAKE=1, EXCITATION IS APPLIED TO THE SUPPORT OF SPRING-DAMPER
C   1 BY A SHAKER OR A SHOCK MACHINE.
C
     431  CONTINUE
C
C   READ EXTERNAL FORCE AND SUPPORT DATA
C
            READ (5,1103) (((FV(J,K,L),J=1,8),K=1,2),L=1,3)
            READ (5,1103) (((SV(J,K,L),J=1,8),K=1,2),L=1,3)
            READ (5,1103) ((FSIN(J,K),J=1,3),K=1,2)
            READ (5,1103) ((SSIN(J,K),J=1,3),K=1,2)
     430  CONTINUE
```

```
      IF(MSHAKE) 410,411,410
  410 CONTINUE
      IF (NRAND) 2411,2410,2411
C
C  WHEN NRAND IS NOT ZERO, A RANDOM EXCITATION IS APPLIED TO THE SUPPORT
C  OF SPRING-DAMPER 1.
C
 2411 READ  (5,1103) (OMEGA(J),J=1,NRAND),PSI
      WRITE (6,2412)
 2412 FORMAT(1H0,26HEXCITATION IS SUM OF TERMS)
      WRITE (6,1103) (OMEGA(J),J=1,NRAND),PSI
      READ (5,1103) (AMP(J),J=1,NRAND)
      WRITE(6,1103) (AMP(J),J=1,NRAND)
      A=0.0
      XX=0.0
      XNEW=0.0
      VELX=0.0
      XPREV=0.0
      GO TO 2413
 2410 CONTINUE
C
C  READ SUPPORT ACCELERATIONS AND FREQUENCY VARIATION
C
      READ(5,1103) TIMONE,TIMTWO,ACCEL1,ACCEL2,FREQ1,FREQ2
      WRITE(6,1210)TIMONE,ACCEL1,FREQ1
      WRITE(6,1210)TIMTWO,ACCEL2,FREQ2
      ZFA=(ACCEL2-ACCEL1)/(TIMTWO-TIMONE)
      ZFF=(FREQ2-FREQ1)/(TIMTWO-TIMONE)
C
C  DETERMINE TYPE OF EXCITATION
C
      IF(ITYPE) 1202,1203,1202
C
C  WHEN ITYPE=1, THE TESTING MACHINE IS A SHOCK MACHINE.
C  WHEN ITYPE=0, THE TESTING MACHINE IS A SHAKER.
C
 1202 WRITE(6,1200)
      A=ACCEL1-TIMONE*ZFA
 1200 FORMAT(1H0,26HEXCITATION IS TIME-HISTORY)
      GO TO 1204
 1203 WRITE(6,1201)
      A=0.0
      DEL2PI=6.2831853*DEL
      WW=0.0
      DEL15=1.5*DEL
 1201 FORMAT(1H0,24HEXCITATION IS SINUSOIDAL)
 1204 CONTINUE
C
C  READ INITIAL SHAKER VELOCITY AND DISPLACEMENT
C
      READ(5,1103) XX,VELX
C
C  THE INITIAL SHAKER DISPLACEMENT IS XX, VELOCITY VELX.
C
      XNEW=XX
      XX=XX-VELX*DEL
      XPREV=XX+XX-XNEW
 2413 CONTINUE
  411 CONTINUE
   45 CONTINUE
```

```
      TCOUNT=-0.000001
      TM=-DEL
   50 TM=TM+DEL
   51 CONTINUE
C
C  SKIP TO STATEMENT 90 IF THIS IS THE FIRST TIME INTERVAL
C
      IF(TM)771,90,771
  771 CONTINUE
C
C  COMPUTE THE GENERALIZED FORCES, FRC
C
      DO 82 K=1,NZ
      NZK=NZZ(K)
      FRC(K)=0.0
  811 DO 81 J=1,NZK
      JJZJ=JZ(K,J)
      INDX=0
  810 INDX=INDX+1
      JZJ=JJZJ
      JJZJ=JJZJ-1000
      IF(JJZJ) 812,812,810
  812 GO TO (30,31,32), INDX
C
C  ADD SPRING FORCE
C
   30 FRC(K)=FRC(K)+AZ(K,J)*FX(JZJ)
      GO TO 81
C
C  ADD FRICTION FORCE
C
   31 FRC(K)=FRC(K)+AZ(K,J)*FY(JZJ)
      GO TO 81
C
C  ADD STOP FORCE
C
   32 FRC(K)=FRC(K)+AZ(K,J)*FW(JZJ)
   81 CONTINUE
   82 CONTINUE
      IF (MSHAKE)  336,337,336
  337 CONTINUE
      IF (MFV-2) 333,331,331
  331 CONTINUE
      DO 33 K=1,NZ
C
C  INTERPOLATE FORCE INPUT TO GET EXTERNALLY APPLIED FORCE
C
      CALL INTRP(FV(1,2,K),FV(1,1,K),A,A,TM-DEL,B,B,B,MFV,1)
C
C  ADD EXTERNAL FORCES TO GENERALIZED FORCES
C
   33 FRC(K)=FRC(K)+B
  333 CONTINUE
      IF (ISINEF) 334,335,334
  334 CONTINUE
C
C  GENERATE SINUSOIDAL FORCE EXCITATION
C
      DO 332 K=1,NZ
      B=FSIN(K,1)*SIN(FSIN(K,2)*6.2832*(TM-DEL))
```

```
C
C   ADD TO GENERALIZED FORCES.
C
  332 FRC(K)=FRC(K)+B
 335   CONTINUE
 336   CONTINUE
C
C   COMPUTE ACCELERATION, F/M
C
      DO 280 K=1,NZ
  280 FRC(K)=FRC(K)/ZM(K)
C
C   COMPUTE AND STORE MAXIMUM ACCELERATION AMPLITUDE
C
      DO 281 K=1,NZ
      ZACC(K)=FRC(K)/GRAV
      ACM=ABS(ZACC(K))
      IF (ACM-ZMACC(K)) 282,282,283
 283   ZMACC(K)=ACM
 282   CONTINUE
 281   CONTINUE
C
C   COMPUTE NEW DISPLACEMENTS
C
      DO 80 K=1,NZ
      DO 93 JJ=1,2
      L=3-JJ
   93 Z(L+1,K)=Z(L,K)
      Z(1,K)=2.*Z(2,K)-Z(3,K)+FRC(K)*DELSQI
      IF(TM.EQ.DEL)Z(1,K)=Z(1,K)-0.5*FRC(K)*DELSQI
      IF(TM.EQ.DEL)Z(3,K)=Z(3,K)+0.5*FRC(K)*DELSQI
   80   CONTINUE
   83   CONTINUE
C
C   COMPUTE NEW VALUES OF EXTENSION OF FORCE ELEMENTS
C
      IF (MX) 8510,851,8510
 8510 DO 85 K=1,MX
      DO 88 J=1,2
      L=3-J
   88 X(L+1,K)=X(L,K)
      X(1,K)=0.0
   85 CONTINUE
  851 IF (MY) 8520,852,8520
 8520 DO 86 K=1,MY
   86 Y(1,K)=0.0
  852 IF(MW) 9000,90,9000
 9000 CALL ZEROM (W,1,MW)
C
   90 CONTINUE
      DO 91 K=1,NZ
      NZN=NZZ(K)
      DO 92 J=1,NZN
      JJZJ=JZ(K,J)
      AZJ=AZ(K,J)*Z(1,K)
      IF (TM-DEL) 94,94,993
   94 AZ2=AZ(K,J)*Z(2,K)
      AZ3=AZ(K,J)*Z(3,K)
  993   CONTINUE
      INDX=0
```

```
  920 INDX=INDX+1
      JZJ=JJZJ
      JJZJ=JJZJ-1000
      IF (JJZJ) 9200,9200,920
C
C   ROUTE TO PROPER ELEMENT
C
 9200 GO TO (921,922,923),INDX
C
C   SPRING-DAMPER ELEMENT EXTENSION
C
  921 X(1,JZJ)=X(1,JZJ)+AZJ
      IF(TM-DEL) 96,96,95
   96 X(2,JZJ)=X(2,JZJ)+AZ2
      X(3,JZJ)=X(3,JZJ)+AZ3
   95 GO TO 92
C
C   FRICTION ELEMENT EXTENSION
C
  922 Y(1,JZJ)=Y(1,JZJ)+AZJ
      IF (TM-DEL) 98,98,97
   98 Y(2,JZJ)=Y(2,JZJ)+AZ2
   97 GO TO 92
C
C   STOP ELEMENT EXTENSION
C
  923 W(JZJ)=W(JZJ)+AZJ
   92 CONTINUE
   91 CONTINUE
C
C   ADD SUPPORT MOTION TO THE EXTENSIONS OF THE FIRST 3 SPRING-DAMPERS
C
      IF (MSHAKE) 931,932,931
  931 CONTINUE
      IF(TM.EQ.0.0)GO TO 9330
      XPREV=XX
      XX=XNEW
      IF (NRAND) 3412,412,3412
 3412 A=0.0
      DO 3413 K=1,NRAND
      OMEGT=OMEGA(K)*(TM-DEL)*6.2832
 3413 A=A+AMP(K)*EXP(-PSI*OMEGT)*(COS(OMEGT)-PSI*SIN(OMEGT))
      GO TO 416
  412 IF(TM-DEL-TIMTWO-.00003) 415,415,414
  414 TIMONE=TIMTWO
      ACCEL1=ACCEL2
      FREQ1=FREQ2
      READ(5,1103) TIMTWO,ACCEL2,FREQ2
      WRITE(6,1210) TIMTWO,ACCEL2,FREQ2
 1210 FORMAT(1H0,4HTIME,F10.4,3X,10HACCEL AMPL,F10.4,3X,4HFREQ,F10.4)
      ZFA=(ACCEL2-ACCEL1)/(TIMTWO-TIMONE)
      ZFF=(FREQ2-FREQ1)/(TIMTWO-TIMONE)
  415 CONTINUE
      A=ACCEL1+(TM-DEL-TIMONE)*ZFA
      IF(ITYPE) 416,417,416
  417 CONTINUE
      IF(TM.NE.DEL)WW=WW+DEL2PI*(FREQ1+(TM-DEL15-TIMONE)*ZFF)
      IF(WW-6.2831853) 418,418,419
  419 WW=WW-6.2831853
  418 CONTINUE
```

```
       A=A*SIN(WW)
  416  CONTINUE
       XNEW=A*DELACC+XX+XX-XPREV
       IF(TM.NE.DEL)GO TO 1416
       CORR=(XNEW+XPREV)*0.5
       XNEW=XNEW-CORR
       XPREV=XPREV+CORR
 1416  CONTINUE
 9330  X(1,1)=X(1,1)+XNEW
       GO TO 933
  932  CONTINUE
       IF (MSV-2)   926,925,925
  925  CONTINUE
       DO 924 K=1,3
       CALL INTRP(SV(1,2,K),SV(1,1,K),A,A,TM,B,B,B,MSV,1)
  924  X(1,K)=X(1,K)+B
  926  CONTINUE
C
C   ADD SINUSOIDAL SUPPORT MOTION
C
       IF (ISINES)   927,928,927
  927  CONTINUE
       DO 1603 K=1,3
       B=SSIN(K,1)*SIN(SSIN(K,2)*6.2832*TM)
 1603  X(1,K)=X(1,K)+B
  928  CONTINUE
  933  CONTINUE
 1060  IF(MX) 6010,601,6010
 6010  DO 60 K=1,MX
       FX(K)=0.
C
C   COMPUTE NEW FORCES IN SPRING-DAMPERS
C
       DO 60 J=1,3
   60  FX(K)=FX(K)+CX(J,K)*X(J,K)
  601  IF (MY) 6020,602,6020
C
C   COMPUTE NEW FORCES IN FRICTION ELEMENTS
C
 6020  DO 61 K=1,MY
       FY(K)=CY(2,K)*(Y(1,K)-Y(2,K))
       FRK=ABS(FY(K))
       IF(FRK-CY(1,K))   61,61,6021
 6021   FY(K)=CY(1,K)*FY(K)/FRK
       Y(2,K)=Y(1,K)-FY(K)/CY(2,K)
   61  CONTINUE
  602  IF (MW)   6030,603,6030
C
C   COMPUTE NEW FORCES IN STOPS
C
 6030  DO 66 K=1,MW
       FW(K)=0.0
       IF ((-W(K))-CW(1,K)) 66,66,661
  661  FW(K)=CW(2,K)*(-W(K)-CW(1,K))
   66  CONTINUE
  603     CONTINUE
C
C   WRITE OUTPUT
C
       IF (TM-TCOUNT) 1095,1096,1096
```

```
1001 FORMAT(1H0,17X,17H**********      .7HTIME = ,F12.5,17H          ****
     C******)
1002 FORMAT(1H0,3HX= ,8F13.8/(4X,8F13.8))
1003 FORMAT(1H0,3HY= ,8F13.8/(4X,8F13.8))
1004 FORMAT(1H0,3HZ= ,8F13.8/(4X,8F13.8))
1005 FORMAT(1H0,3HW= ,8F13.8/(4X,8F13.8))
1006 FORMAT(1H0,4HFX= ,9F12.1/(5X,9F12.1))
1007 FORMAT(1H0,4HFY= ,9F12.1/(5X,9F12.1))
1008 FORMAT(1H0,4HFW= ,9F12.1/(5X,9F12.1))
1011 FORMAT(1H0,17X,31HDISPLACEMENT OF SPRING ELEMENTS)
1012 FORMAT(1H0,17X,33HDISPLACEMENT OF FRICTION CONTACTS)
1013 FORMAT(1H0,17X,34HDISPLACEMENT OF NON-LINEAR SPRINGS)
1014 FORMAT(1H0,17X,25HGENERALIZED DISPLACEMENTS)
1016 FORMAT(1H0,17X,28HSPRING DAMPER ELEMENT FORCES)
1017 FORMAT(1H0,17X,24HFRICTION CONTACTS FORCES)
1018 FORMAT(1H0,17X,24HNON-LINEAR SPRING FORCES)
1019 FORMAT(1H0,14HSPRING ELEMENT,2X,I2,2X,3HC1=,F10.3,2X,3HC2=,F13.3,
     12X/6HNC/CC=,(10(I2,F6.2,2X)))//)
1020 FORMAT(1H0,16HFRICTION ELEMENT,2X,I2,2X,3HC1=,F10.3,2X,3HC2=,F10.3
     1,2X,2X/6HNC/CC=,(10(I2,F6.2,2X)))//)
1021 FORMAT(1H0,17HNON-LINEAR SPRING,2X,I2,2X,3HC1=,F10.3,2X,3HC2=,
     1F13.3,2X/6HNC/CC=,(10(I2,F6.2,2X)))//)
1030 FORMAT(1H0,17X,33HMAXIMUM AND PREVIOUS ACCELERATION)
1031 FORMAT(1H0,3X,(3(F10.4,F10.4,10X))//)
1096 TCOUNT=TCOUNT+TSTP
     WRITE(6,1001)TM
     IF (MX)   6610,6661,6610
6610 WRITE(6,1011)
     WRITE(6,1002)(X(1,K),K=1,MX)
     WRITE(6,1016)
     WRITE(6,1006)(FX(K),K=1,MX)
6661 CONTINUE
     IF(MY)   6630,663,6630
6630 WRITE(6,1012)
     WRITE(6,1003)(Y(1,K),K=1,MY)
     WRITE(6,1017)
     WRITE(6,1007)(FY(K),K=1,MY)
663  CONTINUE
     IF(MW)   6650,665,6650
6650 WRITE(6,1013)
     WRITE(6,1005)(W(K),K=1,MW)
     WRITE(6,1018)
     WRITE(6,1008)(FW(K),K=1,MW)
665  CONTINUE
     WRITE(6,1014)
     WRITE(6,1004) (Z(1,K),K=1,NZ)
     WRITE (6,1030)
     WRITE (6,1031)(ZMACC(K),ZACC(K),K=1,NZ)
     CALL ZEROM(ZMACC,3,1)
1095 IF (TM-TTL) 50,500,500
C
C  LOOP BACK TO STATEMENT 50 FOR NEW TIME STEP
C
  500 GO TO 101
  100 STOP
     END
```

```
CINTRP               INTERPOLATION SUBROUTINE FOR THREE VARIABLES WHICH
CINTRP               ARE FUNCTIONS OF A FOURTH
       SUBROUTINE INTRP(X,Y,W,Z,XX,YY,WW,ZZ,N,M)
       DIMENSION Y(1),X(1),W(1),Z(1)
C      INTERPOLATES LINEARLY BETWEEN POINTS ON A CURVE.
C      X IS A VECTOR OF N VALUES OF ABSCISSA
C      Y,W,Z ARE VECTORS OF N VALUES OF ORDINATES
C      FOR X=XX, Y=YY, W=WW, AND Z=ZZ
C      IF M=3 PROGRAM INTERPOLATES FOR Y,W, AND Z
C      IF M=2 PROGRAM INTERPOLATES FOR Y AND W. IN CALL USE SAME SYMBOL
C      FOR Z AND Y,ZZ AND YY. IF M=1 USE SAME SYMBOL FOR Z,W,Y AND ZZ,WW,
C      YY AND PROGRAM WILL INTERPOLATE FOR Y ONLY.
       DO 10 J=2,N
       K=N+1-J
       IF (X(K+1)-X(K)) 10,10,11
  11   IF (XX-X(K))  10,20,20
  10   CONTINUE
  20   TS1= X(K+1)-X(K)
       TS2=(X(K+1)-XX)/TS1
       TS1=(XX-X(K))/TS1
       IF (M-2)   50,40,30
  30   ZZ= TS1*Z(K+1)+TS2*Z(K)
  40   WW= TS1*W(K+1)+TS2*W(K)
  50   YY= TS1*Y(K+1)+TS2*Y(K)
 100   RETURN
       END
CZEROM     SUBROUTINE ZEROM FOR SETTING A MATRIX TO ZERO(REAL NUMBERS)
C              SUBROUTINE ZEROM
       SUBROUTINE ZEROM(A,I,K)
       DIMENSION A(1)
       II=I*K
       DO 10  J=1,II
  10   A(J)=0.0
       RETURN
       END
```

Problems

3.1   In the computer program described, the restoring force exerted by any spring is linearly related to its extension x, so that $F = -kx$. What changes must be made in the program so that nonlinear springs can be used, in which

$$F = -(k_1 x + k_2 x^2 + k_3 x^3)?$$

3.2   Modify the program so that the elastic–plastic force element described in section 1.10 can be used.

3.3   Instead of the finite difference representation

$$\dot{x}_0 = \frac{(3x_0 - 4x_{-1} + x_{-2})}{2\Delta t}$$

use the representation

$$\dot{x}_0 = \frac{(x_0 - x_{-1})}{\Delta t}$$

in the approximation for the behavior of the damper as given in eq. (3.62). What changes must be made in the computing program? For the three mass system described in the next section (3.14), what changes in response predictions, if any, are observed?

3.4   Modify the program so that it will accept the International System of Units. (*Hint*: Change the value of g used in the programs.)

3.14   EXAMPLE—SYSTEM WITH STOPS†

To show some of the results obtainable by means of the methods of analysis discussed in this chapter, we will discuss the three mass system illustrated in Fig. 3.14. It consists of three identical masses, connected in series by identical springs and dampers. Between each mass, and between the last mass and the support, there are stops with 1/4 in. clearance. We shall describe the response of this system to three types of excitation: a force applied to the first mass as a time history effect; a sinusoidal force; and the case when each mass is given an initial displacement.

Time history force

We consider first a force applied to the first mass which increases linearly from 0 to 100 lb in the initial 0·05 sec. It remains 100 lb for the next 0·05 sec. Then it drops to 0 lb during the next 0·05 sec. The masses are all initially at rest, and the support is stationary. A detailed list of input data as they appear on cards is given in Table 3.4.

The results are plotted in Fig. 3.15. One group of curves gives the deflections Z_1, Z_2, Z_3 at the three masses. It is seen that the deflection at mass 1 reaches about 0·5 in., at $t = 0·12$ sec. At about the same time, the deflection of mass 2 reaches about 0·4 in. and the deflection of mass 3 reaches about 0·2 in. Remembering that the force on mass 1 has dropped to zero at $t = 0·15$ sec, it is not surprising

† The reader is encouraged to apply the computer program of section 3.13 to these examples and all others of this chapter.

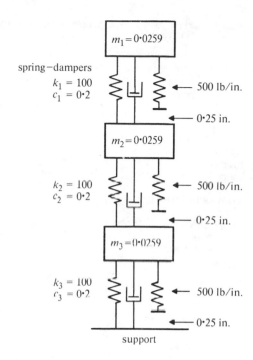

FIGURE 3.14
Three mass system including spring–damper and stop elements (masses in lb sec²/in., springs in lb/in., and dampers in lb sec/in.)

to see the deflections rapidly decrease after that time and go negative at about $t = 0 \cdot 18$ sec.

Another set of curves gives the stretching of the springs X and W connecting the masses. As might be anticipated, the greatest stretch occurs for spring 3 connecting the lowest mass to the support.

The final set of curves gives the compressive forces FX in the springs plus dampers, and FW in the stops. An interesting feature of these curves is that FW remains zero until about $t = 0 \cdot 18$ sec, when the stop closure is $0 \cdot 25$ in. After that time, the stop force rises rapidly. It reaches a peak value of 340 lb for stop 3 at $t = 0 \cdot 21$ sec. The stops are quite effective in preventing much closure of the distance between masses. X_1, X_2, and X_3 are negative for only a short time at about $t = 0 \cdot 21$ sec. Their minimum value is $-0 \cdot 25$ in., as required by the stop setting.

Problems

3.5 Repeat the example computation with the force applied to mass 2 instead of mass 1.
3.6 Repeat the example problem with damping set to zero.
3.7 Repeat the example problem with the stop spring rate reduced from 500 lb/in. to 100 lb/in.

Table 3.4   DATA INPUT FOR THREE MASS SYSTEM OF FIG. 3.14, UNDER TIME HISTORY FORCE EXCITATION

Remarks	Data input							
Control indices	0 0 3 2 5							
Times	0·005	0·010	0·30					
Masses	0·0259	0·0259	0·0259					
Spring 1	1 2 1 2	00·2	100·0	1·0	−1·0			
Spring 2	1 2 2 3	00·2	100·0	1·0	−1·0			
Spring 3	1 1 3 0	00·2	100·0	1·0				
Stop 1	3 2 1 2	00·25	500·0	1·0	−1·0			
Stop 2	3 2 2 3	00·25	500·0	1·0	−1·0			
Stop 3	3 1 3 0	00·25	500·0	1·0				
End of force element data	4							
Initial conditions	0·0	0·0	0·0	0·0	0·0	0·0		
FV, force on mass 1	0·0	100·0	100·0	0·0	0·0	0·0	0·0	0·0
at times specified	0·0	0·05	0·10	0·15	10·0	0·0	0·0	0·0
FV, on mass 2	0·0	0·0	0·0	0·0	0·0	0·0	0·0	0·0
	0·0	0·05	0·10	0·15	10·0	0·0	0·0	0·0
FV, on mass 3	0·0	0·0	0·0	0·0	0·0	0·0	0·0	0·0
	0·0	0·05	0·10	0·15	10·0	0·0	0·0	0·0
No excitation SV	0·0	0·0	0·0	0·0	0·0	0·0	0·0	0·0
on first spring	0·0	10·0	0·0	0·0	0·0	0·0	0·0	0·0
No excitation of	0·0	0·0	0·0	0·0	0·0	0·0	0·0	0·0
second spring	0·0	10·0	0·0	0·0	0·0	0·0	0·0	0·0
No excitation of	0·0	0·0	0·0	0·0	0·0	0·0	0·0	0·0
third spring	0·0	10·0	0·0	0·0	0·0	0·0	0·0	0·0
FSIN is zero	0·0	0·0	0·0	0·0	0·0	0·0		
SSIN is zero	0·0	0·0	0·0	0·0	0·0	0·0		
End of data	1							

3.8 Repeat the example problem with the springs set at 400 lb/in. instead of 100 lb/in. (*Hint:* Reduce *t* to 0·0025 to maintain the same convergence as before. For good accuracy the time step should be approximately $1/3\sqrt{(m/k)_s}$ where $(m/k)_s$ is the smallest ratio of mass to the sum of the springs acting on that mass.)

3.9 Repeat the example problem with the force on mass 1 having values of 0·0, 100, − 100, 100, 0·0, 0·0 at times 0·0, 0·05, 0·15, 0·25, 0·30, 0·50.

Sinusoidal force

We consider next a sinusoidal force applied to mass 1 in Fig. 3.14. We take the amplitude as 100 lb and the frequency as 5 Hz. All masses are initially at rest.

The response, Fig. 3.16, is similar in many respects to that found in Fig. 3.15. Again the stops come into contact briefly after about 0·15 sec. For this forcing input, the masses tend to move in unison, with the top mass moving most. All of the stops come into contact at nearly the same time.

Problems

3.10 Repeat the example problem with the stop clearance increased from 0·25 in. to 1·00 in. (*Note:* The response now has more negative deflection.)

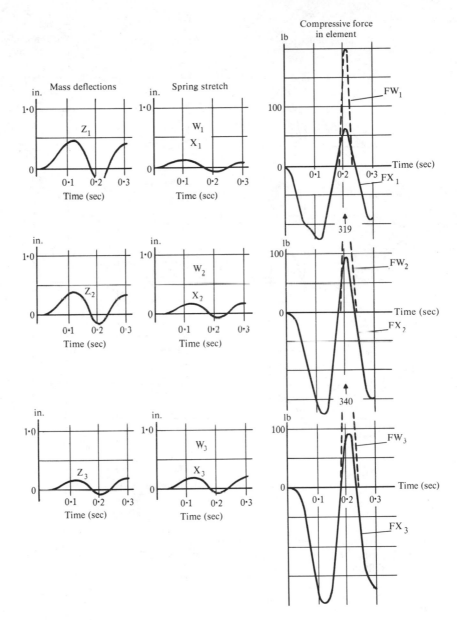

FIGURE 3.15
Response of system in Fig. 3.14 to time history force on mass 1

3.11 Repeat the example problem with the frequency of excitation increased from 5 Hz to 10 Hz. (*Hint:* Take Δt half as great, i.e., 0·0025 sec, to observe the greater detail in the forcing function.)

3.12 Apply the force of the example problem to mass 2 instead of mass 1. (*Note:* The masses do not move in unison to the same extent as before, indicating that the second mode is now oppositely excited.)

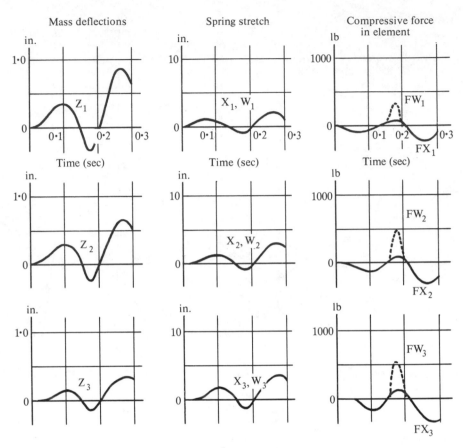

FIGURE 3.16
Response of system in Fig. 3.14 to sinusoidal force on mass 1

3.13   Repeat the example problem with the damping set to zero.
3.14   Repeat the example problem with the stop spring rate reduced from 500 lb/in. to 100 lb/in. (Note the increase in negative deflections.)

Initial deflections

We consider next an initial displacement of the masses in Fig. 3.14. We take the initial position of mass 1 at 1·5 in., mass 2 at 1·0 in., and mass 3 at 0·5 in., while the initial velocities are zero.

The results are plotted in Fig. 3.17. The mass deflections Z give the appearance of a decaying sinusoid. The spring extensions X show some higher frequency effects, particularly when the stops come into contact, W less than $-0·25$. The spring–damper compressive forces FX and the stop compressive forces FW are about what might be expected. They show a perturbation when the stops make contact.

Problems

3.15   Repeat the example problem with the initial deflection 1·0 in. at masses 1 and 3 and 0·0 in. at mass 2. (Note the increased presence of higher frequency response.)

3.16   Repeat the example problem with the stop clearance at 0·5 in. instead of 0·25 in.

3.17   Repeat the example problem with the initial deflections zero and with the initial velocity 10 in./sec at mass 1 and 0·0 at masses 2 and 3.

3.18   Repeat the example problem with the clearance for stops 1 and 2 set to 0·5 and with the stiffness of stop 3 reduced from 500 lb/in. to 100 lb/in.

3.19   Repeat the example problem with mass 2 increased from 0·0259 lb sec²/in. to 0·1036 lb sec²/in.

FIGURE 3.17
Response of system in Fig. 3.14 to initial displacements

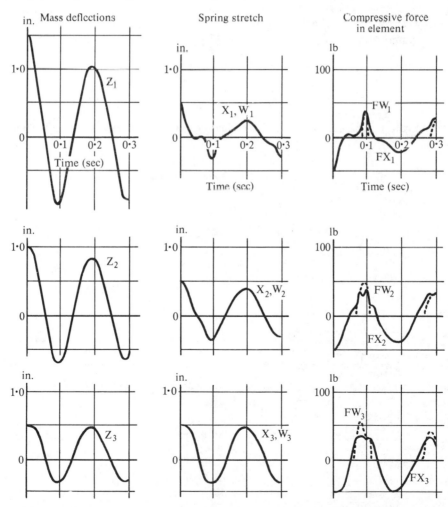

3.15   BEAM BENDING

When a beam is part of a system, it is possible to study the system dynamics with two or three degrees of freedom (thus keeping the amount of effort small) by employing the modal deformation as a generalized coordinate. To illustrate how this is done, consider the system shown in Fig. 3.18. It consists of a beam which is supported at its mid-length by a spring. The spring connects to a second mass, of weight equal to the beam weight, which in turn is connected to a support by another spring and a damper. A stop allows a free 1/4 in. clearance between the second mass and the support.

The beam is of height $h = 0{\cdot}316$ in., breadth $b = 1{\cdot}32$ in., length $l = 80$ in., density $\rho = 0{\cdot}3$ lb/in.^{3}, and has an elastic modulus of 30×10^6 lb/in.^{2}.

We shall model this system with three degrees of freedom consisting of the vertical motion of the rigid mass m_1, the vertical rigid body motion of the beam, and the modal deformation of the beam in its lowest mode of vibration. The quantities required for specifying the system are found as follows. The beam mass is computed from the geometry and weight per unit volume:

$$\text{beam weight} = hbl\rho$$
$$= 0{\cdot}316 \times 1{\cdot}32 \times 80 \times 0{\cdot}3 = 10 \text{ lb}$$

$$\text{beam mass} = m_2 = 10/386 = 0{\cdot}0259 \text{ lb sec}^2/\text{in.}$$

From section 3.11, the modal mass is also equal to the beam mass m_2. The modal spring rate is obtained from eq. (3.41) and Table 3.1 for the free–free mode as

$$k_3 = \omega_n^2 m_n = \kappa^4 EI \, \frac{g}{\rho bh} \cdot \frac{\rho bhl}{g} = \left(\frac{4{\cdot}73}{l}\right)^4 \frac{bh^3 EI}{12}$$

$$= 119{\cdot}2 \text{ lb/in.}$$

FIGURE 3.18
Three degree-of-freedom system including modal deformation as third degree of freedom

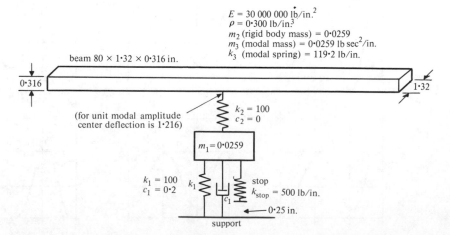

$E = 30\,000\,000 \ \dot{\text{l}}\text{b/in.}^2$
$\rho = 0{\cdot}300$ lb/in.^{3}
m_2 (rigid body mass) $= 0{\cdot}0259$
m_3 (modal mass) $= 0{\cdot}0259$ lb sec^2/in.
k_3 (modal spring) $= 119{\cdot}2$ lb/in.

beam 80 × 1·32 × 0·316 in.

0·316

1·32

(for unit modal amplitude center deflection is 1·216)

$k_2 = 100$
$c_2 = 0$

$m_1 = 0{\cdot}0259$

$k_1 = 100$
$c_1 = 0{\cdot}2$

k_1

c_1

stop
$k_{\text{stop}} = 500$ lb/in.

0·25 in.

support

From Table 3.1, we get the deflection per unit modal amplitude at the midpoint of the beam, where the spring k_2 is attached (we take positive modal deflection at the center and negative at the tips):

$$\text{deflection } (x/l = 0\cdot5) = 1\cdot216$$

This deflection is the coupling ratio between modal amplitude and extension of spring k_2.

The remaining springs, dampers, masses, and stops of this system are also shown in Fig. 3.18. The system specifications are therefore complete.

To illustrate the responses of the system to a number of excitations, we consider three situations: a time history force applied to the rigid mass m_1; a sinusoidal force applied to the rigid mass; and, finally, certain initial displacements applied within the system as a whole.

Time history force

We consider as a first example a force applied to mass 1. The force increases linearly from 0 to 100 lb in the initial 0·05 sec. It remains constant at 100 lb for the next 0·05 sec. Then it drops to zero during the next 0·05 sec and remains zero thereafter. The system is initially at rest. The support remains stationary.

The response of the system is given in Fig. 3.19. The responses Z in the generalized coordinates are all of the order of magnitude of an inch; however, they do not all move in unison.

We interpret the configuration at time $t = 0\cdot230$ sec when $Z_1 = -0\cdot308$, $Z_2 = -1\cdot631$, and $Z_3 = 0\cdot845$ as follows. Referring to Fig. 3.19, mass m_1 has moved downward 0·308 in. The top of spring k_2 has moved downwards $1\cdot631 - 0\cdot845 \times 1\cdot216 = 1\cdot631 - 1\cdot027 = 0\cdot603$ in. Here, the multiplier 1·216 was used to move from the generalized coordinate Z_3 to the actual deflection at that point. The modal bending is such that the center of the beam is higher than the tips. The actual beam shape is given by multiplying the modal deflection given in Table 3.1 by 0·845 in.

Problems

3.20 Repeat the example problem with the stop clearance set to 1·5 in. instead of 0·25 in.

3.21 Repeat the example problem with the spring k_2 increased in stiffness to 200 lb/in.

3.22 Repeat the example problem with two additional springs extending downwards from the beam tips to the support. Take their spring stiffness at 100 lb/in. (*Hint:* Two additional springs must be considered acting on coordinates 2 and 3. Their coupling ratio for coordinate 2 is 1·0, while their coupling ratio for coordinate 3 is the modal tip deflection, −2·0 (see Table 3.1). We have taken positive modal deflection with center up and tips down.)

3.23 Repeat problem 3.22 with stops having a stiffness of 200 lb/in. and a clearance of 0·5 in. added between the beam tips and the support. (*Hint:* Add 2 more stops acting on coordinates 2 and 3. Their coupling ratio for coordinate 2 is 1·0, while their coupling ratio for coordinate 3 is the modal deflection at the tip, −2·0.)

3.24 Repeat the example problem with c_1 set to zero but adding a damper $c_2 = 0\cdot4$ lb sec/in. to go with k_2.

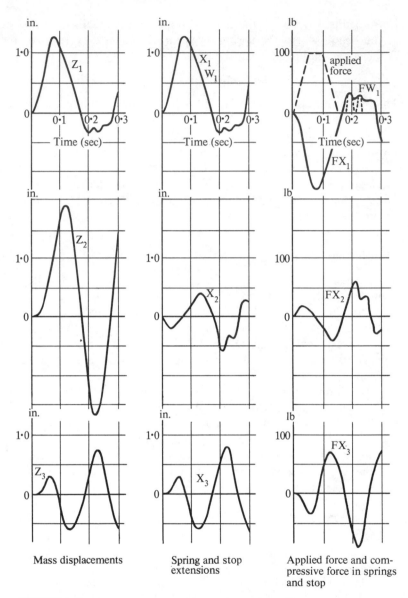

Mass displacements | Spring and stop extensions | Applied force and compressive force in springs and stop

FIGURE 3.19
Response of system in Fig. 3.18 to time history force applied to mass 1

Sinusoidal force

As a second example, we consider a sinusoidal force applied to mass 1. We take the amplitude as 100 lb and the frequency as 5 Hz.

The response, Fig. 3.20, is similar in many respects to that found in Fig. 3.19. Brief contact with the stop occurs between 0·1 and 0·2 sec. The modal motion, coordinate Z_3, has higher frequency components. The rigid body beam motion Z_2 reaches large amplitudes at 0·2 and 0·3 sec. The modal spring force FX_3 is large at the same times.

FIGURE 3.20
Response of beam model in Fig. 3.18 to sinusoidal force applied to mass 1

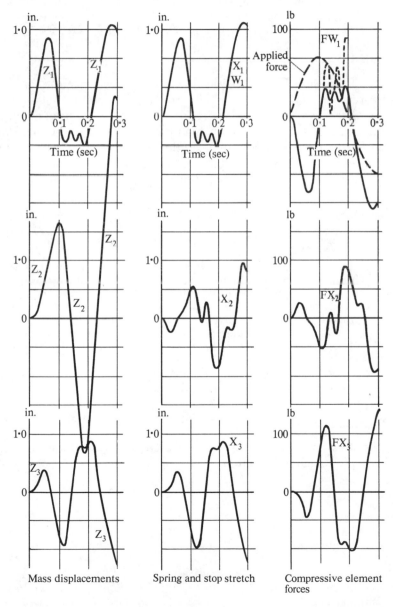

Mass displacements   Spring and stop stretch   Compressive element forces

Problems

3.25   Repeat the example problem with the stop clearance increased to 1·0 in.

3.26   Repeat the example problem with mass 1 increased from 0·0259 to 0·518 lb sec^2/in.

3.27   Repeat the example problem with the applied force considered acting at the center of the beam instead of on mass 1. (*Hint:* The force amplitude on mass 1 goes to 0·0; the force amplitude on mass 2 goes to 100·0; and the force amplitude on coordinate 3 is 1·216 × 100 = 121·6 lb.)

3.28   Repeat the example problem with the applied force consisting of 50 lb at 5 Hz acting upwards at each beam end. (*Hint:* The force amplitude on mass 1 goes to 0·0; the force amplitude on mass 2 goes to 50·0 × 1·0 + 50·0 × 1·0 = 100·0; and the force amplitude on coordinate 3 is 50·0 × (−2·0) + 50·0 × (−2·0) = −200·0.)

Initial displacement

As a final example for the system in Fig. 3.18, we take the initial deflection in the coordinates as $Z_1 = 1·5$, $Z_2 = 1·0$, and $Z_3 = 0·5$, and the initial velocities as zero.

The results are plotted in Fig. 3.21. The stop contacts at about 0·04 and 0·12 sec. The spring k_2 shows a greater high-frequency component than the other springs, indicating relative motion between the mass and beam. The time step used in this computation, 0·005 sec, might have been a little longer than would be desirable to get good accuracy (1/20 of period of highest frequency motion) for X_2.

Problems

3.29   Rerun the example problem with the initial displacements changed to $Z_1 = 0·5, Z_2 = −0·5$, and $Z_3 = −1·0$.

3.30   Rerun the example problem with the initial displacements zero but with the velocities of coordinates 1 and 2 set at −60·0 in./sec. (*Hint:* The initial condition card is changed from (1·5, 1·0, 0·5, 0·0, 0·0, 0·0) to (0·0, 0·0, 0·0, −60·0, −60·0, 0·0).)

3.31   Rerun the example problem with two additional springs attached at the beam tips. Take each spring as having a stiffness of 50 lb/in. (*Hint:* Two additional spring cards are needed. They act on coordinates 2 and 3 with multipliers of 1·0 and −2·0, respectively. Use a value of 0·0025 for the time step since the added springs shorten the vibration periods.)

3.32   Rerun problem 3.31 with the spring 1 removed. (Note that the high frequency content of the motion is substantially less.)

Stress in a beam

The stress in a beam is expressible directly in terms of the bending moment at any particular section. It is given by

$$\sigma = \frac{My}{I} \tag{3.76}$$

where y is the distance measured from the neutral axis of the beam. For a beam of a rectangular section of thickness h and width b, the maximum stress is

$$\sigma_{\max} = \frac{Mh}{2I} = \frac{6M}{bh^2} \tag{3.77}$$

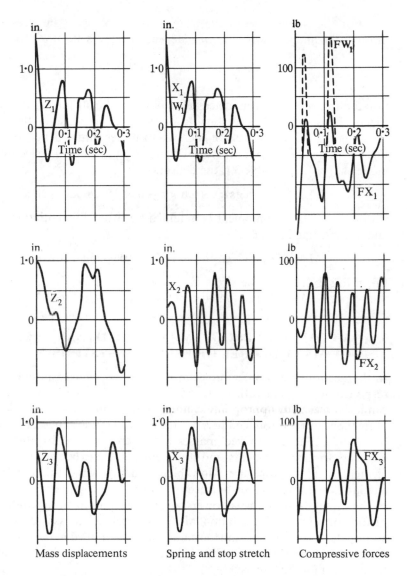

FIGURE 3.21
Results of initial displacements for the beam–mass system in Fig. 3.18

Generally, it is of interest not only to find the dynamic deflections of a structure or system, as we have done so far, but also the stresses. For the most part, indeed, more often than not, in mechanical and civil structures, the actual stresses experienced by the structure are of the greatest importance in assessing the integrity of the structure. Thus, emphasis must also be laid on the determination of the stress as a dynamic response characteristic.

For the simple case of the beam considered here, the stresses can be found through the moment, which in turn depends on the deflection

$$M = EI \frac{d^2 y}{dx^2} \tag{3.78}$$

For the free–free beam modeled here, only the beam mode will produce bending effects. Thus, it is only necessary to observe the response of the third generalized coordinate Z_3. For the case of the system under initial deflections, Z_3 is given in Fig. 3.21. The moment is obtained by noting that the modal deflection is Z_3 times the mode shape. Using Table 3.1, the deflection is

$$y = Z_3(\cos \kappa x + \text{ch } \kappa x - 0.9825 (\sin \kappa x + \text{sh } \kappa x)) \tag{3.79}$$

with $\kappa = 4.370/l$. The moment is obtained by twice differentiating with respect to x and substituting in eq. (3.78):

$$M = EIZ_3\kappa^2(-\cos \kappa x + \text{ch } \kappa x - 0.9825 (-\sin \kappa x + \text{sh } \kappa x)) \tag{3.80}$$

From this expression, the stress in the beam can be computed at any time during the response, since Z_3 is a function of time as given in Fig. 3.21.

3.16   A COMPUTER PROGRAM FOR THE DYNAMIC RESPONSE OF MANY DEGREE-OF-FREEDOM SYSTEMS

In this section, a computer program is described which can compute the dynamic response of a system with up to 65 degrees of freedom. The size of the program can be increased by appropriate changes in the dimensioning statements: 65 degrees of freedom were introduced here to allow a moderate range of sample problems to be studied. The generalized masses are interconnected by linear spring–damper elements, frictional elements, and stop elements. Up to 60 spring–dampers, 80 friction elements, and 24 stops are allowable. In many ways, this computer program is similar to the three degree-of-freedom computer program of section 3.13. It differs in the specification of a narrower range of excitation inputs, allowing either time history or sinusoidal excitation of the generalized masses or of the supports of spring–damper elements. And the data input is simplified in the use of the FORTRAN NAMELIST option, which allows the specification of only data items that need to be changed for that run.

Table 3.5   DESCRIPTION OF INPUT VARIABLES

Variable	Description
DEL	Time step, Δt
NG	Number of generalized coordinates on which a particular force element acts
NC	Identifier for which masses are acted on by a particular force element
CC	Corresponding coupling ratios for force elements ($B_{i,r}$ in eq. 3.55)
C1, C2	For spring–damper elements, C1 is damping, C2 is the spring constant; for friction elements, C1 is the friction coefficient, C2 is the contact local spring constant; for stop elements, C1 is the clearance, C2 is the spring constant

Table 3.5—*continued*

Variable	Description
C3	Used only for stop elements, this is the damping in the element
N1	Identifier for normal force in friction element: numbers less than 100 indicate FU; numbers above 100 indicate stop element force (N1–100); minus values indicate a friction element acting at right angles to the following friction element. Set N1 = 0 for the next element
N2	Not used in program
NZ	Total number of coordinates, up to 65
INDX	Type of force element: 1 = spring–damper, 2 = friction, 3 = stop, 8 = end of element data
ZM	Mass
TSTP	Print time interval
TTL	Total time
IDONE	1, when computation is complete; 0 otherwise
NEWLDS	When 1, same system is run with new loads; 0 otherwise
FV	Time history external forces
JFV	Identifies which masses FV is applied to; several time histories can apply cumulatively to the same mass
NFV	Number of time history force excitations
SV	Time history support displacements (positive displacement stretches support spring)
JSV	Identifies which support springs have displacement excitation; several time histories can apply cumulatively to the same spring
NSV	Number of time history displacement excitations
FSINF	Amplitude and frequency of externally applied sinusoidal forces
JSINF	Identifier of masses to which sinusoidal forces indicated are applied; the same mass can be indicated as being acted on by several different sinusoidal forces
NSINF	An indicator of the number of sinusoidal forces applied; when 0, there are none
SSINS	Amplitude and frequency of sinusoidal support displacements
JSINS	Identifier of support springs to which sinusoidal displacements are applied; the same spring can be indicated as being acted on by several different displacement excitations
NSINS	An indicator of the number of sinusoidal support displacements; when 0, there are none
FU	Normal forces F for friction force μF
MFV	Number of points on time history force curves, FV
MSV	Number of points on time history displacement curves, SV
ZINIT	Initial displacements of coordinates
ZDOT	Initial velocities of coordinates

The principal input variables are described in Table 3.5, in the order in which they appear in the NAMELIST/LIST statement in the computer program. To describe the manner in which the data are specified, it is now necessary to discuss the NAMELIST option and its application.

The NAMELIST input option

This computer program uses an input reading option, known as NAMELIST, which simplifies the specification of input. In the program, all input variables are specified as follows, under the NAMELIST named LIST:

NAMELIST/LIST/DEL, NG, NC, CC, C1, C2, C3, N1, N2, NZ, INDX, ZM, TSTP, TTL, IDONE, NEWLDS, FV, JFV, NFV, SV, JSV, NSV, FSINF, JSINF, NSINF, SSINS, JSINS, NSINS, FU, MFV, MSV, ZINIT, ZDOT

For each problem to be computed with the program, not all of these variables need to be specified—only those which are needed in the problem description need be given. The NAMELIST option facilitates this specification. Thus, if out of the list at a particular instant we want to read values of DEL, NZ, ZM(1), ZM(2), ZM(3), we might set in the program a read statement

$$\text{READ (5, LIST)}$$

and, in the input card deck, the input data

$$\text{\$LIST/DEL} = 0{\cdot}01, \text{NZ} = 15, \text{ZM}(1) = 2, 4, 6\text{\$}$$

This card would cause the storage of the values

$$
\begin{aligned}
\text{DEL} &= 0{\cdot}01 \\
\text{NZ} &= 15 \\
\text{ZM}(1) &= 2{\cdot}0 \\
\text{ZM}(2) &= 4{\cdot}0 \\
\text{ZM}(3) &= 6{\cdot}0
\end{aligned}
$$

Note that all the variables assigned to the NAMELIST named LIST were not read— only those identified on the data card itself. (However, at least one of the variables assigned to the NAMELIST named LIST must be on the data card.) The data items are separated by commas. If more than one card is needed, the last item on each card must be a constant followed by a comma. A \$ following the last data item signals the end of the data. Data may appear in only columns 2 through 72. Integer numbers are automatically converted to real numbers where required.

A chief advantage of using the NAMELIST option for reading in data is that new values are read in only for the data whose values change. This is chiefly useful when one has a series of cards to be read, in which the values of only one or two variables need to be changed on each card.

Computer program description

A flow chart of the computer program is given in Fig. 3.22. The time step integration loop is the chief feature of the program. The input is read by two READ statements. They actually simply call the NAMELIST, allowing, at that point in the logical program flow, a change in any NAMELIST variable. We have found it most useful to use the first READ at statement 101 to obtain the controlling time intervals, the excitation description, and the initial values of the generalized coordinates. The second READ at statement 2 is used to specify each force element (spring–damper, frictional, or stop) individually. The recommended order in which the data can be read is shown in Table 3.6.

The program is written for any consistent set of units. For example, both pound, inch, second units or newton, metre, second units may be used.

The computer program listing is now given.

Computer program listing

```
C         DYNAMICS OF A MANY DEGREE OF FREEDOM SYSTEM
C
      COMMON ZM(65),Z(3,65),FRC(65),NZ,JZ(1000),AZ(1000),JMZ(67)
C
C   THE GENERALIZED MASSES ARE STORED IN ZM.
C   Z CONTAINS THE PRESENT AND PREVIOUS TWO GENERALIZED DISPLACEMENTS.
C   FRC PROVIDES FOR THE 65 GENERALIZED FORCES. NZ IS THE NUMBER OF
C   GENERALIZED COORDINATES USED. JZ IS A SELF-GENERATED MAP OF THE
C   MASSES ON WHICH THE SPRING, DAMPER, STOP ETC, FORCES ACT
C   AZ CONTAINS MULTIPLIERS FOR THE FORCES USED IN COMPUTING THE
C   GENERALIZED FORCES.
C   SELF GENERATED JMZ(N) CONTAINS THE STARTING LOCATION IN AZ AND JZ FOR
C   NUMBERS RELATED TO GENERALIZED COORDINATE N.
C
      COMMON INDX,MX,MY,MW,FX(60),X(3,60),CX(3,60),FY(80),Y(2,80),
     CCY(2,80),JY(80),FW(24),W(3,24),CW(5,24),FV(10,2,5),JFV(5),
     CSV(10,2,5),JSV(5),FU(18),FSINF(2,5),JSINF(5),NSINF,SSINS(2,5),
     CJSINS(5),NSINS,MFV,MSV,ZINIT(65),ZDOT(65)
C
C   INDX OF 1,2,3 CORRESPONDS WITH X,Y,W TYPE FORCES.
C   INDX OF 8 IS END OF FORCE DATA. MX,MY,MW ARE THE SELF GENERATED
C   NUMBER OF ELEMENTS OF EACH TYPE. FX,FY,FW ARE THE FORCES OF EACH
C   TYPE. CX,CY,CW ARE CORRESPONDING COEFFICIENTS.
C   JY GIVES THE APPROPRIATE NORMAL FORCE NUMBER FOR FRICTION. FV GIVES
C   THE EXTERNAL FORCE-TIME HISTORY AT MASSES JFV. SV GIVES THE SUPPORT
C   DISPLACEMENT UNDER SPRING-DAMPER ELEMENTS JSV.
C   FOR SV AND FV 10 POSITION (OR FORCE) VALUES ARE STORED WITH 10
C   CORRESPONDING TIME VALUES FOR 5 SPRING ENDS (OR MASSES) IDENTIFIED
C   IN JSV (OR JFV). SSINS AND FSINS STORE THE AMPLITUDE AND FREQ OF
C   SINUSOIDAL DISPLACEMENT (OR FORCE) FOR 5 SPRING ENDS (OR MASSES)
C   IDENTIFIED IN JSINS (OR JSINF). NSINS IS THE NUMBER OF VALUES IN
C   JSINS, ETC.
C   WHEN READING X (SPRING-DAMPER) ELEMENTS C1 IS DAMPING AND C2 IS
C   SPRING RATE. WHEN READING Y ELEMENTS, C1 IS FRICTION COEFFICIENT,
C   C2 IS CONTACT SPRINGBACK SPRING RATE,
C   AND N1 IS NORMAL FORCE IDENTIFIER. VALUES OF N1 LESS THAN 100 REFER
C   TO FU(N1),READ-IN NORMAL FORCES.FOR N1 ABOVE 100 THEY REFER TO
C   FW(N1-100). GIVING N1 A MINUS VALUE MEANS IT IS A PAIR, THAT IS AT
C   RIGHT ANGLES TO, THE FOLLOWING Y-TYPE FORCE. (SET N1=0 FOR THE FOLLOW-
C   ING FORCE.) WHEN READING W, STOP ELEMENTS, C1 IS THE FREE SPACE,
C   C2 IS THE SPRING RATE, AND C3 IS THE DAMPING.
C
      COMMON NG,NC(10),CC(10),DEL,TM,TTL,TSTP,C1,C2,C3,N1,N2
C
C   NG IS THE NUMBER OF GENERALIZED COORDINATES AFFECTED BY A PARTICULAR
C   FORCE
C   NC IS THE COORDINATE NUMBER AND CC IS THE DISPLACEMENT AT THE FORCE
C   PER UNIT GENERALIZED COORDINATE (1 FOR TRANSLATION, LEVER ARM FOR
C   ROTATION, DISPLACEMENT IN MODE FOR NORMAL MODE, ETC.)
C   DEL IS THE TIME STEP, TM THE TIME, TTL THE TOTAL TIME OF INTEREST,
C   TSTP THE TIME BETWEEN OUTPUT PRINTS.
C
C   INPUT VARIABLES
C
      NAMELIST/LIST/DEL,NG,NC,CC,C1,C2,C3,N1,N2,NZ,INDX,ZM,TSTP,TTL,
     CIDONE,NEWLDS,FV,JFV,NFV,SV,JSV,NSV,FSINF,JSINF,NSINF,SSINS,JSINS,
     C NSINS,FU,MFV,MSV,ZINIT,ZDOT
C
      NAMELIST/LISTGO/IDONE,NZ,NEWLDS,DEL,TSTP,TTL,ZM,FU,FV,SV,JFV,JSV,
     1FSINF,SSINS,JSINF,JSINS,NFV,NSV,NSINF,NSINS,MFV,MSV,ZINIT,ZDOT
```

```
C
  101   READ(5,LIST)
        IF(IDONE.EQ.1) GO TO 100
        IF (NEWLDS.EQ.1) GO TO 40
C
C  SET UP INITIAL VALUES OF SOME PARAMETERS
C
    1 K=0
      J=-9
    4 J=J+10
      K=K+1
      JMZ(K)=J
      IF(K .LT. 66) GO TO 4
      INC=0
      MX=0
      MY=0
      MW=0
      CALL ZEROM(CX,3,60)
      CALL ZEROM(CY,2,80)
      CALL ZEROM(CW,5,24)
      CALL IZEROM(JY,1,80)
      CALL ZEROM(AZ,1,1000)
      CALL IZEROM(JZ,1,1000)
C
    2 READ (5,LIST)
      IF(INC.EQ.1) GO TO 3
      DELSQI=DEL**2
      DELSQ=1./DELSQI
      DELI=0.5/DEL
      INC=1
    3 CONTINUE
C
C  SET UP COEFFICIENTS IN FORCE ELEMENTS
C
      GO TO (10,12,14,40,40,40,40,40),INDX
C
C  SPRING-DAMPER COEFFICIENTS
C
   10 MX=MX+1
      WRITE(6,1019) (MX,C1,C2,((NC(J),CC(J)),J=1,NG))
      CX(1,MX)=-DELI*3.*C1-C2
      CX(2,MX)=DELI*4.*C1
      CX(3,MX)=-DELI*C1
      DO 11 K=1,NG
   11   CALL STORE (MX,NC(K),CC(K),AZ,JZ,JMZ)
      GO TO 2
C
C  FRICTION ELEMENT COEFFICIENTS
C
   12 MY=MY+1
      WRITE(6,1020) (MY,N1,C1,C2,((NC(J),CC(J)),J=1,NG))
      CY(1,MY)=C1
      CY(2,MY)=C2
      JY(MY)=N1
      DO 13 K=1,NG
   13 CALL STORE((MY+1000),NC(K),CC(K),AZ,JZ,JMZ)
      GO TO 2
C
C  STOP ELEMENT COEFFICIENTS
C
```

```
   14 MW=MW+1
      WRITE(6,1021) (MW,C1,C2,C3,((NC(J),CC(J)),J=1,NG))
      CW(1,MW)=C1
      CW(2,MW)=-DELI*3.*C3-C2
      CW(3,MW)=DELI*4.*C3
      CW(4,MW)=-DELI*C3
      CW(5,MW)=-C1*C2
      DO 15 K=1,NG
   15 CALL STORE((MW+2000),NC(K),CC(K),AZ,JZ,JMZ)
      GO TO 2
   40 CONTINUE
C
C     PRINT ALL ITEMS IN LISTGO FOR DIAGNOSIS
C
      WRITE(6,LISTGO)
C
      CALL ZEROM(X,3,60)
      CALL ZEROM(Y,2,80)
      CALL ZEROM(W,3,24)
C
C COMPUTE INITIAL DISPLACEMENTS Z(-1), Z(-2) FROM INITIAL CONDITIONS
C
      DO 41 J=1,NZ
      Z(1,J)=ZINIT(J)
      Z(2,J)=Z(1,J)-DEL*ZDOT(J)
   41 Z(3,J)=Z(2,J)-DEL*ZDOT(J)
   45 CONTINUE
      TCOUNT=-0.000001
C
C  SET INITIAL TIME
C
      TM=-DEL
C
C  START TIME STEP INTEGRATION LOOP
C
   50 TM=TM+DEL
   51 CONTINUE
      IF (TM.EQ.0.0) GO TO 90
C
C COMPUTE CONTRIBUTIONS TO GENERALIZED FORCES FROM EACH FORCE ELEMENT
C
  771 CONTINUE
      DO 82 K=1,NZ
      L=JMZ(K)
      LL=JMZ(K+1)-1
      FRC(K)=0.0
      DO 81 J=L,LL
      JJZJ=JZ(J)
      INDX=0
  810 INDX=INDX+1
      JZJ=JJZJ
      JJZJ=JJZJ-1000
      IF(JJZJ.GT.0) GO TO 810
      IF (JZJ.EQ.0) GO TO 82
C
C  SPRING-DAMPER ELEMENT
C
      GO TO (30,31,32), INDX
   30 FRC(K)=FRC(K)+AZ(J)*FX(JZJ)
      GO TO 81
```

```
C
C   FRICTION ELEMENT
C
   31 FRC(K)=FRC(K)+AZ(J)*FY(JZJ)
      GO TO 81
C
C   STOP ELEMENT
C
   32 FRC(K)=FRC(K)+AZ(J)*FW(JZJ)
   81 CONTINUE
   82 CONTINUE
C
      IF (NFV.EQ.0) GO TO 330
C
C   ADD CONTRIBUTION OF EXTERNAL FORCES
C
      DO 33 K=1,NFV
      JFVK=JFV(K)
      CALL INTRP(FV(1,2,K),FV(1,1,K),A,A,TM,B,B,B,MFV,1)
   33 FRC(JFVK)=FRC(JFVK)+B
  330 IF (NSINF.EQ.0) GO TO 331
      DO 332 K=1,NSINF
      JFVK=JSINF(K)
      B=FSINF(1,K)*SIN(FSINF(2,K)*6.2832*TM)
  332 FRC(JFVK)=FRC(JFVK)+B
  331 CONTINUE
C
C   ACCELERATIONS OF EACH MASS
C
      DO 280 K=1,NZ
  280 FRC(K)=FRC(K)/ZM(K)
C
C   COMPUTE NEW GENERALIZED DISPLACEMENTS AT END OF TIME STEP
C
      DO 80 K=1,NZ
      DO 93 JJ=1,2
      L=3-JJ
   93 Z(L+1,K)=Z(L,K)
      Z(1,K)=2.*Z(2,K)-Z(3,K)+FRC(K)*DELSQI
      IF(TM.EQ.DEL)Z(1,K)=Z(1,K)-0.5*FRC(K)*DELSQI
   80 CONTINUE
   83 CONTINUE
C
C   COMPUTE CONTRIBUTION OF GENERALIZED DISPLACENENTS TO THE
C   FORCE ELEMENT EXTENSIONS
C
      IF (MX.EQ.0) GO TO 851
      DO 85 K=1,MX
      DO 88 J=1,2
      L=3-J
   88 X(L+1,K)=X(L,K)
   85 X(1,K)=0.0
  851 IF (MY.EQ.0) GO TO 852
      DO 86 K=1,MY
   86 Y(1,K)=0.0
  852 IF (MW.EQ.0) GO TO 90
      DO 853 K=1,MW
      W(3,K)=W(2,K)
      W(2,K)=W(1,K)
  853 W(1,K)=0.0
```

```
   90 CONTINUE
      DO 91 K=1,NZ
      L=JMZ(K)
      LL=JMZ(K+1)-1
      DO 92 J=L,LL
      JJZJ=JZ(J)
      IF (JJZJ.LE.0) GO TO 92
      AZJ=AZ(J)*Z(1,K)
      IF(TM.NE.0.0) GO TO 993
      AZ2=AZ(J)*Z(2,K)
      AZ3=AZ(J)*Z(3,K)
  993 CONTINUE
      INDX=0
  920 INDX=INDX+1
      JZJ=JJZJ
      JJZJ=JJZJ-1000
      IF (JJZJ.GT.0) GO TO 920
      GO TO (921,922,923),INDX
C
C   SPRING-DAMPER ELEMENT EXTENSION
C
  921 X(1,JZJ)=X(1,JZJ)+AZJ
      IF (TM.NE.0.0) GO TO 92
      X(2,JZJ)=X(2,JZJ)+AZ2
      X(3,JZJ)=X(3,JZJ)+AZ3
      GO TO 92
C
C   FRICTION ELEMENT EXTENSION
C
  922 Y(1,JZJ)=Y(1,JZJ)+AZJ
      IF(TM.NE.0.0) GO TO 92
      Y(2,JZJ)=Y(2,JZJ)+AZ2
      GO TO 92
C
C   STOP ELEMENT EXTENSION
C
  923 W(1,JZJ)=W(1,JZJ)+AZJ
      IF (TM.NE.0.0) GO TO 92
      W(2,JZJ)=W(2,JZJ)+AZ2
      W(3,JZJ)=W(3,JZJ)+AZ3
   92 CONTINUE
   91 CONTINUE
      IF (NSV.EQ.0) GO TO 10601
C
C   CONTRIBUTION OF SUPPORT DISPLACEMENTS TO FORCE ELEMENT EXTENSIONS
C
      DO 924 K=1,NSV
      JSVK=JSV(K)
      CALL INTRP(SV(1,2,K),SV(1,1,K),A,A,TM,B,B,B,MSV,1)
  924 X(1,JSVK)=X(1,JSVK)+B
10601 IF (NSINS.EQ.0) GO TO 10602
      DO 10603 K=1,NSINS
      JSVK=JSINS(K)
      B=SSINS(1,K)*SIN(SSINS(2,K)*6.2832*TM)
10603 X(1,JSVK)=X(1,JSVK)+B
10602 CONTINUE
C
C   COMPUTE NEW FORCES IN ELEMENTS DUE TO THE ELEMENT EXTENSIONS
C
 1060 IF (MX.EQ.0) GO TO 602
```

```
          DO 60 K=1,MX
          FX(K)=0.
          DO 60 J=1,3
   60 FX(K)=FX(K)+CX(J,K)*X(J,K)
  602 IF (MW.EQ.0) GO TO 603
          DO 66 K=1,MW
          FW(K)=0.0
          IF((-W(1,K)).LE.CW(1,K)) GO TO 66
          DO 669 J=1,3
  669 FW(K)=FW(K)+CW(J+1,K)*W(J,K)
          FW(K)=FW(K)+CW(5,K)
          IF(FW(K).LT.0.0)  FW(K)=0.0
   66 CONTINUE
  603   CONTINUE
  601 IF (MY.EQ.0) GO TO 604
          DO 61 K=1,MY
          IF(JY(K))  6001,61,6002
 6001 JYK=-JY(K)
          YSQ=((Y(1,K)-Y(2,K))**2+(Y(1,K+1)-Y(2,K+1))**2)**0.5
          GO TO 6003
 6002 JYK=JY(K)
          YSQ=ABS(Y(1,K)-Y(2,K))
 6003 IF(JYK.GT.100) GO TO 63
          FYK=FU(JYK)
          GO TO 65
   63 FYK=FW(JYK-100)
   65 IF (YSQ.LT.0.000001) YSQ=.000001
          YF1=CY(1,K)*FYK
          YF2=CY(2,K)*(Y(1,K)-Y(2,K))
          IF(JY(K).LT.0) YF3=CY(2,K+1)*(Y(1,K+1)-Y(2,K+1))
          YFK=ABS(YF2)
          IF(JY(K).LT.0) YFK=(YF2**2+YF3**2)**0.5
          IF(YFK-YF1) 651,651,652
  651 YSQI=YFK/YSQ
          GO TO 653
  652 YSQI=YF1/YSQ
  653 CONTINUE
          FY(K)=YSQI*(Y(2,K)-Y(1,K))
          IF(JY(K).LT.0)  FY(K+1)=YSQI*(Y(2,K+1)-Y(1,K+1))
          IF(YFK-YF1)  61,61,654
  654 YF1=YF1/YFK
          Y(2,K)=Y(1,K)-YF1*(Y(1,K)-Y(2,K))
          IF (JY(K).LT.0) Y(2,K+1)=Y(1,K+1)-YF1*(Y(1,K+1)-Y(2,K+1))
   61 CONTINUE
  604 CONTINUE
C
C  PRINT OUTPUT
C
          IF (TM.LT.TCOUNT) GO TO 1095
 1001 FORMAT(1H0,17X,17H*********         ,7HTIME = ,F12.5,17H          ****
     C******)
 1002 FORMAT(1H0,3HX= ,8F13.8/(4X,8F13.8))
 1003 FORMAT(1H0,3HY= ,8F13.8/(4X,8F13.8))
 1004 FORMAT(1H0,3HZ= ,8F13.8/(4X,8F13.8))
 1005 FORMAT(1H0,3HW= ,8F13.8/(4X,8F13.8))
 1006 FORMAT(1H0,4HFX= ,9F12.1/(5X,9F12.1))
 1007 FORMAT(1H0,4HFY= ,9F12.1/(5X,9F12.1))
 1008 FORMAT(1H0,4HFW= ,9F12.1/(5X,9F12.1))
 1011 FORMAT(1H0,17X,31HDISPLACEMENT OF SPRING ELEMENTS)
 1012 FORMAT(1H0,17X,33HDISPLACEMENT OF FRICTION CONTACTS)
```

```
1013 FORMAT(1H0,17X,34HDISPLACEMENT OF NON-LINEAR SPRINGS)
1014 FORMAT(1H0,17X,25HGENERALIZED DISPLACEMENTS)
1016 FORMAT(1H0,17X,28HSPRING DAMPER ELEMENT FORCES)
1017 FORMAT(1H0,17X,24HFRICTION CONTACTS FORCES)
1018 FORMAT(1H0,17X,24HNON-LINEAR SPRING FORCES)
1019 FORMAT(1H0,14HSPRING ELEMENT,2X,I2,2X,3HC1=,F10.3,2X,3HC2=,F13.3,
    12X/6HNC/CC=,(10(I2,F6.2,2X))//)
1020 FORMAT(1H0,16HFRICTION ELEMENT,2X,I2,2X,3HN1=,I4,2X,3HC1=,F10.8,
    12X,3HC2=,F13.3,2X/6HNC/CC=,(10(I2,F6.2,2X))//)
1021 FORMAT(1H0,17HNON-LINEAR SPRING,2X,I2,2X,3HC1=,F10.3,2X,3HC2=,
    1F13.3,2X,3HC3=,F10.3,2X/6HNC/CC=,(10(I2,F6.2,2X))//)
     TCOUNT=TCOUNT+TSTP
     WRITE(6,1001)TM
      IF(MX.EQ.0) GO TO 661
     WRITE(6,1011)
     WRITE(6,1002)(X(1,K),K=1,MX)
     WRITE(6,1016)
     WRITE(6,1006)(FX(K),K=1,MX)
 661 CONTINUE
      IF(MY.EQ.0) GO TO 663
     WRITE(6,1012)
     WRITE(6,1003)(Y(1,K),K=1,MY)
     WRITE(6,1017)
     WRITE(6,1007)(FY(K),K=1,MY)
 663 CONTINUE
      IF(MW.EQ.0) GO TO 665
     WRITE(6,1013)
     WRITE(6,1005)(W(1,K),K=1,MW)
     WRITE(6,1018)
     WRITE(6,1008)(FW(K),K=1,MW)
 665 CONTINUE
     WRITE(6,1014)
     WRITE(6,1004) (Z(1,K),K=1,NZ)
C
C  LOOP TO STATEMENT 50 FOR NEW TIME STEP
C
1095 IF (IM.LI.ITL) GO TO 50
     GO TO 101
 100 STOP
     END
```

```
      SUBROUTINE STORE(M,NN,C,AZ,JZ,JMZ)
C  NN IS THE MASS NUMBER
C  M IS THE ELEMENT NUMBER
C  JZ STORES ELEMENT NUMBERS
C  AZ STORES COUPLING RATIOS
C  JMZ IS A MAP OF JZ AND AZ
      DIMENSION AZ(1),JZ(1),JMZ(1)
      IF(NN.EQ.0) GO TO 100
      N=IABS(NN)
      L=ISIGN(M,NN)
   10 K=JMZ(N)-1
   24 K=K+1
C  SUBROUTINE SHIFT REARRANGES STORAGE IF NECESSARY
      IF(K.EQ.JMZ(N+1)) CALL SHIFT(K-1,N+1,JMZ,JZ,AZ,1000,66)
C  LOOKS FOR AN EMPTY STORAGE LOCATION
      IF (JZ(K).NE.0) GO TO 24
C  STORES ELEMENT NUMBER IN JZ
      JZ(K)=L
C  STORES COUPLING RATIO IN JZ
      AZ(K)=C
  100 RETURN
      END
      SUBROUTINE SHIFT(KK,JJ,JMZ,JZ,AZ,M,N)
C  KK+1 IS LOCATION IN JZ TO BE EMPTIED
C  JJ IS THE NEXT MASS NUMBER
C  M IS THE SIZE OF JZ(1000 IN THIS PROGRAM)
C  N IS THE NUMBER OF DEGREES OF FREEDOM PLUS 1(66 IN THIS PROGRAM)
      DIMENSION JMZ(1),JZ(1),AZ(1)
C  LOOKS FOR AN EMPTY STORAGE LOCATION
      DO 10 J=JJ,N
      JMZJ=JMZ(J)-1
      IF (JZ(JMZJ).EQ.0) GO TO 11
      JMZ(J)=JMZJ+2
   10 CONTINUE
      JMZJ=JMZJ+1
   11 CONTINUE
C  CHECKS THAT THE SIZE OF JZ HAS NOT BEEN EXCEEDED
      IF (JMZJ.LT.(M+1)) GO TO 14
      WRITE(6,1001)(JMZ(J),J=1,N)
      STOP
 1001 FORMAT (20I6)
C  MOVES ITEMS IN JZ AND AZ UP TO PROVIDE MORE SPACE FOR MASS (JJ-1)
   14 K=JMZJ
   12 K=K-1
      IF (K.EQ.KK) GO TO 13
      JZ(K+1)=JZ(K)
      AZ(K+1)=AZ(K)
      GO TO 12
C  SETS EMPTIED SPACE TO ZERO
   13 JZ(K+1)=0
      AZ(K+1)=0.0
  100 RETURN
      END
```

```
CINTRP              INTERPOLATION SUBROUTINE FOR THREE VARIABLES WHICH
CINTRP                      ARE FUNCTIONS OF A FOURTH
        SUBROUTINE INTRP(X,Y,W,Z,XX,YY,WW,ZZ,N,M)
        DIMENSION Y(1),X(1),W(1),Z(1)
C       INTERPOLATES LINEARLY BETWEEN POINTS ON A CURVE.
C       X IS A VECTOR OF N VALUES OF ABSCISSA
C       Y,W,Z ARE VECTORS OF N VALUES OF ORDINATES
C       FOR X=XX, Y=YY,W=WW, AND Z=ZZ
C       IF M=3 PROGRAM INTERPOLATES FOR Y,W, AND Z
C       IF M=2 PROGRAM INTERPOLATES FOR Y AND W. IN CALL USE SAME SYMBOL
C       FOR Z AND Y,ZZ AND YY. IF M=1 USE SAME SYMBOL FOR Z,W,Y AND ZZ,WW,
C       YY AND PROGRAM WILL INTERPOLATE FOR Y ONLY.
        DO 10 J=2,N
        K=N+1-J
        IF (X(K+1)-X(K)) 10,10,11
   11   IF (XX-X(K))   10,20,20
   10   CONTINUE
   20   TS1= X(K+1)-X(K)
        TS2=(X(K+1)-XX)/TS1
        TS1=(XX-X(K))/TS1
        IF (M-2)    50,40,30
   30   ZZ= TS1*Z(K+1)+TS2*Z(K)
   40   WW= TS1*W(K+1)+TS2*W(K)
   50   YY= TS1*Y(K+1)+TS2*Y(K)
  100   RETURN
        END
        SUBROUTINE IZEROM(M,I,K)
        DIMENSION  M(1)
        II=I*K
        DO 10  J=1,II
   10 M(J)=0
        RETURN
        END
CZEROM     SUBROUTINE ZEROM FOR SETTING A MATRIX TO ZERO(REAL NUMBERS)
C              SUBROUTINE ZEROM
        SUBROUTINE  ZEROM(A,I,K)
        DIMENSION  A(1)
        II=I*K
        DO 10  J=1,II
   10 A(J)=0.0
        RETURN
        END
```

Table 3.6   ORDER OF INPUT VARIABLES

Function	Variables
Mass description	ZM, NZ
Time specification	TSTP, TTL, DEL
Control indices	IDONE, NEWLDS
Excitation description	FV, JFV, NFV, MFV, SV, JSV, NSV, MSV, FSINF, JSINF, NSINF, SSINS, JSINS, NSINS, FU
Initial conditions	ZINIT, ZDOT
Force element description	INDX, NG, NC, CC, C1, C2, C3, N1

Storage and retrieval of information

The program for the three degree-of-freedom system stored the force element coupling ratios B in the array AZ and the force element identifiers in the array JZ. The space in AZ and JZ used for each mass was preassigned as 20 memory locations. In a system with up to 65 mass coordinates, it is likely that some masses may be acted on by more than 20 force elements, and others by fewer than 20. For this reason, the present program provides initially for 10 spaces for each coordinate. The starting point for the storage pertaining to each coordinate in the arrays JZ and AZ is stored in the array JMZ. The storage is accomplished by the subroutine STORE. This subroutine checks the availability of storage space before storing data. If there is not enough, it calls on subroutine SHIFT to provide more space. SHIFT at the same time changes the starting points of the array JMZ.

Input specification

To illustrate the way in which input data can be specified to the computer program, consider the eight examples shown in Fig. 3.23. There are actually two systems: one, a single mass system with one spring support; another, a three mass system with three springs. They are both excited by different force or displacement conditions in each of the eight examples. The following remarks show how the data are read in by the NAMELIST feature.

For example (1), consisting of a single mass and spring, excited at the support by a unit displacement:

$LIST/DEL = 0.01, NZ = 1, ZM(1) = 10, TSTP = 0.1, TTL = 1, IDONE = 0,
NSV = 1, JSV(1) = 1, SV(1, 1, 1) = 1, 1, SV(1, 2, 1) = 0, 20, MSV = 2,
NEWLDS = 0, NFV = 0, NSINF = 0, NSINS = 0$

Now, the data for the spring–damper element are needed:

$LIST/INDX = 1, C1 = 0, C2 = 10, NG = 1, NC(1) = 1, CC(1) = 1$

Note that the NAMELIST routine stores $SV(1, 1, 1) = 1$ and $SV(2, 1, 1) = 1$. In general, only the first number in a sequence need be identified. Now the index INDX is set equal to 8, indicating that there are no further force elements:

$LIST/INDX = 8$

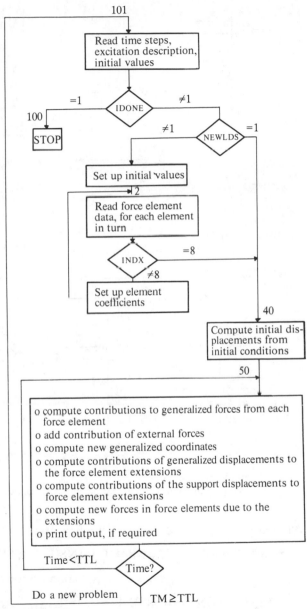

FIGURE 3.22
Flow diagram for 65 degree-of-freedom program

For example (2), much of the same data are still operative, and only the force descriptions need be changed:

$LIST/NEWLDS = 1, NFV = 1, NSV = 0, JFV(1) = 1, FV(1, 1, 1) = 10, 10, FV(1, 2, 1) = 0, 20, MFV = 2$

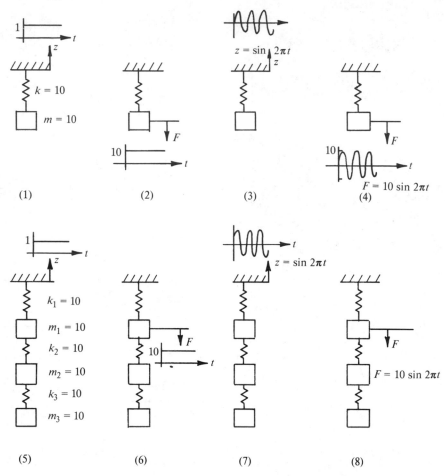

FIGURE 3.23
Showing the eight examples for the discussion of input specification

Similarly, example (3) only requires a new excitation:

$LIST/NFV = 0, NSINS = 1, JSINS(1) = 1, SSINS(1, 1) = 1, 1$

and example (4) likewise has the input specification:

$LIST/NSINS = 0, NSINF = 1, JSINF(1) = 1, FSINF(1, 1) = 10, 1$

For example (5), which now contains three masses and springs, we set

$LIST/NZ = 3, ZM(1) = 3*10, NEWLDS = 0, NSINF = 0, NSV = 1$
$LIST/INDX = 1, C1 = 0, C2 = 10, NG = 1, NC(1) = 1, CC(1) = 1$
$LIST/NG = 2, NC(1) = 1, 2, CC(1) = -1, 1$
$LIST/NC(1) = 2, 3$
$LIST/INDX = 8$

Finally, for each remaining example, we change only the excitations. The data are:

$LIST/NEWLDS = 1, NSV = 0, NFV = 1$
$LIST/NFV = 0, NSINS = 1$
$LIST/NSINS = 0, NSINF = 1$

We then add a card to stop the computation:

$LIST/IDONE = 1$

Problems

3.33 What is the input for example (6) where the force is applied to mass 2 instead of mass 1? (*Hint:* Set JFV(1) = 2.)

3.34 Repeat example (8) but apply the force to mass 3 instead of mass 1. (*Hint:* Set JSINF(1) = 3.)

3.35 Combine the action of the forces in problems 3.33 and 3.34. (*Hint:* Set JFV(1) = 2, JSINF(1) = 3, NFV = 1, and NSINF = 1.)

3.36 Repeat example (8) but apply the same force to masses 2 and 3 as is applied to mass 1. (*Hint:* Set FSINF(1, 1) $-$ 10, 1, 10, 1, 10, 1, NSINF $=$ 3, and JSINF(1) $=$ 1, 2, 3.)

3.37 Repeat example (1) with force of example (2) applied simultaneously. (*Hint:* Set SV(1, 1, 1) = 1, 1, SV(1, 2, 1) = 0, 20, MSV = 2, FV(1, 1, 1) = 10, 10, FV(1, 2, 1) = 0, 20, MFV = 2, NSV = 1, and NFV = 1.)

3.17 MASS STRIKING A BEAM

Consider now how the impact of a ball striking a beam can be modeled. The ideas developed here on modeling the forces between the ball and beam during the impact will be useful in a later chapter of this book, when we come to the study of more general impacts, as in chapter 7. When the ball and beam are in contact, the contact forces can be induced by a spring between the mass of the ball and that of the beam. And when the ball loses contact, or bounces away from the beam, the spring force is zero. This situation can be modeled by a stop spring whose force is zero during the periods of noncontact. Before the initial contact, a certain clearance is assumed between the ball and the beam. The stop spring contact, therefore, can be illustrated as in Fig. 3.24. The precise choice of the force–deflection curve during contact will depend on the problem at hand—in this case, it is assumed that the spring rate k is constant for all values of contact closure. (Nonlinearity can be modeled by using several stops.) In general, the value of k will depend on the elastic constants of the material in the ball. The stop element can include damping, and to that extent could simulate inelastic effects during contact.

The motion of the beam is modeled by expressing it in terms of its normal modes, which are used as generalized coordinates. Let us take, as an example, the case of a ball falling freely and striking a simply supported beam at its midpoint. Then, because of the symmetry of the situation, we need only use the symmetric

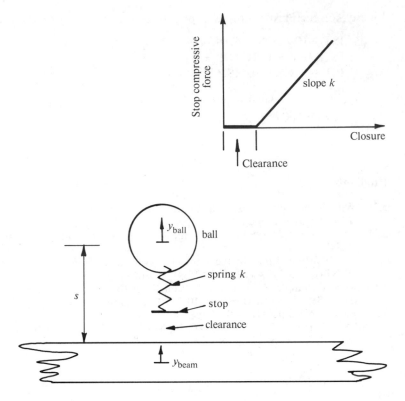

FIGURE 3.24
Illustrating how an impact situation can be modeled

modes of the beam as its generalized coordinates. Thus, we have the following information available:

$$\text{mode shape} = y = \sin(n\pi x/l) \quad (n = 1, 3, 5, 7, \ldots)$$

$$\text{frequency} \quad = \omega = 2{\cdot}85\, hn^2/l^2 (Eg/\rho)^{1/2}$$

where E is modulus, g is acceleration of gravity, ρ is weight per unit volume, l is beam length, and h is beam thickness. The beam width is b, so the beam weight is ρbhl.

As a result, we can compute the generalized mass and the generalized spring, which are:

$$\text{generalized mass} \ = bhl\rho/2g$$

$$\text{generalized spring} = 4{\cdot}06\, n^4 bE(h/l)^3$$

The coupling ratio of the stop element is $+1$ to the ball mass coordinate, and to the beam mass coordinates is given by: -1 for $n = 1$; $+1$ for $n = 3$; -1 for $n = 5$; $+1$ for $n = 7$; etc.

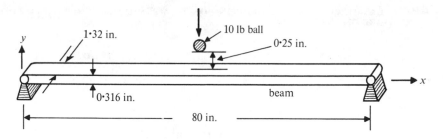

Generalized coordinate 1, ball displacement
Generalized coordinate 2, modal deflection, $n = 1$
Generalized coordinate 3, modal deflection, $n = 3$
Generalized coordinate 4, modal deflection, $n = 5$
Generalized coordinate 5, modal deflection, $n = 7$

Mode shape, $n = 1$, $y_1 = \sin(\pi x/80)$
Mode shape, $n = 3$, $y_3 = \sin(3\pi x/80)$
Mode shape, $n = 5$, $y_5 = \sin(5\pi x/80)$
Mode shape, $n = 7$, $y_7 = \sin(7\pi x/80)$

FIGURE 3.25
Weight of 10 lb dropping 0·25 in. on to a flexible steel beam simply supported at the ends

For this example, we take the beam as 80 in. long, 1·32 in. wide, and 0·316 in. thick, as shown in Fig. 3.25. We will include only the first four symmetric modes. The material is steel with $E = 30\,000\,000$ lb/in.^{2} and $\rho = 0.3$ lb/in.^{3}. We then have:

$b = 1.32$ in.
$h = 0.316$ in.
$l = 80.0$ in.
$g = 386.0$ in./sec^2
$\rho = 0.3$ lb/in.^{3}
beam mass $= m = 0.0259$ lb sec^2/in.
generalized mass, all modes $= 0.01295$ lb sec^2/in.
$k_1 = $ (generalized spring, $n = 1$) $= 9.9$ lb/in.
$k_3 = 81 \times 9.9 = 802$ lb/in.
$k_5 = 625 \times 9.9 = 6180$ lb/in.
$k_7 = 2401 \times 9.9 = 23\,750$ lb/in.

We will take the ball as having a weight of 10 lb (mass $= 0.0259$ lb sec^2/in.). The stiffness of the local contact between ball and beam is assumed to be 500 lb/in. We model the contact with a stop having an initial rate of 500 lb/in. The pull of gravity on the ball is modeled by applying to it a constant force of -10 lb.

Since the frequency when $n = 7$ is

$$\left(\frac{2.85}{2\pi}\right)\left(\frac{0.316 \times 7^2}{80^2}\right)\sqrt{\frac{30\,000\,000 \times 386}{0.3}} = 216 \text{ Hz}$$

and good accuracy requires a time step of about $1/20$ of the shortest period, Δt can be taken as $0.000\,25$ sec. The input data required is shown in Table 3.7.

Table 3.7 DATA INPUT FOR BALL DROPPING ON TO A BEAM. THE GENERALIZED COORDINATES ARE GIVEN IN FIG. 3.25

```
$LIST/DEL=0.00025,NZ=5,ZM(1)=0.0259,4*0.01295,TSTP=0.001,TTL=0.2,

IDONE=0,NEWLDS=0,JFV(1)=1,FV(1,1,1)=-10,-10,FV(1,2,1)=.0,.5,MFV=2,

 NFV=1$

$LIST/INDX=1,C1=0,C2=9.9,NG=1,NC(1)=2,CC(1)=1$

$LIST/NC(1)=3,C2=   802$

$LIST/NC(1)=4,C2=  6180$

$LIST/NC(1)=5,C2=23750$

$LIST/INDX=3,NG=5,C1=0.25,C2=500,NC(1)=1,2,3,4,5,CC(1)=1,-1,1,-1,1$

$LIST/INDX=8$

$LIST/IDONE=1$
```

The results of the calculations done with the computer program of section 3.16 are shown in Fig. 3.26.

It is interesting to note that the ball first contacts the beam after about 0·036 sec, when it has fallen the 0·25 in. initial clearance between itself and the beam. The contact of the ball with the beam is not steady. There is separation at $t = 0·06$ sec, recontact at $t = 0·085$ sec, separation again at $t = 0·112$ sec, recontact at $t = 0·13$ sec, etc. The deflection of the ball (coordinate 1 in Fig. 3.25) and the deflection in the first mode (coordinate 2) differ by about 0·25 in., as they should. The higher modes, coordinates 3, 4, 5, have small deflections. Their contributions to the force are significant, however, and cause the higher frequency components in the force curve.

Problems

3.38 In the example problem, the stop spring rate is 500 lb/in., while the modal spring rate in the highest frequency mode is 23 750 lb/in. It is likely, therefore, that the highest frequency mode has little effect on the answer. We might, therefore, consider using a value of Δt meeting the convergence criterion but not the 'good accuracy' criterion. For convergence, Δt must be less than $(1/2\pi f)$, where f is the highest frequency. In the example, f is 216 Hz so $\Delta t < 0·00147$ sec. Rerun the example problem with $\Delta t = 0·001$ sec (four times bigger time step) and compare the deflection results with those in the example.

3.39 Change the stop spring rate in the example problem to 5000 lb/in. What effect does this have on the impact process? What is the duration of contacts with the beam?

3.40 In the example problem, the beam is considered to have no damping. Recompute the response for a damping of 20 per cent of critical in each mode. (*Hint*: Critical damping is $2\sqrt{mk}$. The damping for $n = 1$ is therefore $0·20 \times 2 \times \sqrt{0·01295 \times 9·9} = 0·1434$ lb sec/in.

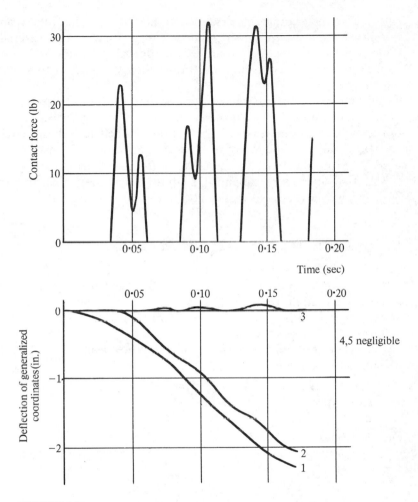

FIGURE 3.26
Contact force of ball on beam and deflections in generalized coordinates

Similarly, for $n = 3, 5, 7$, the damping is 1·290, 3·59, and 7·04 lb sec/in., respectively.) Does the damping smoothen the contact–force curve? Does it affect the maximum contact force?

3.41 In the example problem, the stop spring is considered to have no damping. Recompute the response for 20 per cent damping in the stop element. (*Hint:* Critical damping is $2\sqrt{mk} = 2\sqrt{0·0259 \times 500} = 3·6$. The stop damping is therefore $0·2 \times 3·6 = 0·72$.)

3.18   BEAMS AS SPRING–DAMPER ELEMENTS

Beams are used in structures to such an extent that it is desirable to discuss how they may be modeled as spring–damper elements in such a way as to be compatible with the computer programs described in this chapter. The stiffness and damping of the spring–damper elements are related to the geometric and material properties of the beams themselves. As a result, we first describe how the stiffness of such beam springs can be derived, before illustrating their application to structural dynamic modeling.

Two sorts of beam element are to be described. In this section, we discuss the modeling of a beam by means of two orthogonal springs, which allow the inclusion of rotary inertia in the analysis. Such spring models have been developed by Wen and Beylerian[4] for the study of elastoplastic beam deformation, and have been used by Moreadith and his associates[5,6] for the structural analysis of pipe whip restraints. For a more simplified analysis, where only displacements are to be modeled, we show how springs can be derived in section 3.19.

Derivation of the basic stiffness of each beam spring

To begin with, let us consider the four beams shown in Fig. 3.27 which are distorted in four ways: in bending, in shear with balanced bending, in extension, and in torsion. We shall derive the stiffness of each beam component in turn. The beams have a length l, elastic modulus E, shear modulus G, area of cross-section A, bending stiffness EI, and torsional stiffness GJ.

Consider the beam in pure bending, as shown in Fig. 3.27(a). The governing differential equation is

$$EI \frac{d^2 y}{dx^2} = +M \tag{3.81}$$

Direct integration yields

$$EIy = +\frac{Mx^2}{2} + c_1 x + c_2 \tag{3.82}$$

Since the boundary conditions at the beam ends demand that $y = 0$ at $x = 0$ and $x = l$, it is easily shown that

$$c_1 = -\frac{Ml}{2}, \quad c_2 = 0 \tag{3.83}$$

We take θ as the change in slope (in the direction M) from $x = 0$ to $x = l$. Thus, since the beam slope at the ends is equal but opposite in sign,

$$\frac{\theta}{2} = \left(\frac{dy}{dx}\right)_{x=0} = \frac{Ml}{2EI} \tag{3.84}$$

we have

$$\theta = \frac{Ml}{EI} \tag{3.85}$$

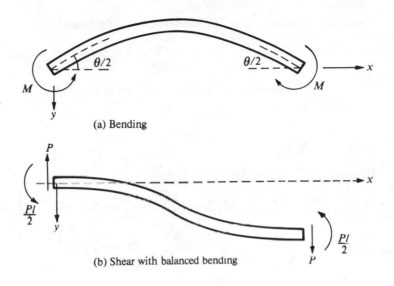

(a) Bending

(b) Shear with balanced bending

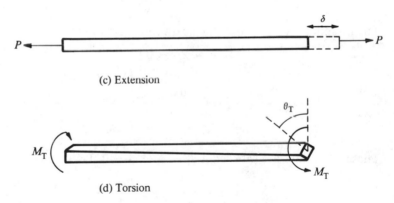

(c) Extension

(d) Torsion

FIGURE 3.27
Four component beam springs

The bending stiffness k_B of this configuration is given by

$$k_B = \frac{M}{\theta} = \frac{EI}{l} \tag{3.86}$$

In the case of the shear configuration of Fig. 3.27(b), we observe that the displacement y is made up of two parts: a portion y_b due to bending, and a portion y_s due to shear. Thus,

$$y = y_b + y_s \tag{3.87}$$

The shear displacement y_s is governed by the equation

$$\frac{dy_s}{dx} = +\frac{P\alpha}{AG} \tag{3.88}$$

where α is the shear correction factor. The bending displacement is governed by the equation

$$EI\frac{d^2 y_b}{dx^2} = -Px + \frac{Pl}{2} \tag{3.89}$$

Integrating the two equations, we obtain

$$\left.\begin{array}{l} EIy_b = -\dfrac{Px^3}{6} + \dfrac{Plx^2}{4} + c_1 x + c_2 \\[2ex] y_s = +\dfrac{P\alpha x}{AG} + c_3 \end{array}\right\} \tag{3.90}$$

so that

$$EIy = -\frac{Px^3}{6} + \frac{Plx^2}{4} + \left(c_1 + \frac{P\alpha EI}{AG}\right)x + c_2 + EIc_3 \tag{3.91}$$

The boundary conditions at the beam ends demand that

$$\left.\begin{array}{l} \left.\begin{array}{l} y = 0 \\[1ex] \dfrac{dy_b}{dx} = 0 \end{array}\right\} (x = 0) \\[4ex] \dfrac{dy_b}{dx} = 0 \quad (x = l) \end{array}\right\} \tag{3.92}$$

These conditions are satisfied if $c_1 = c_2 = c_3 = 0$. Thus,

$$EIy = -\frac{Px^3}{6} + \frac{Plx^2}{4} + \frac{P\alpha EIx}{AG} \tag{3.93}$$

and the value of y at $x = l$ is given by

$$y_l = +\frac{Pl^3}{12EI}(1 + \phi) \tag{3.94}$$

where $\phi = 12EI\alpha/AGl^2$. The shearing stiffness k_S of this beam is therefore given by

$$k_S = \left(\frac{P}{y_l}\right) = \frac{12EI}{l^3(1 + \phi)} \tag{3.95}$$

For the beam in the extensional mode shown in Fig. 3.27(c), the extension δ is given by

$$\delta = \frac{Pl}{AE} \tag{3.96}$$

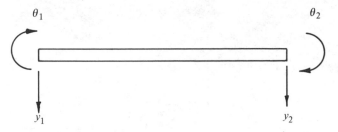

FIGURE 3.28
Showing the end displacements y_1, y_2, θ_1, θ_2

Consequently, the stiffness k_E is

$$k_E = \frac{P}{\delta} = \frac{EA}{l} \tag{3.97}$$

Similarly, the beam in a torsional mode shown in Fig. 3.27(d) has a stiffness

$$k_T = \frac{GJ}{l} \tag{3.98}$$

Beam springs

Now consider the beam shown in Fig. 3.28, which is distorted under loads at its ends and has end displacements y_1, y_2, θ_1, and θ_2 which describe this distortion. One now asks how the extensions y and θ, which describe the two bending distortion modes of the beam in Fig. 3.27(a) and Fig. 3.27(b), are related to these end displacements. We denote the bending mode by y_B and the shear mode by y_S. In addition, we define the two rigid body motions y_D and y_R. y_D represents lateral translation perpendicular to the beam length, while y_R represents a rotation about the beam midpoint. Unit values of the modes y_B, y_S, y_D, y_R have the external coordinates shown in Fig. 3.29. The relationship between the modes and the external coordinates is given in Table 3.8.

Table 3.8 MODES AND EXTERNAL COORDINATES FOR A BEAM SPRING

		Component		
Mode	y_1	y_2	θ_1	θ_2
B	0	0	$-1/2$	$1/2$
S	$-1/2$	$1/2$	0	0
D	1	1	0	0
R	$-l/2$	$l/2$	1	1

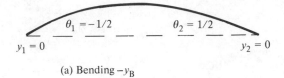

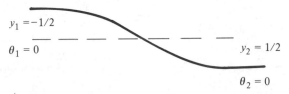

(a) Bending $-y_B$

(b) Shear with balanced bending $-y_S$

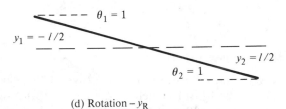

(c) Translation $-y_D$

(d) Rotation $-y_R$

FIGURE 3.29
Definition of the basic deformation modes

If, now, the amplitudes of displacement in the modes are represented by the symbols y_B, y_S, y_D, y_R, then we can write in matrix form:

$$\begin{Bmatrix} y_1 \\ y_2 \\ \theta_1 \\ \theta_2 \end{Bmatrix} = \begin{bmatrix} 0 & -1/2 & 1 & -l/2 \\ 0 & 1/2 & 1 & l/2 \\ -1/2 & 0 & 0 & 1 \\ 1/2 & 0 & 0 & 1 \end{bmatrix} \begin{Bmatrix} y_B \\ y_S \\ y_D \\ y_R \end{Bmatrix} \qquad (3.99)$$

Solving this set of equations gives:

$$\begin{Bmatrix} y_B \\ y_S \\ y_D \\ y_R \end{Bmatrix} = \begin{bmatrix} 0 & 0 & -1 & 1 \\ -1 & 1 & -l/2 & -l/2 \\ 1/2 & 1/2 & 0 & 0 \\ 0 & 0 & 1/2 & 1/2 \end{bmatrix} \begin{Bmatrix} y_1 \\ y_2 \\ \theta_1 \\ \theta_2 \end{Bmatrix} \qquad (3.100)$$

The displacements in modes D and R represent rigid body translation and rotation, respectively. They do not contribute to any potential energy storage during the deformation. If we eliminate those modes, we have

$$\begin{Bmatrix} y_B \\ y_S \end{Bmatrix} = \begin{bmatrix} 0 & 0 & -1 & 1 \\ -1 & 1 & -l/2 & -l/2 \end{bmatrix} \begin{Bmatrix} y_1 \\ y_2 \\ \theta_1 \\ \theta_2 \end{Bmatrix} \tag{3.101}$$

The coupling matrix $[B]$ between the internal extensions y_B and y_S of springs k_B and k_S, eqs. (3.86) and (3.95) respectively, and the external displacements y_1, y_2, θ_1, θ_2, is a basic feature of the component element method. It is used to assemble the simple component elements into a complex geometric assembly. In the case of the beam, then,

$$[B] = \begin{bmatrix} 0 & 0 & -1 & 1 \\ 1 & 1 & l/2 & l/2 \end{bmatrix} \tag{3.102}$$

The first line in matrix $[B]$ couples the beam end displacements to spring k_B and the second line couples the beam end displacements to spring k_S.

Beam–mass systems

Systems consisting of beams connecting masses can now be modeled. As examples, consider the two systems illustrated in Fig. 3.30 and Fig. 3.31. First, in Fig. 3.30, we show two beams of length l_1 and l_2 connecting two masses. The left-hand mass has also another spring which connects it to a support. The masses each have two degrees of freedom: they can translate vertically and rotate about their center of gravity. The generalized masses are therefore the masses m_1, m_2, and the mass moments of inertia I_1 and I_2. The corresponding degrees of freedom are denoted by ϕ_1, ϕ_3 in translation and by ϕ_2, ϕ_4 in rotation, as shown in the figure. Each

FIGURE 3.30
Structural idealization of a cantilever beam system

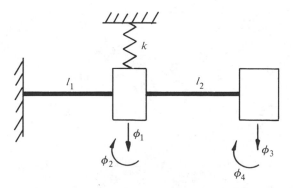

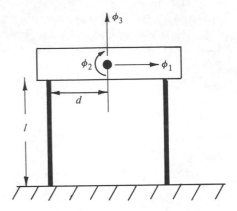

FIGURE 3.31
Structural idealization of a single-story
building with rigid roof, as two cantilever
beams supporting a mass

beam is now modeled by two springs of stiffness, k_B and k_S, respectively. The spring
stiffnesses are given by eqs. (3.86) and (3.95). Spring k_B controls the bending movement
of the beam, while k_S controls the shear movement with balanced bending. There
are therefore a total of five spring elements, four for the beams and one for spring k,
and four mass elements comprising the system.

The computer program described in section 3.16 can be used to assemble the
system and subsequently to calculate its dynamic response to external stimuli. The
input data necessary for specifying the spring–damper elements are shown in Table 3.9.
The springs are numbered, and the spring constants are given in one column of the

**Table 3.9 INPUT DATA SPECIFYING SPRING MODEL FOR CANTILEVER BEAM OF
FIG. 3.30**

Spring number	Spring stiffness	NG	NC	$\dot{C}$C
1	$K_1 = k_{B1} = \dfrac{EI_1}{l_1}$	1	2	1
2	$K_2 = k_{S1} = \dfrac{12EI_1}{l_1^3(1 + \phi_1)}$	2	1, 2	$1, -\dfrac{l_1}{2}$
3	$K_3 = k_{B2} = \dfrac{EI_2}{l_2}$	2	2, 4	$-1, 1$
4	$K_4 = k_{S2} = \dfrac{12EI_2}{l_2^3(1 + \phi_2)}$	4	1, 2, 3, 4	$-1, -\dfrac{l_2}{2},$
				$1, -\dfrac{l_2}{2}$
5	$K_5 = k$	1	1	1

table. We recall that the computer program requires the specification of the three parameters NG, NC, and CC. NG specifies the number of generalized coordinates to which the particular force element is coupled. For spring 1 (having stiffness k_{B1}), which connects mass 1 to the support, there is connection with a single coordinate ϕ_2, the mass rotation. (The coupling ratio to ϕ_1 is zero, as will subsequently be shown.) Thus, NG $= 1$ for spring 1. NC identifies which coordinate is acted on by the force element. In this case, since the force element acts on coordinate ϕ_2, we set NC $= 2$. CC is the appropriate coupling coefficient for the spring. Referring to eq. (3.101), we observe that for spring B:

$$y_B = 0{\cdot}0\, y_1 + 0{\cdot}0\, y_2 - \theta_1 + \theta_2$$

Now because the beam is cantilevered, $y_1 = 0$ and $\theta_1 = 0$. The local coordinates y_2 and θ_2 are identified with the global coordinates ϕ_1 and ϕ_2. Since the coefficient of y_2 is $0{\cdot}0$ and that of θ_2 is $1{\cdot}0$,

$$y_{B1} = \phi_2$$

so that the coupling coordinate which is the coefficient of ϕ_2 is CC $= 1$. In a similar manner, the second spring of stiffness k_{S1} is coupled to the two coordinates ϕ_1 and ϕ_2 so that NG $- 2$. The coordinate numbers are NC $= 1, 2$. The coupling coefficients are obtained once more from eq. (3.101):

$$y_{S1} = -y_1 + y_2 - (l/2)\theta_1 - (l/2)\theta_2$$

Since in terms of the element local coordinates, $y_1 = \theta_1 = 0$ for this cantilevered beam, we have, in terms of the global coordinates that describe the system:

$$y_{S1} = \phi_1 - (l/2)\phi_2$$

Consequently, CC $= 1, -l/2$ for this spring.

Turning now to the second beam, we first inspect the bending spring of stiffness k_{B2}. This spring stretches between two masses and is coupled to the coordinates ϕ_2 and ϕ_4 (the coupling is zero to ϕ_1 and ϕ_3). Referring to eq. (3.101), the coupling coefficients in the local coordinates are given by

$$y_{B2} = -\theta_1 + \theta_2$$

and in global coordinates this equation becomes

$$y_{B2} = -\phi_2 + \phi_4$$

so that in this case CC $= -1, 1$. Similarly, the shearing spring having stiffness k_{S2} is coupled to four coordinates $\phi_1, \phi_2, \phi_3, \phi_4$, according to the requirements of eq. (3.101):

$$y_{S2} = -y_1 + y_2 - (l/2)\theta_1 - (l/2)\theta_2$$

Transforming from local to global coordinates, we have:

$$y_{S2} = -\phi_1 + \phi_3 - (l/2)\phi_2 - (l/2)\phi_4$$
$$= -\phi_1 - (l/2)\phi_2 + \phi_3 - (l/2)\phi_4$$

Thus, for this spring, $CC = -1, -l/2, 1, -l/2$. Finally, for the fifth spring, which stretches from a support to mass m_1 and is coupled to only the coordinate ϕ_1, we have $NG = 1$, $NC = 1$, $CC = 1$. The specification of the system of springs is now complete and the remaining inputs of mass and external forces can proceed in the usual way.

Consider now the structural idealization of a single-story building with a rigid roof, shown in Fig. 3.31. The floor supports are idealized as two vertical beams, and the rigid roof is allowed three degrees of freedom; two translatory coordinates ϕ_1 and ϕ_3 in the horizontal and vertical directions, respectively, and a rotation ϕ_2 about the center of mass. We identify six springs in this system. Each vertical beam has three spring components. Two of these spring components are k_B and k_S, and control the bending and shearing action of the beam. The other spring component is $k_E = EA/l$, and controls the elongation of the beam. The necessary input data for the computer programs of either section 3.13 or section 3.16 are shown in Table 3.10. The specification of each spring coupling coefficient follows from eq. (3.101) as before. Thus, the first spring, of stiffness k_B, is coupled to only the rotational coordinate ϕ_2, and from eq. (3.101) the appropriate coupling coefficient is 1. Similarly, the second spring, of stiffness k_S, is coupled to both the horizontal translation ϕ_1 and the rotation ϕ_2. Consequently, $NG = 2$, $NC = 1, 2$. From eq. (3.101), we find that, since the spring is a cantilever and therefore fixed at its support, $CC = 1, -l/2$. The third spring is coupled to the vertical movement ϕ_3 and the rotation ϕ_2 (through the lever arm d). Thus, $NG = 2$, $NC = 2, 3$ and $CC = d, 1$. The remaining three springs for the right-hand beam are similarly derived.

The ability to specify a system of such springs opens up a wide range of possible modeling applications, particularly in the area of civil structures such as buildings, nuclear power generation plants and the like, in machinery support structures, and in other specialized design areas. Figure 3.32 illustrates some likely applications. In Fig. 3.32(a), a building is modeled as a cantilever beam. The mass

Table 3.10   INPUT DATA SPECIFYING SPRING MODEL FOR SINGLE-STORY BUILDING OF FIG. 3.31

Spring number	Spring stiffness	NG	NC	CC
1	$K_1 = k_B = \dfrac{EI}{l}$	1	2	1
2	$K_2 = k_S = \dfrac{12EI}{l^3(1+\phi)}$	2	1, 2	$1, -\dfrac{l}{2}$
3	$K_3 = k_E$	2	2, 3	$d, 1$
4	$K_4 = k_B = \dfrac{EI}{l}$	1	2	1
5	$K_5 = k_S = \dfrac{12EI}{l^3(1+\phi)}$	2	1, 2	$1, -\dfrac{l}{2}$
6	$K_6 = k_E$	2	2, 3	$-d, 1$

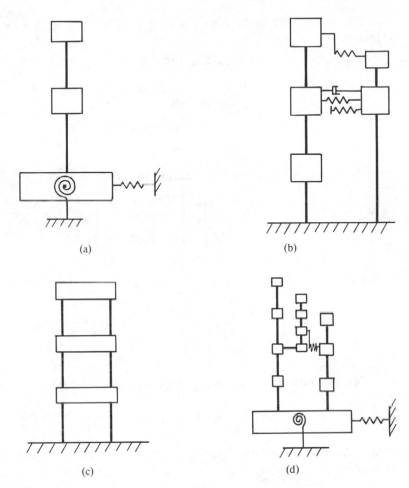

(a)

(b)

(c)

(d)

FIGURE 3.32
Numerous other structural idealizations are possible for representing civil structures and buildings

is lumped at two points along the beam (usually at the floors), and at the foundation. The soil is modeled as translational and rotational springs, as shown. In Fig. 3.32(b), two beams are joined by springs, and by a damping element and a stop element. The stop element can be used to model the case where the beams may strike one another when large displacements occur. This type of model can be useful in analyzing the vibrational aspects of pipe dynamics in heat exchangers. Further more complex models of buildings are shown in Fig. 3.32(c) and Fig. 3.32(d).

Consistent mass matrix for a beam

When the distributed beam mass is to be accounted for, rather than the lumped masses in Fig. 3.30, it is desirable to represent the beam by a mass matrix.

The beam deformation modes in Fig. 3.28 involve a cubic displacement expression

$$y = a + b(x/l) + c(x/l)^2 + d(x/l)^3$$

where l is the beam length and a, b, c, d are coefficients to be determined. Taking the origin of coordinates at the left end,

$$y_1 = a, \quad y_2 = a + b + c + d$$
$$l\theta_1 = b, \quad l\theta_2 = b + 2c + 3d$$

In matrix form we can write these relationships as

$$\begin{Bmatrix} y_1 \\ y_2 \\ l\theta_1 \\ l\theta_2 \end{Bmatrix} = \begin{bmatrix} 1 & 0 & 0 & 0 \\ 1 & 1 & 1 & 1 \\ 0 & 1 & 0 & 0 \\ 0 & 1 & 2 & 3 \end{bmatrix} \begin{Bmatrix} a \\ b \\ c \\ d \end{Bmatrix}$$

so that, upon inversion,

$$\begin{Bmatrix} a \\ b \\ c \\ d \end{Bmatrix} = \begin{bmatrix} 1 & 0 & 0 & 0 \\ 0 & 0 & 1 & 0 \\ -3 & 3 & -2 & 1 \\ 2 & -2 & 1 & 1 \end{bmatrix} \begin{Bmatrix} y_1 \\ y_2 \\ l\theta_1 \\ l\theta_2 \end{Bmatrix}$$

Consider now the kinetic energy expression for the beam,

$$\text{K.E.} = \frac{\rho}{2} \int_0^l [\dot{a} + \dot{b}(x/l) + \dot{c}(x/l)^2 + \dot{d}(x/l)^3]^2 \, dx$$

where ρ is the mass per unit length. Integrating and expressing the result in matrix form gives

$$\text{K.E.} = \frac{\rho l}{2} [\dot{a}, \dot{b}, \dot{c}, \dot{d}] \begin{bmatrix} 1 & 1/2 & 1/3 & 1/4 \\ 1/2 & 1/3 & 1/4 & 1/5 \\ 1/3 & 1/4 & 1/5 & 1/6 \\ 1/4 & 1/5 & 1/6 & 1/7 \end{bmatrix} \begin{Bmatrix} \dot{a} \\ \dot{b} \\ \dot{c} \\ \dot{d} \end{Bmatrix}$$

Substituting for $[\dot{a}, \dot{b}, \dot{c}, \dot{d}]$ in terms of $[\dot{y}_1, \dot{y}_2, l\dot{\theta}_1, l\dot{\theta}_2]$ gives

$$\text{K.E.} = \frac{\rho l}{2} [\dot{y}_1, \dot{y}_2, \dot{\theta}_1, \dot{\theta}_2] \begin{bmatrix} \dfrac{13}{35} & \dfrac{9}{70} & \dfrac{11l}{210} & -\dfrac{13l}{420} \\ \dfrac{9}{70} & \dfrac{13}{35} & \dfrac{13l}{420} & -\dfrac{11l}{210} \\ \dfrac{11l}{210} & \dfrac{13l}{420} & \dfrac{l^2}{105} & -\dfrac{l^2}{140} \\ -\dfrac{13l}{420} & -\dfrac{11l}{210} & -\dfrac{l^2}{140} & \dfrac{l^2}{105} \end{bmatrix} \begin{Bmatrix} \dot{y}_1 \\ \dot{y}_2 \\ \dot{\theta}_1 \\ \dot{\theta}_2 \end{Bmatrix}$$

The consistent mass matrix for the beam in Fig. 3.28 is therefore given by

$$[m_\mathrm{e}] = \rho l \begin{bmatrix} \dfrac{13}{35} & \dfrac{9}{70} & \dfrac{11l}{210} & -\dfrac{13l}{420} \\[2ex] \dfrac{9}{70} & \dfrac{13}{35} & \dfrac{13l}{420} & -\dfrac{11l}{210} \\[2ex] \dfrac{11l}{210} & \dfrac{13l}{420} & \dfrac{l^2}{105} & -\dfrac{l^2}{140} \\[2ex] -\dfrac{13l}{420} & -\dfrac{11l}{210} & -\dfrac{l^2}{140} & \dfrac{l^2}{105} \end{bmatrix}$$

Having obtained the mass matrix $[m_\mathrm{e}]$ for a single beam element, an assembly process can be used to write the contributions from each beam. The equations of motion therefore can be written in matrix form as

$$[m]\{\ddot{x}\} = \{F\}$$

where $\{x\}$ represents the generalized coordinates and $\{F\}$ is the force on the co ordinates. It is desirable, in practice, that the fully assembled mass matrix has a narrow band, so that the system is more easily solvable. Solution yields the accelerations $\{\ddot{x}\}$, written symbolically as

$$\{\ddot{x}\} = [m]^{-1}\{F\}$$

The matrix $[m]$ is known as the consistent mass matrix, because it expresses more closely than a diagonal matrix (which appears in the case of a lumped mass model) how the distribution of kinetic energy is effected in the system. If the consistent mass matrix for the beam springs is to be used in conjunction with the computer program of section 3.16, the program must be modified to account for the matrix form.

Connection between component springs and the beam stiffness matrix

There is an intimate connection between the coupling coefficients and the stiffness of a beam system shown in Fig. 3.33. There, the beam is under external forces at its ends P_1, P_2, M_1, M_2 and experiences the end deflections y_1, y_2, θ_1, θ_2. The stiffness matrix $[K]$ for this system is the relationship between force and displacement:

$$\begin{Bmatrix} P_1 \\ P_2 \\ M_1 \\ M_2 \end{Bmatrix} = [K] \begin{Bmatrix} y_1 \\ y_2 \\ \theta_1 \\ \theta_2 \end{Bmatrix} \qquad (3.103)$$

(a) Displacements

(b) Forces

FIGURE 3.33
Forces causing end displacements in a bent beam

It can be shown that

$$[K] = [B]^T \begin{bmatrix} k_B & & & \\ & k_S & & \\ & & 0 & \\ & & & 0 \end{bmatrix} [B] \tag{3.104}$$

or, using for $[B]$ the matrix in eq. (3.100),

$$[K] = \begin{bmatrix} 0 & -1 & 1/2 & 0 \\ 0 & 1 & 1/2 & 0 \\ -1 & -1/2 & 0 & 1/2 \\ 1 & -1/2 & 0 & 1/2 \end{bmatrix} \begin{bmatrix} k_B & & & \\ & k_S & & \\ & & 0 & \\ & & & 0 \end{bmatrix} \begin{bmatrix} 0 & 0 & -1 & 1 \\ -1 & 1 & -1/2 & -1/2 \\ 1/2 & 1/2 & 0 & 0 \\ 0 & 0 & 1/2 & 1/2 \end{bmatrix}$$

$$\tag{3.105}$$

The lower two rows in the coupling matrix $[B]$ are those elements associated with the rigid body displacements. Carrying out the multiplication, we find that

$$[K] = \begin{bmatrix} k_S & -k_S & k_S l/2 & k_S l/2 \\ -k_S & k_S & -k_S l/2 & -k_S l/2 \\ k_S l/2 & -k_S l/2 & (k_B + k_S l^2/4) & (-k_B + k_S l^2/4) \\ k_S l/2 & -k_S l/2 & (-k_B + k_S l^2/4) & (k_B + k_S l^2/4) \end{bmatrix}$$

and, using eqs. (3.86) and (3.95),

$$[K] = \frac{EI}{(1+\phi)l^3} \begin{bmatrix} 12 & -12 & \vdots & 6l & 6l \\ -12 & 12 & \vdots & -6l & -6l \\ \cdots & \cdots & \cdots & \cdots & \cdots \\ 6l & -6l & \vdots & (4+\phi)l^2 & (2-\phi)l^2 \\ 6l & -6l & \vdots & (2-\phi)l^2 & (4+\phi)l^2 \end{bmatrix} \quad (3.106)$$

This is the stiffness matrix for such a beam.

We now observe that the inverse process will lead to the determination of component element stiffnesses and coupling ratios. For, if we can find a way of diagonalizing a given, known, stiffness matrix, we will obtain these component descriptions by this means. Thus, we can generalize the results discussed previously.

General springs

A more general spring has its stiffness described by a symmetric matrix whose dimensions are equal to the number of external coordinates. Such a matrix can be expressed as

$$[K] = [B]^T[\lambda][B] \quad (3.107)$$

where $[\lambda]$ is a diagonal matrix of the eigenvalues and $[B]$ contains the corresponding coupling ratios of springs to global coordinates in rows. There will be as many zero values of the λs as there are rigid body displacement modes for the spring. The remaining λs represent orthogonal springs analogous to k_B and k_S in eqs. (3.86) and (3.95), while the coupling ratios provide the terms in a matrix analogous with that in eq. (3.101). The diagonalization process can be carried out by the Jacobi method, as described in section 6.5.

3.19 SIMPLIFIED BEAM SPRINGS

In this section, we derive some further beam spring models, which are applicable to situations where the rotary coordinate is not of interest, but where the displacement coordinate is dominant. In many cases, the rotational inertia contributes very little to the total kinetic energy. In addition, it complicates a subsequent dynamic analysis by adding rotational degrees of freedom to the problem and by adding high frequency terms to the response computation. The latter are particularly undesirable if the subsequent dynamic analysis is done using the step-by-step central difference approximations, since the time steps used then become very small. As a result, it is advantageous in certain cases to use the beam springs described in this section.

Another beam spring

In some cases, it is desired to model a beam spring by specifying only displacements perpendicular to the beam. This can readily be done if use is made of the fact that the elastic potential energy of a portion of structure is given by

$$\text{P.E.} = 1/2[\delta][k][\delta]^T \tag{3.108}$$

where δ is the displacement vector and k is the stiffness contribution of the portion of structure.

Straight beam

Consider the beam shown in Fig. 3.34(a). Location l is midway between the co-ordinates 1 and 2, while location r is midway between the coordinates 2 and 3. We approximate the change in slope between locations l and r as

$$\text{slope change} = \left(\frac{y_3 - y_2}{b}\right) - \left(\frac{y_2 - y_1}{a}\right) \tag{3.109}$$

We take the average curvature between location l and location r as

$$\left(\frac{d^2y}{dx^2}\right)_{av} = \frac{1}{\left(\frac{a}{2} + \frac{b}{2}\right)}\left[\left(\frac{y_3 - y_2}{b}\right) - \left(\frac{y_2 - y_1}{a}\right)\right]$$

$$= \left[\frac{2}{ab(a + b)}\right][by_1 - (a + b)y_2 + ay_3] \tag{3.110}$$

The potential energy stored between location l and location r is

$$\text{P.E.} = \left(\frac{EI}{2}\right)\left(\frac{a + b}{2}\right)\left(\frac{d^2y}{dx^2}\right)^2_{av}$$

$$= \tfrac{1}{2}[y_1, y_2, y_3]\begin{bmatrix} b \\ -(a + b) \\ a \end{bmatrix}\left[\frac{2EI}{a^2b^2(a + b)}\right][b, -(a + b), a]\begin{bmatrix} y_1 \\ y_2 \\ y_3 \end{bmatrix} \tag{3.111}$$

We recognize that the beam spring representation of eq. (3.111) follows eq. (3.107), if we use a spring stiffness of

$$k = \frac{2EI}{a^2b^2(a + b)} \tag{3.112}$$

with coupling ratios of

$$(b, -(a + b), a) \text{ to } (y_1, y_2, y_3)$$

Right angle beam

Another beam spring involves the case where a beam makes a right-angled turn as shown in Fig. 3.34(b). Moments along the beam are continuous, so the concepts are quite similar to the previous case. We approximate the change in slope between

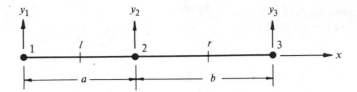

(a) A straight beam

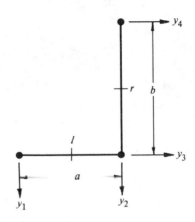

(b) A right-angled beam

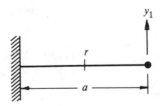

(c) Built-in beam end

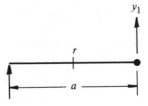

(d) Simply supported beam end

FIGURE 3.34
Beam elements

locations l and r as

$$\text{slope change} = \left(\frac{y_4 - y_3}{b}\right) - \left(\frac{y_2 - y_1}{a}\right) \tag{3.113}$$

The average curvature between location l and location r is

$$\left(\frac{d^2 y}{dx^2}\right)_{\text{av}} = \frac{1}{\left(\dfrac{a}{2} + \dfrac{b}{2}\right)}\left[\left(\frac{y_4 - y_3}{b}\right) - \left(\frac{y_2 - y_1}{a}\right)\right]$$

$$= \left[\frac{2}{ab(a + b)}\right][by_1 - by_2 - ay_3 + ay_4] \tag{3.114}$$

Again, the potential energy is given by

$$\text{P.E.} = \left(\frac{EI}{2}\right)\left(\frac{a + b}{2}\right)\left(\frac{d^2 y}{dx^2}\right)_{\text{av}}^2$$

$$= 1/2[y_1,\, y_2,\, y_3,\, y_4]\begin{bmatrix} b \\ -b \\ -a \\ a \end{bmatrix}\left[\frac{2EI}{a^2 b^2 (a + b)}\right][b,\, -b,\, -a,\, a]\begin{bmatrix} y_1 \\ y_2 \\ y_3 \\ y_4 \end{bmatrix} \tag{3.115}$$

We recognize that the beam stiffness is the same as given in eq. (3.112) with coupling ratios of

$$(b,\, -b,\, -a,\, a) \quad \text{to} \quad (y_1,\, y_2,\, y_3,\, y_4) \tag{3.116}$$

Built-in beam end
Here we consider a portion of beam structure from a built-in end to the midpoint r, as illustrated in Fig. 3.34(c). We approximate the slope change between the built-in end and location r as

$$\text{slope change} = y_1/a \tag{3.117}$$

We take the average curvature between the built-in end and location r as

$$\left(\frac{d^2 y}{dx^2}\right)_{\text{av}} = \frac{1}{(a/2)}(y_1/a) = \frac{2}{a^3}(ay_1) \tag{3.118}$$

The potential energy stored in the beam between the built-in end and location r is

$$\text{P.E.} = \frac{EI}{2}\left(\frac{a}{2}\right)\left(\frac{d^2 y}{dx^2}\right)_{\text{av}}^2$$

$$= 1/2(y_1)(a)\left(\frac{2EI}{a^5}\right)(a)(y_1) \tag{3.119}$$

and we recognize that for this portion of beam between the built-in end and location r the spring stiffness is

$$k = \frac{2EI}{a^5} \tag{3.120}$$

and the coupling ratio to the displacement y_1 is a.

Simply supported beam end

The potential energy stored in the portion of beam between location r and the simply supported end in Fig. 3.34(d) is taken as zero, since the bending moment at the supported end is zero. The slope at r is, to the same order of approximation as the previous cases, the same as that at the simply supported end. Therefore, the curvature in the region between r and the end is zero. As a result, there is a negligible amount of potential energy stored in this element. Thus, no spring is required to represent it.

Total beam

The total beam stiffness is represented by the assembly of the individual springs in the computer program by the usual process that has been described previously for other elements.

Convergence

To obtain an insight into the convergence of this representation of beam springs, we consider the deflection at the tip of the cantilever beam shown in Fig. 3.35(a). Beam theory gives

$$\text{tip deflection} = \frac{Pl^3}{3EI} = \frac{1}{3}$$

We first model this beam, as shown in Fig. 3.35(b), with two beam segments. For the contribution of part g, the force at the first node due to the contribution of segment g is given, through eq. (3.120), by

$$F_{1g} = \left(\frac{1}{2}\right) \frac{2EI}{(0.5)^5} \left(\frac{1}{2}\right) y_1$$

$$F_{1g} = 16y_1$$

Similarly, for the contribution of part h, the forces at the nodes 1 and 2 are, through eq. (3.112),

$$\begin{Bmatrix} F_{1h} \\ F_{2h} \end{Bmatrix} = \begin{bmatrix} -1.0 \\ 0.5 \end{bmatrix} \begin{bmatrix} \frac{2EI}{(0.5)^4(1)} \end{bmatrix} [-1, 0.5] \begin{Bmatrix} y_1 \\ y_2 \end{Bmatrix}$$

$$\begin{Bmatrix} F_{1h} \\ F_{2h} \end{Bmatrix} = \begin{bmatrix} 32 & -16 \\ -16 & 8 \end{bmatrix} \begin{Bmatrix} y_1 \\ y_2 \end{Bmatrix}$$

The total load is the sum of the parts, or,

$$F_1 = F_{1g} + F_{1h} = 16y_1 + 32y_1 - 16y_2 = 0$$

$$F_2 = F_{2h} = -16y_1 + 8y_2 = 1$$

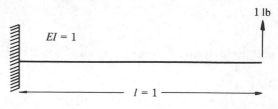

(a) Beam under static load

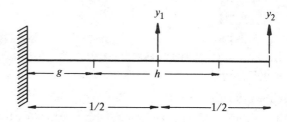

(b) Two-element beam model

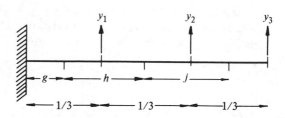

(c) Three-element beam model

FIGURE 3.35
Model of a cantilever beam under static load

Solving gives $y_2 = 3/8$, an error of 12 per cent.

Next, we model the beam in three beam segments, as shown in Fig. 3.35(c). The contribution of part g is now

$$F_{1g} = \left(\frac{1}{3}\right)\frac{2EI}{(1/3)^5}\left(\frac{1}{3}\right)y_1 = 54y_1$$

For the contribution of part h, we have

$$\begin{Bmatrix} F_{1h} \\ F_{2h} \end{Bmatrix} = \begin{bmatrix} -2/3 \\ 1/3 \end{bmatrix} \frac{2EI}{(1/3)^4(2/3)} \begin{bmatrix} -\frac{2}{3}, \frac{1}{3} \end{bmatrix} \begin{Bmatrix} y_1 \\ y_2 \end{Bmatrix}$$

$$\begin{Bmatrix} F_{1h} \\ F_{2h} \end{Bmatrix} = \begin{bmatrix} 108 & -54 \\ -54 & 27 \end{bmatrix} \begin{Bmatrix} y_1 \\ y_2 \end{Bmatrix}$$

For the contribution of part j, we have

$$\begin{Bmatrix} F_{1j} \\ F_{2j} \\ F_{3j} \end{Bmatrix} = \begin{bmatrix} 1/3 \\ -2/3 \\ 1/3 \end{bmatrix} \frac{2EI}{(1/3)^4(2/3)} [1/3, -2/3, 1/3] \begin{Bmatrix} y_1 \\ y_2 \\ y_3 \end{Bmatrix}$$

$$\begin{Bmatrix} F_{1j} \\ F_{2j} \\ F_{3j} \end{Bmatrix} = \begin{bmatrix} 27 & -54 & 27 \\ -54 & 108 & -54 \\ 27 & -54 & 27 \end{bmatrix} \begin{Bmatrix} y_1 \\ y_2 \\ y_3 \end{Bmatrix}$$

Again, the total load is the sum of the parts, or,

$$F_1 = F_{1g} + F_{1h} + F_{1j} = 189y_1 - 108y_2 + 27y_3 = 0$$

$$F_2 = F_{2h} + F_{2j} = -108y_1 + 135y_2 - 54y_3 = 0$$

$$F_3 = F_{3j} = 27y_1 - 54y_2 + 27y_3 = 1$$

Solving gives $y_3 = 0.356$, an error of 7 per cent.

It is apparent that increasing the number of beam segments reduces the error, and that even with as few as three segments, for a cantilever beam where the moment varies continually, the error is below 7 per cent.

Application to pipe dynamics

An important problem in the analysis of piping systems is the prediction of the pipe dynamics. We can analyze such a system using the type of beam spring just described and using the computer program of section 3.16. The following example illustrates how this is done. This example not only illustrates how the beam springs described in this section can be used for studying the elastic response of a system, but is also applied to the event where a plastic hinge is allowed to form in the pipe. This event is modeled through the application of the concepts embodied in the friction element that has been discussed previously. In addition, a piping restraint is applied in the example in the form of a stop element situated toward the end of the pipe, as shown in Fig. 3.36(a).

Figure 3.36(a) shows the overall dimensions of a piping system. The pipe is a 26 in. line with an inner diameter of 23·358 in. The pipe material is elastic up to an extreme fiber stress of 39 000 lb/in.^{2} and then becomes fully plastic. The modulus is 30 000 000 lb/in.^{2}. The pipe weight is 358·5 lb/ft. The pipe restraint is considered to provide a clearance of 5·995 in., after which it resists with a spring force of 523 175 lb/in. The force input was 445 650 lb from 0 to 0·00203 sec, 311 995 lb from

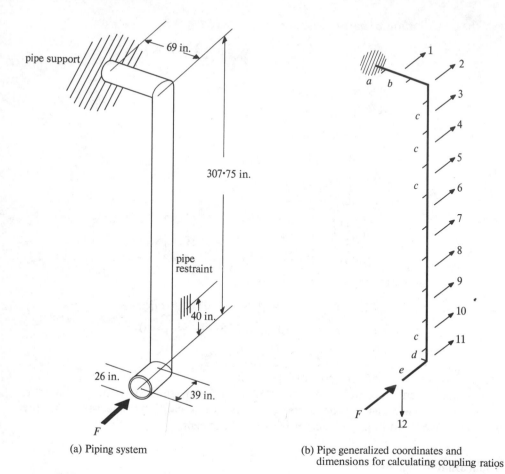

(a) Piping system

(b) Pipe generalized coordinates and dimensions for calculating coupling ratios

FIGURE 3.36
Model of piping system

0·00203 sec to 0·004516 sec, and 475 508 lb thereafter. Modeling of the piping system was done with the 12 coordinates shown in Fig. 3.36(b). The coupling ratios and pipe stiffnesses appropriate to this model are shown in Table 3.11, and can be calculated from the formulas in eqs. (3.112), (3.116), and (3.120).

It was desired to model the beam springs as elastic–plastic elements. This could readily be done in the program described in section 3.16 by using the friction element, defined by INDX = 2. To use this element, the identifier N1 was set to 1 and a value of 1 000 000 was read in for FU(1). The variable C1 was then read in as (1/1 000 000) of the fully plastic spring force, and the variable C2 was read in as the spring stiffness. It can be shown that the maximum spring force for a beam element like a in Fig. 3.36(b) is $\sigma_{max}(I/c)(1/a^2)$, where σ_{max} is the extreme fiber stress at which full plasticity starts, c is the extreme fiber distance, I is the moment of inertia, and a is the distance from the built-in end to coordinate 1. For the beam element a, the

Table 3.11   STIFFNESS AND COUPLING RATIOS OF PIPING ELEMENTS

Segment	Spring stiffness (lb/in.³)	Coupling ratio (in.)	Coordinates coupled
a	9700	34·5	1
b	4840	− 69, 34·5	1, 2
c	2830	38·4, − 76·8, 38·4	2, 3, 4
			3, 4, 5
			⋮
			8, 9, 10
d	2560	40, − 78·4, 38·4	9, 10, 11
e	2465	39, − 39, 40	10, 11, 12

maximum spring force is therefore 19 900 lb/in. The maximum spring force for beam portions like b to e in Fig. 3.36(b) is $\sigma_{max}(I/c)(1/ab)$. This formula gives the following values for the maximum spring force in each segment:

$$\text{segment } b, \text{ maximum force} = 19\,900 \text{ lb/in.}$$
$$\text{segment } c, \text{ maximum force} = 16\,060 \text{ lb/in.}$$
$$\text{segment } d, \text{ maximum force} = 15\,420 \text{ lb/in.}$$
$$\text{segment } e, \text{ maximum force} = 15\,200 \text{ lb/in.}$$

These values were used in the analysis.

Modeling the pipe restraint was done using a stop element defined in the computer program by INDX = 3, with the input variables NG = 1, NC(1) = 10, CC(1) = − 1·0, C1 = 5·995, and C2 = 523 175.

The pipe mass could have been modeled as uncoupled masses, as required by the program of section 3.16. However, in this case, the program was modified so that each beam portion between two coordinates was modeled by a 2 × 2 matrix. When this was done for the portion between coordinates 1 and 2, for example, the inertia forces F_I contributed by this portion are given by

$$\begin{Bmatrix} F_{I1} \\ F_{I2} \end{Bmatrix} = m \begin{bmatrix} 1/3 & 1/6 \\ 1/6 & 1/3 \end{bmatrix} \begin{Bmatrix} \ddot{y}_1 \\ \ddot{y}_2 \end{Bmatrix}$$

where m is the mass of the portion of beam between coordinates 1 and 2. In addition, the mass of the portion of beam between coordinates 11 and 12 was considered to act fully with coordinate 11. A time step of 0·0001 sec was used.

Some results of interest are shown in Fig. 3.37. There, the displacements of the various locations on the pipe are shown as a function of time. The point at which the pipe first strikes the pipe restraint is also shown. In Fig. 3.37(b), the pipe moment is shown at three locations on the pipe. Note how the pipe yields plastically when the maximum moment of $23·5 \times 10^6$ lb in. is reached. This event occurs at various times and at various locations in the pipe.

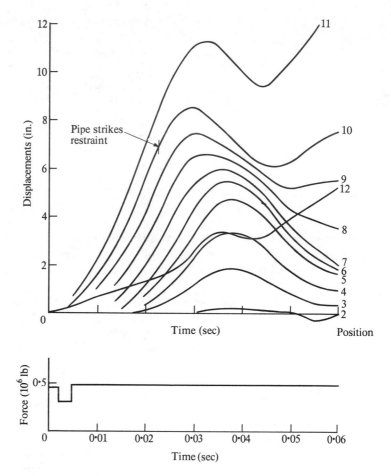

FIGURE 3.37(a)
Pipe displacements and applied force

3.20 SUMMARY

The basic purpose of this chapter has been to develop a logical and systematic means of computing the response of a system having many degrees of freedom. The description of the systems by means of generalized coordinates has enabled the study of a variety of systems, some of which consist of discrete masses, others of beams and continuous structures, or a combination of discrete and continuous portions.

To further amplify the power of the computer programs described in this chapter, we shall pursue the study of systems with many degrees of freedom in the following chapter. There, the examples are all selected from the general area of vehicle dynamics.

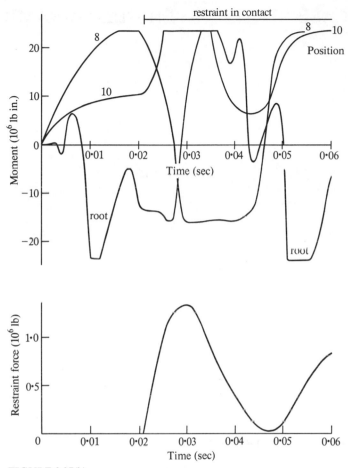

FIGURE 3.37(b)
Pipe moments and restraint force. Note that the pipe yields plastically at certain times

3.21 REFERENCES

1. S. H. Crandall, *Engineering Analysis,* Ch. 2, McGraw-Hill Book Company, Inc., New York, 1956.
2. W. F. Stokey, 'Vibrations of systems having distributed mass and elasticity,' in C. M. Harris and C. E. Crede (Eds.), *Shock and Vibration Handbook,* vol. 1, pp. 7.11–7.16, McGraw-Hill Book Company, Inc., New York, 1961.
3. D. Young and R. P. Felgar, 'Tables of characteristic functions representing normal modes of a vibrating beam,' The University of Texas Publication No. 4913, July 1949.
4. R. K. Wen and N. Beylerian, 'Elasto–plastic response of Timoshenko beams,' *J. Struct. Div., Proc. ASCE,* **93**, No. ST3, 131–146, 1967.
5. F. L. Moreadith, G. J. Patterson, C. R. Angstadt, and J. F. Glova, 'Structural analysis and design of pipe whip restraints,' *Structural Design of Nuclear Plant Facilities,* vol. II, pp. 201–342 (particularly pp. 282–283), American Society of Civil Engineers, New York, 1974.
6. G. J. Patterson, 'Lumped-parameter model of nonlinear dynamic analysis of pipe-whip restraints,' *Proc. 2nd ASCE Specialty Conference on Structural Design of Nuclear Power Plant Facilities,* vol. 1-B, pp. 1444–1484, American Society of Civil Engineers, New York, 1975.

4

CASE STUDIES—VEHICLE DYNAMICS

4.1 INTRODUCTION

Vehicles present some challenging problems in terms of computing their responses to dynamic situations. Their suspension systems are varied and contain many of the elements that we have already studied in previous chapters: nonlinear springs, stops, friction, and dampers. Their interactions with tracks and guideways provide further stimuli for dynamic behavior.

The purpose of this chapter is two-fold. First, we wish to illustrate the power of the computer program described in the previous chapters and, to do this, we have chosen four examples of vehicle dynamics to which to apply the programs. Second, we wish to show the types of problem that can be solved by applying the programs. The first two examples, concerning an automobile and an aircraft landing, are stated in simplified terms. They are intended to show certain points in a clear fashion. Thus, the displacements of the passengers and the g-levels they experience as the car travels over a bump are illustrated. However, the elements comprising the car are minimal to permit an easy discussion. Similarly, the aircraft is modeled in a very simple manner. Certain features of the results are clearly visible, however. For example, the aircraft's wheels are seen to skid in a realistic manner during the landing.

The third and fourth examples concern the dynamics of locomotives and of tracked air cushion vehicles. In these examples, a larger number of degrees of freedom are involved. In addition, the modeling of the restoring forces in the vehicle suspension is done in a more realistic manner. A disadvantage accrues, however, in modeling with such complexity. Computation costs go up, as does the volume of output which must be examined. In practice, it is usually possible to focus on one particular aspect of the vehicle design. Thus, in a locomotive, one may want to know what forces are imposed on the bolster, or between the wheel and track. Otherwise, one may wish to find out under what circumstances the wheel lifts off the track. In the examples discussed, only a small number of such output responses are given. Actually, the computer programs give considerably more information than is discussed in this chapter. It should be noted that additional force elements such as air cushions, creep force, etc., must be added to the programs of section 3.16 if they are to be used for the locomotive and air cushion vehicle examples. While the authors have actually made the program modifications and computations involved, only some results from these examples are given here to illustrate the variety of calculations possible with this basic method.

The program arrangement is specifically designed for ease of modification. To include additional force elements in the programs of section 3.16 the procedure followed is (using the spring, stop, and friction elements as examples):

(a) Additional elements can use INDX values of 4, 5, 6, 7. In this case, the GO TO after statement 3 becomes

$$\text{GO TO (10, 12, 14, 16, 18, 20, 22, 40), INDX}$$

(b) Following statement 15 would be a group of statements descriptive of each additional force element.
(c) Following statement 32 would be a group of statements converting force element force to generalized force.
(d) Following statement 90 would be a group of statements resetting element extensions.
(e) Following statement 923 would be extension statements for the new elements.
(f) Following statement 604 would be force computation statements for the new elements.
(g) Following statement 665 would come WRITE statements for the new elements.

4.2 AUTOMOBILE DYNAMICS

To show the scope of the component element method, and to illustrate the use of a computer program of the type discussed in section 3.16, we will model an automobile traveling over a bumpy road. The reader is urged to make use of the computer program of section 3.16 to reproduce and extend the case studies described in this chapter. The model will be simplified, but will nevertheless include the 11 degrees of freedom, which are shown in Fig. 4.1, as follows:

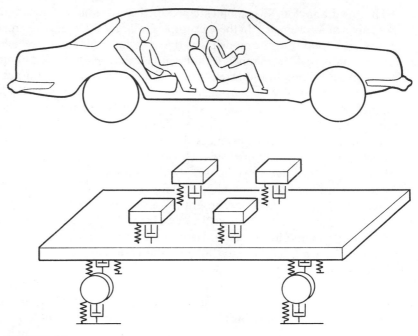

FIGURE 4.1
Model of an automobile

3 for the car body—vertical, pitch and roll
2 for the front wheels—vertical at each wheel
2 for the rear wheel set—vertical and roll of rear axle
4 for the passengers—vertical

The model will include 12 spring–dampers as follows:

4 between the wheels and the road, representing the tires
2 between the front wheels and the car body, representing springs and shocks
2 between the rear axle and the car body, representing springs and shocks
4 between the four passengers and the car body, representing the car seats

The model also contains four stops to limit the upward travel of the wheels towards the car body; these represent the snubbers.

This model is far from using the full capability of the computer program of section 3.16. The model is also quite simplified in comparison with an actual automobile. Lateral and yawing motions, for example, are not included.

In Fig. 4.2, we show the car body and the points of attachment of the springs to the wheels and to the passengers. Note how the passenger and seat mass has been lumped together. It is connected to the car body by a spring–damper. The front wheels have spring–dampers representing the tires, and a spring–damper–stop system that represents the suspension. The rear wheels and axle are modeled as a rigid

○ − denotes position of spring elements
1, 2 − front wheel locations, front springs
3, 4 − rear wheel locations
5,...,8 − passenger locations
9, 10 − rear suspension springs
(all dimensions in inches)

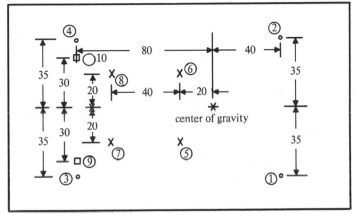

FIGURE 4.2
Automobile body mass and geometry

body with both mass and inertia properties. Tire and suspension springs at the rear are modeled in a manner similar to those of the front wheels. These elements are illustrated in Fig. 4.3. Now, in order to further discuss the vehicle dynamics, we will use the following numerical values. The car body weighs 2500 lb. The center of gravity is 40 in. behind the front wheels. The pitching radius of gyration is 50 in. and the rolling radius of gyration is 31·6 in. The wheel base is 120 in. The front car seats are 20 in. in back of the center of gravity and 20 in. to either side of the car's longitudinal centerline. The back car seats are 40 in. in back of the front seats. The passengers and seats are taken to weigh 175 lb each, and the seats have a spring rate of 30 lb/in. and damping of 8 lb sec/in.

We will consider the loading to result from running the right wheels of the car through a pothole at 10 miles per hour. The hole depth increases to 3 in. in the first 5 in. of travel, stays at 3 in. for the next 10 in. of travel, and returns to zero in the final 5 in. of travel.

The generalized coordinates are defined as follows:

coordinate 1, upward displacement of car body center of gravity
coordinate 2, pitching of car body, positive nose up
coordinate 3, rolling of car body, positive right side up
coordinate 4, right front wheel upward
coordinate 5, left front wheel upward
coordinate 6, upward displacement of rear axle center of gravity
coordinate 7, rolling of rear axle, positive right side up

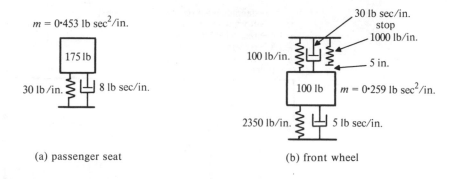

(a) passenger seat

(b) front wheel

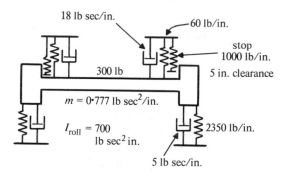

(c) rear axle and wheels

FIGURE 4.3
Constants for springs, dampers, stops and masses of passengers, front wheels, and rear axle

coordinate 8, right front passenger upward
coordinate 9, left front passenger upward
coordinate 10, right rear passenger upward
coordinate 11, left rear passenger upward

The time increment for the application of the finite difference equations was chosen as $\Delta t = 0.0025$ sec. This increment was chosen since the period of the wheel suspension is about 0.050 sec, and it is desired to follow the wheel dynamics. In addition, it will be noted that the time required to travel 20 in. (the hole width) at 10 miles per hour is 0.1136 sec. The time increment gives 45 time steps in this time interval.

Selected portions of the results obtained are plotted in Fig. 4.4. In the upper figure, 4.4(a), are given displacements. Curve (1) is the road beneath the right front wheel. Curve (2) is the wheel. Note the lag in response of the wheel, due to the tire springiness. Nevertheless, the wheel follows the road fairly closely. Curve (3) is the right front wheel tire extension. For the first few hundredths of a second, the tire extends as the road drops away. As the hole is passed, at about $t = 0.10$ sec, the road rises and the maximum tire compression occurs at about $t = 0.115$ sec. Curve (4) gives displacements of the left front wheel. These movements are negligible. Curve (5) gives the displacement of the right front passenger. It reaches a maximum after the hole is passed, and is about 0.8 in. at about $t = 0.15$ sec. Comparing curves (1) and (5), it is evident that the car suspension substantially reduced the amplitude of passenger displacement and made it much more gradual. Both of these factors help passenger comfort.

FIGURE 4.4
Response of auto hitting a bump at 10 miles per hour

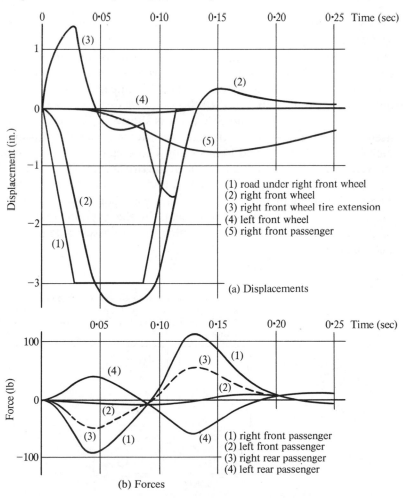

(1) road under right front wheel
(2) right front wheel
(3) right front wheel tire extension
(4) left front wheel
(5) right front passenger

(a) Displacements

(1) right front passenger
(2) left front passenger
(3) right rear passenger
(4) left rear passenger

(b) Forces

In the lower figure, 4.4(b), are given the forces between each passenger and the car. Force (1), for the right front passenger, is initially downward; then upward after the bump is passed. The peak value down is about 95 lb and up is about 115 lb. The passenger and seat weight is about 175 lb, so these forces cause accelerations in excess of $\frac{1}{2}g$. The left rear passenger, curve (4), receives forces upwards when the front right passenger feels forces downwards. The magnitude is somewhat smaller. The forces on the left front passenger, curve (2), are the smallest. The right rear passenger, curve (3), feels forces like the right front passenger, but smaller.

Problems

4.1 Repeat the example problem with one-tenth as much damping in the springs that join the car to the wheels. What effect does this have on the response after the bump is passed?
4.2 Repeat the example problem with one-tenth as much damping in the seat springs. What is the effect on the passenger seat forces?

A very interesting dynamic study of a semitrailer truck has been made by Potts and Walker.[1] They have modeled the truck suspension by nonlinear elements which allow damping to vary nonlinearly as a function of velocity, and wherein the force deflection relationship in the rear axle spring is different in loading and unloading, thus allowing hysteresis to occur. The nonlinear element descriptions must be obtained experimentally.

The model of the semitrailer truck is shown in Fig. 4.5, where the coordinates used and the spring–damper elements are shown in detail. Potts and Walker use the Newmark method, referred to in section 1.8, to calculate the dynamic displacements to excitations applied to the supports of the tire springs. They compared their results with a 1/8 scale model experiment and found rather good agreement in displacements. The accelerations of the centers of mass were somewhat harder to predict, there being more high-frequency components in the experimental data. However, the maximum accelerations, which are often used to assess ride quality, compared well between experiment and analysis.

FIGURE 4.5
Model of a truck showing nine generalized coordinates (after Potts and Walker,[1] reproduced from *Journal of Engineering for Industry, Trans. ASME*, **96**, 1974, American Society of Mechanical Engineers, with permission)

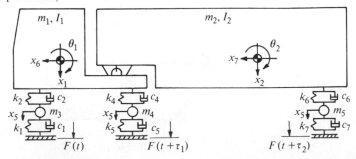

In using a model such as illustrated in Fig. 4.5, the application of the computer programs described in this book would require that the truck–trailer pivot be modeled by a stiff spring representing the pivot flexibility. Potts and Walker derived specific differential equations for the model in Fig. 4.5, where the pivot is assumed to form a pin joint.

4.3   AIRCRAFT LANDING

The landing impact of aircraft is another interesting problem which can be analyzed by means of the concepts we have outlined in chapter 3. A number of features which are typical of landing dynamics can be studied. They include:

      (a) Wheel spin-up.
      (b) Friction between tire and runway.
      (c) Shock strut vertical forces.
      (d) Bending backwards of landing gear.
      (e) Modal vibration of wings.

As a simple example, consider the landing of a 100 000 lb aircraft. The landing speed will be taken as 2000 in./sec (114 miles per hour), while the sinking velocity is 100 in./sec (8·33 ft/sec). The wheels are 95 in. to either side of the center-line of the aircraft at the fore and aft center of gravity. The aerodynamic lift is assumed equal to the weight during the initial impact. The nose gear does not come into contact during the initial impact.

The coordinates used will be as follows:

      1, vertical motion of center of gravity, positive upwards
      2, forward motion of center of gravity
      3, yaw, positive, nose to right
      4, roll, positive, right wing up
      5, modal motion (fuselage stiff, wing flexing, free–free vibration)
      6, forward motion of right wheel
      7, forward motion of left wheel
      8, angular rotation of right wheel, positive counterclockwise
      9, angular rotation of left wheel, positive counterclockwise

The landing gears have a fore-and-aft spring constant of 20 000 lb/in. and a damping of 30 lb sec/in. They have a vertical spring constant of 10 000 lb/in. with a damping of 1000 lb sec/in., and behave like stops, i.e., they only develop force after the wheels contact the ground.

The coefficient of friction between the tire and the runway is 0·5. A local spring stiffness of 2000 lb/in. exists at the contact point to simulate the tires. The wheel radius is 12 in.

The lowest symmetric vibration mode is included. For this mode, the modal amplitude at the wheel is 0·1 in. when the modal mass is 300 lb sec^2/in. The modal spring rate is 7500 lb/in., corresponding to a modal frequency just below 1 Hz.

The modal damping is taken at 3000 lb sec/in.

The inertial values are:

coordinate 1, 259 lb sec²/in. (mass of airplane)

coordinate 2, 259 lb sec²/in. (mass of airplane)

coordinate 3, 14·8 × 10⁶ lb sec² in. (yaw moment of inertia)

coordinate 4, 8·4 × 10⁶ lb sec² in. (roll moment of inertia)

coordinate 5, 300 lb sec²/in. (modal mass)

coordinate 6, 1·295 lb sec²/in. (mass of wheel, 500 lb wheel weight)

coordinate 7, 1·295 lb sec²/in. (mass of wheel)

coordinate 8, 129·5 lb sec² in. (wheel moment of inertia, radius of gyration of wheel 10 in.)

coordinate 9, 129·5 lb sec² in. (wheel moment of inertia)

As a first example, we consider a symmetric landing. At the initial instant of time, it is assumed that both wheels are 1 in. above the runway.

The results are shown in Fig. 4.6. The forces develop after 0·01 sec when the wheels touch down. The vertical force FW at the wheel is quite high (100 000 lb). This is primarily because of the high damping in the shock strut. The vertical force has dropped to about 40 000 lb after 0·2 sec.

The tractive friction force FY on the wheels builds up more gradually to about 50 000 lb, at which point skidding takes place while the wheel is spun up. After about 0·06 sec, the wheel is up to speed and FY decreases. When the strut springs back, the wheel speed is excessive. This results in negative skidding from about 0·11 sec to 0·14 sec. Positive skidding takes place again from about 0·17 sec to 0·20 sec.

During the initial phase of aircraft landing, while the wheels are spun up, the landing gear is bent backwards. The forces developed are denoted by FX in Fig. 4.6. The peak value is over 80 000 lb. It is apparent why fatigue can occur in a landing gear, since FX goes from positive to negative values several times during a landing.

Displacements during landing are shown in Fig. 4.7. The maximum downward displacement of the aircraft center of gravity is about 7·5 in. The wheel rotation is also shown. For a 2000 in./sec speed and a 12 in. radius, an average angular velocity of 167 rad/sec might be expected after the initial start-up phase. The values oscillate around a line having a slope of 167 rad/sec. The oscillation is due to skidding and to the fore-and-aft vibration of the shock strut, which is also shown in the figure.

Problems

4.3   Repeat the example problem with the damping in the shock strut in a vertical direction reduced from 1000 to 500 lb sec/in., and the spring rate increased from 10 000 to 20 000 lb/in. What effect does this have on the maximum forces developed? Is the maximum vertical displacement at the center of gravity increased by this change?

4.4   Repeat the example problem with a local tire spring stiffness at the friction contact of 20 000 lb/in. rather than the 2000 lb/in. value used in the example. Does this change the wheel dynamics significantly? How much is the maximum fore-and-aft force in the shock strut affected?

4.5   Repeat the example problem with a higher frequency vibration mode having lower damping by letting the modal spring rate be 67 500 lb/in. rather than 7500 lb/in., and by lowering the

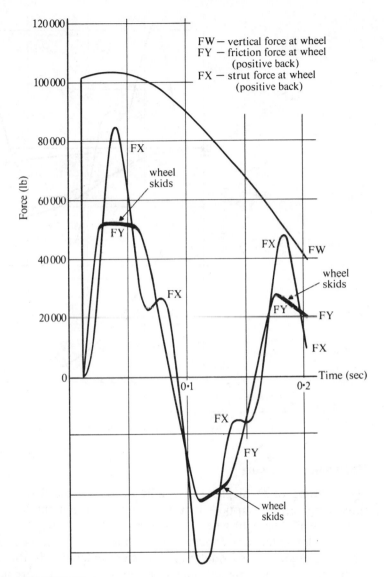

FIGURE 4.6
Forces during symmetrical airplane landing

damping to 500 lb sec/in. from the 3000 lb sec/in. value in the example. Does this affect the modal response appreciably? Are the maximum shock strut forces changed more than 10 per cent?

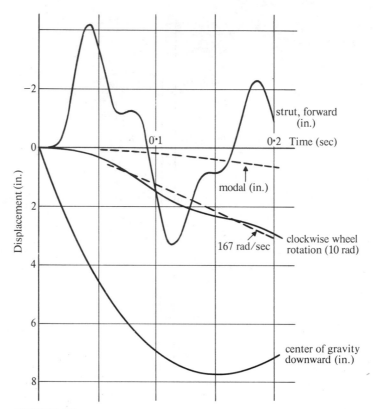

FIGURE 4.7
Displacements during symmetrical landing of airplane

Unsymmetric landing

The landing of an airplane is usually on one wheel first. In the example of the preceding section, suppose that at the initial instant the left wheel is just making contact while the right wheel is still 3 in. off the runway.

The forces that develop are shown in Fig. 4.8. Comparison of Fig. 4.8 and Fig. 4.7 shows a great deal of similarity. The chief difference is the time lag of 0·03 sec until the right wheel makes contact. In addition, the forces are now about 10 per cent higher on the wheel striking first and about 10 per cent lower on the wheel striking second.

Problems

4.6   Add the first unsymmetric mode to the modeling of the example problem. Let the modal mass be 150 lb sec^2/in., the modal spring rate be 75 000 lb/in., and the modal damping be 1000 lb sec/in. Let the modal amplitude be 0·8 in. at the right landing gear and −0·8 in. at the left landing gear. Does the addition of the unsymmetric mode have much effect on the shock strut forces?

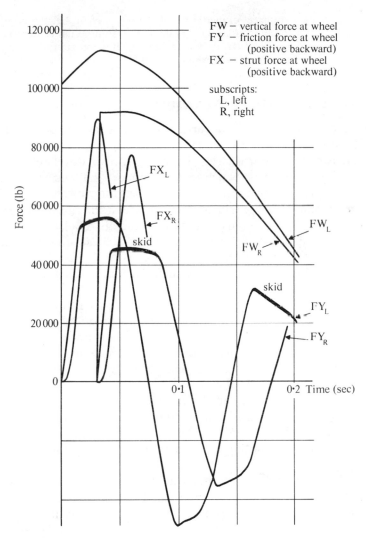

FIGURE 4.8
Forces during unsymmetrical airplane landing

4.7   Add a nose gear to the example problem. Let its vertical stiffness and damping be 5000 lb/in. and 500 lb sec/in., respectively. Let its stiffness be 10 000 lb/in., and the fore-and-aft damping be 15 lb sec/in. Let the nose wheel have a mass of 0·6475 lb sec²/in., a moment of inertia of 64·75 lb sec² in., and a radius of 12 in. Let the nose wheel be 190 in. forward of the center of gravity and an inch from the ground at time $t = 0$. Let the aircraft pitching moment of inertia be 6 400 000 lb sec^2 in. Let the nose wheel friction coefficient be 0·5, and the contact spring rate be 1000 lb/in. What is the maximum nose wheel vertical and drag force? (*Hint:* Include coordinates for aircraft pitch, nose wheel fore-and-aft displacement, and nose wheel rotation.)

4.4   LOCOMOTIVE DYNAMICS

Railway locomotives are carried on an assortment of suspension arrangements. A typical bolster and truck combination is shown in Fig. 4.9. The locomotive is carried on two of these truck sets. In this type, two wheel sets are vertically loaded by two equalizer bars, which in turn support the truck frame. The truck frame has pedestal guides that restrain the wheel sets fore and aft and laterally. The connections are somewhat loose and contact is made at the rub plates. Swing hangers are supported by the truck and in turn support the gib. A bolster is placed on the gibs through a leaf spring support. Finally, the locomotive rests on the bolster and can pivot on it. The whole arrangement, therefore, consists of masses connected by a number of springs. Forces are also transmitted by friction on rub plates. The rub plates also act as stops.

The dynamics of locomotives on tracks is the subject of a large body of literature.[2−7] Cain[2] has described how locomotives sometimes oscillate, or hunt, on rough sections of track, and even on straight track sections. The literature on this subject brings out the mechanical interactions resulting from rail roughness, creepage,[6] and car dynamics, but generally solves rather simplified versions of the governing equations of motion.

Meacham and Ahlbeck[7] have studied, with an analog computer, some rather complex dynamic situations in railroad vehicles, particularly the phenomenon of the rocking of freight cars and the manner in which the car wheels interact with the track. A model of the freight car used by them is shown in Fig. 4.10. The vehicle suspension modeling is clearly illustrated there. In their study, they pay close attention to how the track deformation can play an important role in the dynamics of the car.

The locomotive dynamics can be inspected in some detail by the method presented in this book, provided the following quantities are known:

Mass

The mass, moments of inertia (in pitch, yaw, roll), and center of gravity are needed for each of the following:

(a) The car.
(b) The bolsters.
(c) The truck frames.
(d) Each wheel set.
(e) The traction motors.

Tractive force

The pull of the entire locomotive on the draw bar and the traction of each wheel set.

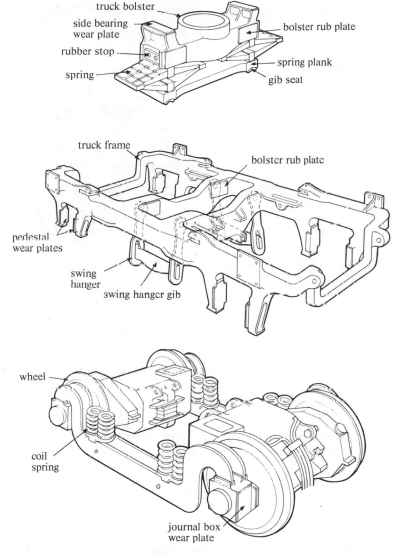

FIGURE 4.9
A typical bolster and truck combination

Secondary suspension and lateral restraint of bolster

(a) The vertical spring stiffness and damping.
(b) Lateral stiffness and damping.
(c) The lateral and fore-and-aft stiffness when stopped, and clearance before stopping.
(d) The friction coefficient between frame and bolster at all contact points.

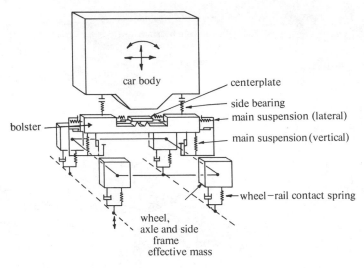

FIGURE 4.10
Freight car model used for computer studies of car-rocking phenomenon (from Meacham and Ahlbeck,[7] reproduced from *Journal of Engineering for Industry, Trans. ASME,* **91,** 1969, American Society of Mechanical Engineers, with permission)

In the examples discussed here, the stop elements have a compressive force of

$$
\left.
\begin{array}{ll}
F_x = -k(x + R) & (-x > R) \\
F_x = 0 & (-x < R)
\end{array}
\right\}
\tag{4.1}
$$

where k is the stop stiffness, R is the stop initial clearance, and x is the extension of the stop element.

Primary suspension
Stiffness and damping are modeled as usual in this suspension.

Motor reaction springs
The stiffness and damping vertically are required for the restraining cushion between the driving motor and the frame.

Wheel set stops
The clearance and stiffness of the wheel set stops and the coefficient of friction at all contacts between frame and wheel set.

Rail
The vertical and lateral alignment of each rail is needed as a function of position along the track. The stiffness of the rail vertically and laterally as a function of position is also needed, as is information on rail curvature and superelevation.

Wheel

For the wheel, creep coefficients, coning angle, radius, and nominal flange clearance are required.

Lateral creep in railroad wheels is the displacement transverse to the direction of rolling. It is a flexible interaction between the wheel and the track. Both the wheel and track are elastic. In general, pairs of wheels are joined rigidly by an axle. The coning angle on the tread makes it possible for the wheel pair to turn with the track on a curved track, if the wheel pair takes a suitable transverse position. If the wheel pair is not in a suitable transverse position, the wheel flanges will soon contact the rail, and, before that occurs, forces are induced which cause elastic deformation and corresponding fore-and-aft creep of the wheel. The force at the point of contact, which is required to produce a given creep, is related through a creep coefficient to the wheel velocity and the overall velocity of the locomotive. The following formulas give the creep forces for both lateral and fore-and-aft motion.

Creep force element laterally

$$F_x = -f_x(\dot{x}/V - \theta) \tag{4.2}$$

where f_x = lateral creep coefficient (see, for example, Newland[6]). It should be noted, however, that an actual wheel is distorted generally as well as by Hertzian deformations, so the actual value of f_x may be significantly lower than that determined from these references

x = displacement of wheel (positive, laterally outward)

V = locomotive velocity

θ = rolling direction angle of wheel from straight ahead, positive outward when facing forward

Creep force element fore and aft

$$F_\theta = -f_y(d\dot{\theta}/V + \alpha x/r)d \tag{4.3}$$

where f_y = fore-and-aft creep coefficient

d = half the span from wheel to wheel

α = coning angle of wheel

r = radius of wheel

Car dynamics

With the data provided as described above, we will now show the types of result that can be obtained concerning the movements of the locomotive cab, truck frames, and individual wheels. Also, important from the design viewpoint, forces between the wheels and track, and within the wheel supports, can be obtained as illustrated in typical instances. Some additions to the force elements of the program in section 3.16 were required for these solutions. They used values of INDX between 3 and 8.

First, we show results which concern a locomotive which rests on two of the truck sets which were illustrated in Fig. 4.9. Figure 4.11(a) shows displacements which result from a one per cent grade change at a velocity of 100 ft/sec. The frame and cab pitch angles are followed in Fig. 4.11(b), as are the spring forces in the wheel sets in Fig. 4.11(c). The effect of a one per cent track yaw on the lateral displacements

FIGURE 4.11
The track is level up to position $x = 0$ and then has a one per cent downward grade. At the 50 ft station the right and left tracks are down 0·5 ft. The rail stiffness is taken constant at 2 000 000 lb/ft vertically and 500 000 lb/ft laterally

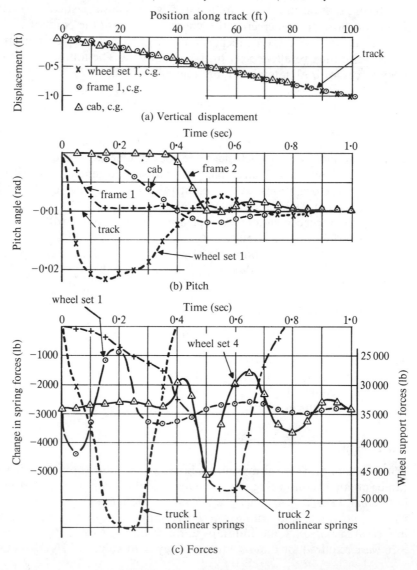

(a) Vertical displacement

(b) Pitch

(c) Forces

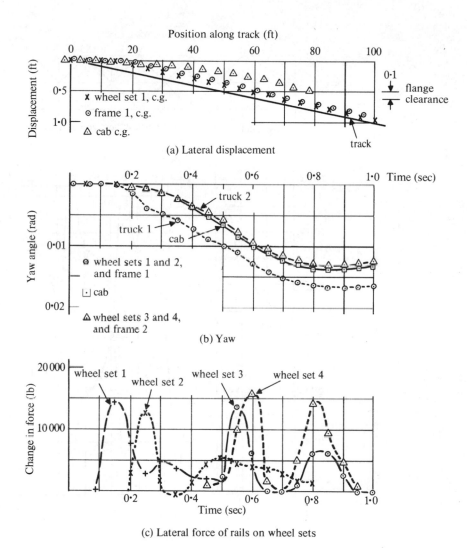

(a) Lateral displacement

(b) Yaw

(c) Lateral force of rails on wheel sets

FIGURE 4.12
Effect of track yaw of one per cent with an initial flange clearance of 0·1 ft. The track is considered straight up to $x = 0$ and then veers sidewise 1 ft every 100 ft, with the same stiffness as in Fig. 4.11. A significant item in this case is the flange contact forces

is shown in Fig. 4.12(a). The effect on the truck yaw angles is shown in Fig. 4.12(b), and the effect on the force of the rails on the wheels is shown in Fig. 4.12(c). If the track has vertical oscillations on it, illustrated by the sawtooth profile in Fig. 4.13(a), the wheel loads oscillate in a nonlinear fashion, and forces in opposite wheels are nearly symmetrically opposed, as shown in Fig. 4.13(b). Similar behavior is observed when the rail stiffness varies along the track, as shown in Fig. 4.14. More realistic track deflections and stiffness variations can also be studied. Figure 4.15 shows

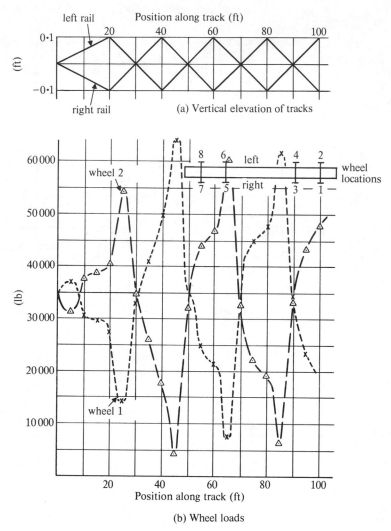

(a) Vertical elevation of tracks

(b) Wheel loads

FIGURE 4.13
Effect of track roll—up on the right and down on the left when facing forward

measured variations along sections of a test track. The effect of this track on the car roll and the bolster spring forces is shown in Fig. 4.16.

A locomotive with two trucks containing three wheel sets each is shown diagrammatically in Fig. 4.17. There, the generalized coordinates are indicated by arrows, and consist of uncoupled angular and linear displacements of each item in the locomotive assembly. There are 63 degrees of freedom in all. Some of the necessary springs, stops, and friction elements are illustrated in Fig. 4.18. When contact is made, each rub plate must allow frictional movement in two orthogonal directions in its plane, whereas perpendicular to its plane it acts as a stop. (Two friction elements and one stop element are used at each rub plate.)

The dynamics of this locomotive as it runs about a curved track can be studied. The additional necessary input is the tractive force moment on the car due to the coupler swing. The appropriate geometric quantities are defined in Fig. 4.19. The moment on the car due to the tractive force, Fig. 4.19, is given by

$$\text{moment} = -L(\alpha - \beta)P_{\text{traction}} \tag{4.4}$$

where   L = distance from cab center of gravity to coupler pivot
α = angular direction of car
β = angular direction of draw bar pull
P_{traction} = draw bar pull

FIGURE 4.14
Effect of variation of track stiffness (velocity 100 ft/sec). The track stiffness is considered to drop from 2 000 000 lb/ft to 300 000 lb/ft alternately on the right and then the left track, simulating rail joints

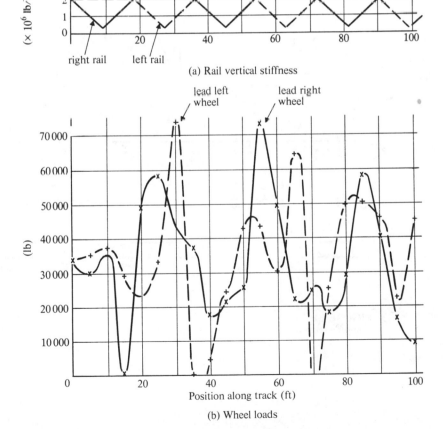

(a) Rail vertical stiffness

(b) Wheel loads

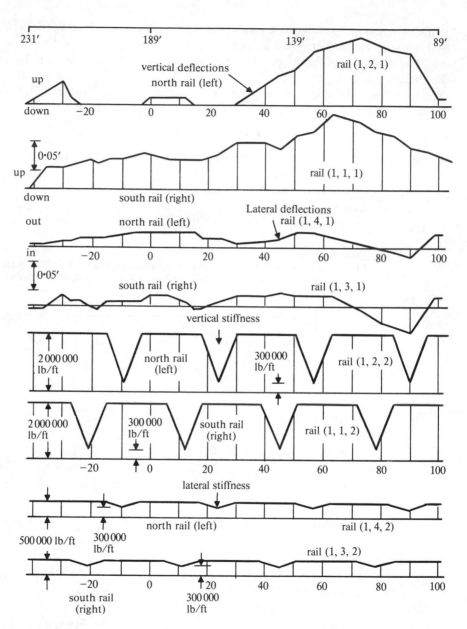

FIGURE 4.15
Run on a test track—test track parameters, showing variation in rail deflection and stiffness.

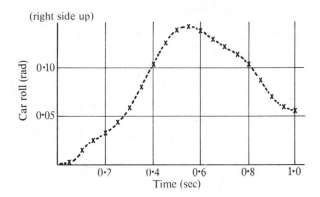

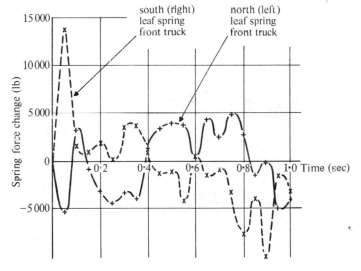

FIGURE 4.16
Car roll and spring force change. Car velocity 100 ft/sec.

The track curvature is assumed to increase linearly as a curve is entered, and then to hold constant. It can readily be shown that in the portion where the curvature is increasing:

$$y = z^3/6SR \qquad (z < S)$$

where z = distance along track from start of curve
y = lateral track position from initial straight line
R = final radius of curvature
S = value of z at which curvature becomes constant

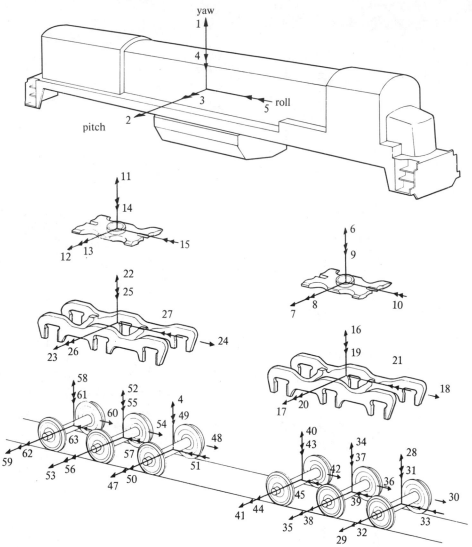

FIGURE 4.17
Generalized coordinates for locomotive dynamics

In the region where the curvature is constant,

$$y = \frac{[z(z - S) + S^2/3]}{2R} \qquad (z > S)$$

The centrifugal force that develops in going around a curve is proportional to the curvature (at constant speed). The superelevation is therefore taken to increase linearly from zero at $z = 0$ to E at $z = S$. For larger values of z, the superelevation

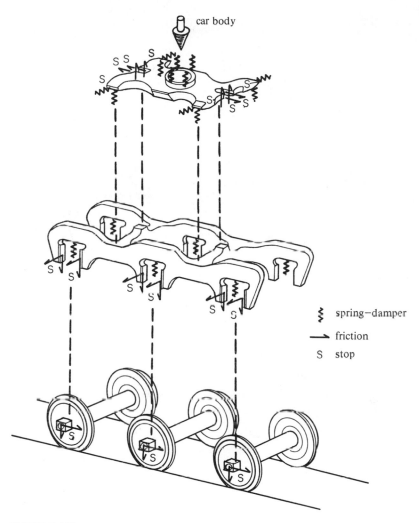

FIGURE 4.18
Force elements active on the suspension system

is constant at the value E. The wheel support forces have a component transverse to the track surface. Account is taken of this in computing the lateral creep of the wheel.

Some results for the wheel support forces and the lateral forces of the wheel flange against the rail are shown in Fig. 4.20 and Fig. 4.21 for two wheel sets: the leading wheel set of the first truck, and the middle wheel set of the second truck. Oscillations will be observed in the forces, particularly in the lateral force components. In this calculation, the radius of the track changed to a final value of 1910 ft at a distance of 350 ft along the curve. Beyond that point, the radius holds constant. The corresponding superelevation at that radius is 0·375 ft.

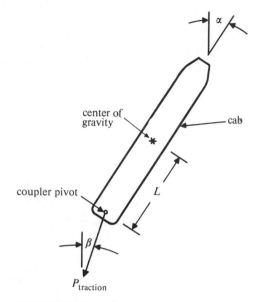

FIGURE 4.19
Restoring moment due to traction load

FIGURE 4.20
Some forces on the leading wheel set of the first truck

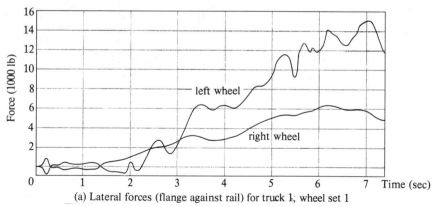

(a) Lateral forces (flange against rail) for truck 1, wheel set 1

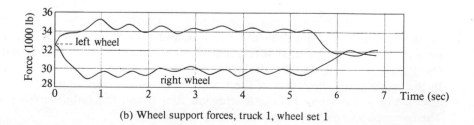

(b) Wheel support forces, truck 1, wheel set 1

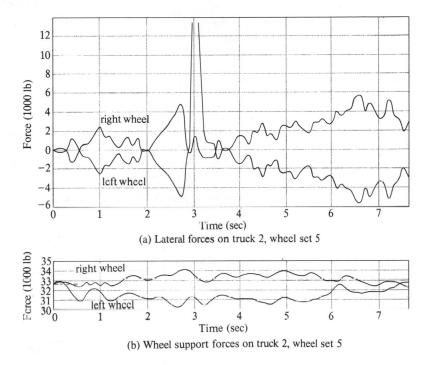

(a) Lateral forces on truck 2, wheel set 5

(b) Wheel support forces on truck 2, wheel set 5

FIGURE 4.21
Some forces on the middle wheel set of second truck

4.5 MASS MOVING ACROSS A BEAM—TIME-VARYING COUPLING RATIOS

Consider now the simple problem of a mass that runs with velocity v across a simply supported flexible beam of length l, as shown in Fig. 4.22(a). A constant force mg acts downward on the mass, due to gravity. The mass is assumed to be in contact with the beam through a spring–damper–stop system, such as, for example, a rubber tire. The force–deflection characteristic of the system is shown in Fig. 4.22(b). When the mass moves downwards, the spring k is compressed. If the mass moves upward, the spring is assumed to lose contact with the beam—this situation is modeled by using a stop element in the computer programs, but with a negative initial clearance (mg/k) to express the fact that it statically supports the mass.

The purpose of introducing this example is to show that situations arise where the coupling ratios may vary in time. At any instant in time, the position of the mass along the beam is given by vt. As shown in Fig. 4.22(c), the mode shapes of the beam are given by $\sin(n\pi x/l)$, $n = 1, 2, 3, \ldots$. The coupling ratios for the modal masses are given by the deflection of that mode at the point of contact of the moving mass. Thus, they are given in each nth mode by $\sin(n\pi vt/l)$. As a result, these coupling ratios vary in time. The response of the mass as it travels across the beam can be found by applying the computer programs of sections 3.13 or 3.16

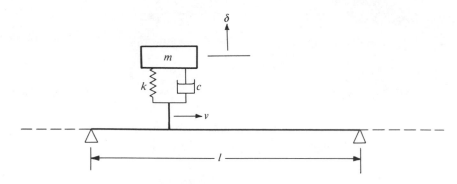

(a) Model of mass moving across a beam

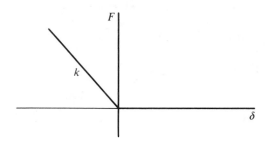

(b) Force−deflection curve for spring−stop system

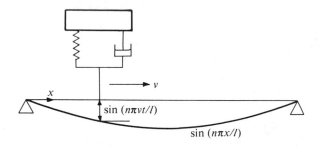

(c) Coupling coefficient between beam and mass

FIGURE 4.22
Modeling a mass moving across a beam

(depending on the number of degrees of freedom included in modeling the beam dynamics). A change in the programs must be made, however, to allow the coupling ratios CC for the beam modal masses to change at each time step.

Problems

4.8   Modify the three degree-of-freedom system computer program to allow for time-varying coupling ratios of the type described for the beam.

4.9   Calculate the motion of a mass of 10 lb moving at a speed of 10 ft/sec across a beam of length 60 in., whose width is 3 in. and thickness is 1 in. Assume its density to be $0 \cdot 1$ lb/in.^{3}, its modulus to be 10 000 000 lb/in.^{2}, and that the spring constant between the mass and beam is 1000 lb/in. Damping is to be two per cent of critical. Use the first two beam modes.

4.10   The beam has a second single support also at its midpoint. Repeat the calculations for the mass dynamic motion. Does the stop element lose contact with the beam after crossing the support at its midpoint? (*Hint:* Model the beam midpoint support by a spring with a stiffness of 10 000 lb/in.)

Having discussed briefly the concept of time-varying coupling ratios, we will now move on to discuss a more complex example concerning an air cushion vehicle moving along a flexible guideway. In this case, also, the coupling ratios of the guideway modes are time-varying. The concept of time-varying coupling ratios will also occur in chapter 7, where we will discuss the problem of an oblique impact of a mass on a flexible blade of an aircraft engine.

4.6 AIR CUSHION VEHICLES

One of the modes of high speed transportation under active study for the future is an air cushion vehicle on an elevated guideway. The use of air cushions in place of wheels may better meet the needs for high speed ground transportation. However, in themselves, they present new challenges in vehicle dynamics. A recent survey of the dynamics of vehicles on elevated guideways has been made by Richardson and Wormley.[8]

To provide an adequate guideway for the vehicle to run on also presents problems. The smoothest ride would undoubtedly come from a heavy, stiff guideway. Such a guideway would be expensive, however. Going to a lighter, more flexible guideway introduces traveling waves in the guideway, which are not always desirable. Thus, it becomes important to adequately understand how these traveling waves are propagated and how the vehicle dynamics act to cause them.

The guideway dynamics can be described in terms of equations of motion, with the amplitudes in the normal modes as generalized coordinates. Damping in the guideway system is considered distributed in proportion to the mass, and thus does not affect the orthogonality of the modes.

To obtain the normal modes of the guideway system, the lateral, vertical, and torsional motion at the center of twist is described with a 15 term sine series for each motion. An expression is then written for the kinetic and potential energy of the system in terms of amplitudes of the terms in the sine series expansions. The kinetic energy includes the energy due to coupling of the lateral and torsional velocities when the center of gravity is not at the center of twist.

Natural frequencies and modes of guideway

We express the kinetic and potential energy of the guideway in terms of the generalized coordinates A_m, B_m, C_m. We take

$$
\left.
\begin{aligned}
y_A &= \sum_m A_m \sin \frac{m\pi x}{L} \quad \text{(downward motion)} \\[2mm]
y_B &= \sum_m B_m \sin \frac{m\pi x}{L} \quad \text{(lateral motion at center of twist)} \\[2mm]
y_C &= \sum_m C_m \sin \frac{m\pi x}{L} \quad \text{(rotation)}
\end{aligned}
\right\}
\tag{4.5}
$$

where x is the coordinate giving position along the guideway and L is its length. The guideway stiffness is expressed in terms of bending and torsional components:

$$
\begin{aligned}
(EI_A) &= \text{bending stiffness in vertical plane} \\
(EI_B) &= \text{bending stiffness in lateral plane} \\
(GJ) &= \text{torsional stiffness}
\end{aligned}
$$

The guideway mass is taken as:

$$
\begin{aligned}
(M) &= \text{mass per unit length} \\
(J) &= \text{rotary inertia about the center of gravity}
\end{aligned}
$$

The offset between the center of gravity and the center of twist is D.

The kinetic energy T is given by

$$T = \int 1/2 \rho \dot{y}^2 \, dx$$

$$
T = 1/2 \int_0^L dx \left\{ (M) \left[\sum_m \dot{A}_m \sin \frac{m\pi x}{L} \right]^2 + (M) \left[\sum_m (\dot{B}_m + D\dot{C}_m) \sin \frac{m\pi x}{L} \right]^2 \right.
$$

$$
\left. + (J) \left[\sum_m \dot{C}_m \sin \frac{m\pi x}{L} \right]^2 \right\}
$$

$$
T = \frac{L(M)}{4} \sum_m (\dot{A}_m^2 + \dot{B}_m^2 + 2\dot{B}_m \dot{C}_m D + \dot{C}_m^2 D^2) + \frac{L(J)}{4} \sum_m \dot{C}_m^2
\tag{4.6}
$$

The potential energy stored due to bending is

$$
U_B = \frac{(EI_A)\pi^4}{4L^3} \sum_m (m^4 A_m^2) + \frac{(EI_B)\pi^4}{4L^3} \sum_m (m^4 B_m^2)
\tag{4.7}
$$

and that due to twisting is

$$
U_T = 1/2 \int_0^L dx \, (GJ) \left[\sum_m \left(\frac{m\pi}{L} \right) C_m \cos \frac{m\pi x}{L} \right]^2
$$

$$
= \frac{(GJ)\pi^2}{4L} \sum_m (m^2 C_m^2)
\tag{4.8}
$$

The total potential energy is

$$U = U_B + U_T \tag{4.9}$$

Lagrange's equation states that for each value of m (in the absence of external force):

$$
\left.
\begin{aligned}
\frac{d}{dt}\left(\frac{\partial T}{\partial \dot{A}_m}\right) + \frac{\partial U}{\partial A_m} &= 0 \\[2mm]
\frac{d}{dt}\left(\frac{\partial T}{\partial \dot{B}_m}\right) + \frac{\partial U}{\partial B_m} &= 0 \\[2mm]
\frac{d}{dt}\left(\frac{\partial T}{\partial \dot{C}_m}\right) + \frac{\partial U}{\partial C_m} &= 0
\end{aligned}
\right\} \tag{4.10}
$$

Substituting eqs. (4.6) to (4.9) into (4.10), we obtain

$$
\left.
\begin{aligned}
\frac{(M)L}{2}\ddot{A}_m + \frac{(EI_A)\pi^4 m^4}{2L^3} A_m &= 0 \\[2mm]
\frac{(M)L}{2}(\ddot{B}_m + D\ddot{C}_m) + \frac{(EI_B)\pi^4 m^4}{2L^3} B_m &= 0 \\[2mm]
\frac{(M)L}{2}(D\ddot{B}_m + D^2\ddot{C}_m) + \frac{(J)L}{2}\ddot{C}_m + \frac{(GJ)\pi^2 m^2}{2L} C_m &= 0
\end{aligned}
\right\} \tag{4.11}
$$

When vibrating in the nth normal mode at a frequency ω_n:

$$
\begin{aligned}
\ddot{A}_m &= -\omega_n^2 A_m \\
\ddot{B}_m &= -\omega_n^2 B_m \\
\ddot{C}_m &= -\omega_n^2 C_m
\end{aligned}
\tag{4.12}
$$

Substituting these values into eq. (4.11) gives

$$\omega_n^2 = \left(\frac{\pi^4 m^4}{L^4}\right)\left(\frac{(EI_A)}{(M)}\right) \qquad (A_m \neq 0,\ B_m = 0,\ C_m = 0) \tag{4.13}$$

and when $A_m = 0$, the pair of equations:

$$
\left.
\begin{aligned}
\left[\left(\frac{\pi^4 m^4}{L^4}\right)\left(\frac{(EI_B)}{(M)}\right) - \omega_n^2\right] B_m - D\omega_n^2 C_m &= 0 \\[2mm]
[-D\omega_n^2]B_m + \left[\left(\frac{\pi^2 m^2}{L^2}\right)\left(\frac{(GJ)}{(M)}\right) - \omega_n^2\left(\frac{(J)}{(M)} + D^2\right)\right]C_m &= 0
\end{aligned}
\right\} \tag{4.14}
$$

Thus, the remaining natural frequencies ω_n can be determined by obtaining values of ω_n for which eqs. (4.14) have solutions other than zero. In the process the absolute values of A_m, B_m, C_m can also be calculated from the condition that the modal mass be unity, i.e.,

$$\int \rho y^2 \, dx = 1 \tag{4.15}$$

The mode shapes are then given by eq. (4.5). The specific values of A_m, B_m, and C_m appropriate to each mode of frequency ω_n are henceforth denoted by $A_{n,m}$, $B_{n,m}$, $C_{n,m}$. There are 45 modes in all.

Guideway equations of motion

A unique feature of the normal modes of a system is that, when the amplitudes in the normal modes are used as generalized coordinates, the equations of motion are uncoupled for systems having damping distributed in proportion to mass.

Denoting by z_n the amplitude in the nth normal mode,

$$
\left.
\begin{aligned}
y_A &= \sum_n z_n \sum_m A_{n,m} \sin \frac{m\pi x}{L} \quad \text{(downward motion)} \\[2mm]
y_B &= \sum_n z_n \sum_m B_{n,m} \sin \frac{m\pi x}{L} \quad \text{(lateral motion at center of twist)} \\[2mm]
y_C &= \sum_n z_n \sum_m C_{n,m} \sin \frac{m\pi x}{L} \quad \text{(rotation)}
\end{aligned}
\right\} \qquad (4.16)
$$

The dynamic equation for the guideway displacement in the nth mode (remembering we have normalized the modes to have unit modal mass) is

$$
\ddot{z}_n + 2\omega_n (c_n/c_c)\,\dot{z}_n + \omega_n^2 z_n = Q_n \qquad (4.17)
$$

where (c_n/c_c) is the ratio of modal damping to the critical value. The term Q_n is the generalized force in the mode, and is defined as the work that would be done by the actual forces during a unit displacement in mode n:

$$
Q_n = \int f y_n \, dx \qquad (4.18)
$$

where f is the force applied at location x and y_n is the modal amplitude at x.

In the dynamic analysis the force f is applied to the guideway by the air cushion elements, and by the intermediate support columns of the guideway.

Car equations of motion

A diagram of the air cushion vehicle, its suspension system, and the guideway is shown in Fig. 4.23. The car rests on two trucks which are connected to it by springs, and by an active element. The force in the active element between car and trucks is proportional to both the velocity and acceleration of its point of attachment to the car, so that the force can be expressed as

$$
F = c_1 \dot{x} + c_2 \ddot{x} \qquad (4.19)
$$

where x is the displacement of the car in inertial space. This equation essentially allows feedback to control the car behavior. Each truck is separated from the guideway by two air cushions. One cushion provides lateral restraint, while the other gives vertical support. The force generated by an air cushion depends not only on

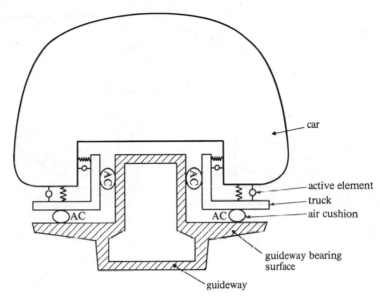

FIGURE 4.23
Diagram of an air cushion vehicle on a guideway

the separation between the truck and guideway and the rate at which this separation changes, but also on the rate of change of force in the air cushion. In order to allow as generalized a force description as necessary, it is quite feasible to set

$$F + T_2\dot{F} = -k(T_1\dot{y} + y) \tag{4.20}$$

where F is the air cushion force and y is the extension of the air cushion. The truck can also be allowed to provide additional car support forces. For example, one can use as an element a spring–damper in parallel, so that

$$F = -kw - c\dot{w} \tag{4.21}$$

where w is the extension. Or one can use as an element a spring–damper in series so that

$$F = -k(u - \zeta) = -c\dot{\zeta} \tag{4.22}$$

Here u is the total extension of the spring and damper and ζ is the extension of the damper.

The car is given five degrees of freedom: vertical and lateral motion, and pitch, yaw, and roll. Each of the four undercarriage trucks is given two degrees of freedom that describe the vertical and lateral motion. Thus, the car is described by 13 generalized coordinates.

The equation governing each car degree of freedom is

$$M_n\ddot{z}_n = Q_n \tag{4.23}$$

where M_n is the generalized mass in that coordinate and Q_n the generalized force.

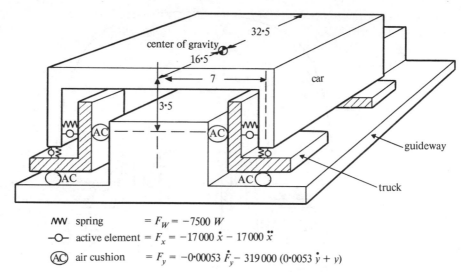

WW   spring $= F_W = -7500\ W$

—O— active element $= F_x = -17\,000\ \dot{x} - 17\,000\ \ddot{x}$

(AC) air cushion $= F_y = -0{\cdot}00053\ \dot{F}_y - 319\,000\ (0{\cdot}0053\ \dot{y} + y)$

FIGURE 4.24
Geometry of air cushion vehicle (dimensions in feet)

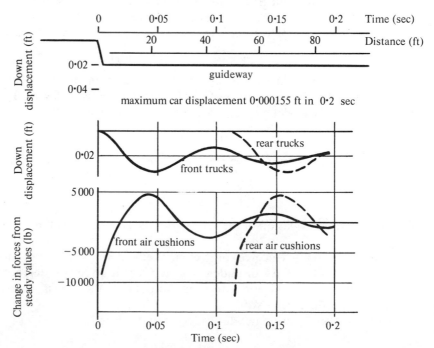

FIGURE 4.25
Effect of guideway down-drop

For example, for car translation, M_n is the mass, while for car pitch it is the pitching moment of inertia. The generalized car displacement is z_n. For example, for car translation, it is the center of gravity movement, and for car pitch it is the pitching rotation. Q_n is the generalized force formed by summing the products obtained by multiplying the actual element forces by the corresponding coupling ratios. For example, for vertical car displacement, the force from a front upper active element contributes directly to the vertical generalized force, so its coupling ratio is 1·0.

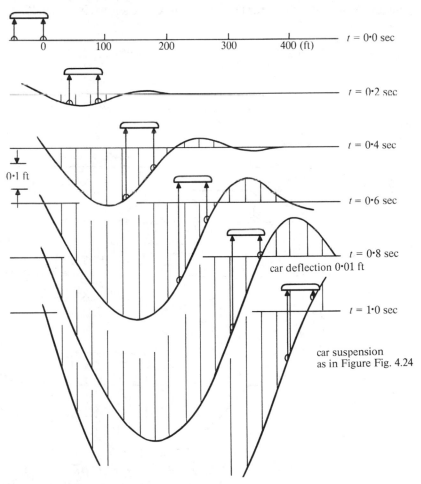

run down flexible guideway, 300 miles per hour, 400 ft span,
$EI = 2\cdot3 \times 10^{10}$ lb ft^2, guideway mass 103·2 lb sec^2/ft^2,
live load deflection 4·4 ft = (span/91),
car weight 76 500 lb

FIGURE 4.26
Effect of guideway flexibility—long span guideway

Those coupling ratios associated with the guideway modal masses vary with time, just as described in the previous section for the case of the mass traversing a flexible beam.

Wind forces, both upward, lateral, and in pitch, can be modeled in the same manner. In a computing program similar to that in section 3.16, all of these force elements were modeled. They used values of INDX between 3 and 8.

Some typical results are described here for a particular vehicle in which each truck mass is 80 lb sec^2/ft, and the car mass is 1740 lb sec^2/ft. The car has moments of inertia of 670 000 lb sec^2 ft in pitch, 662 000 lb sec^2 ft in yaw, and 18 500 lb sec^2 ft in roll. The geometrical relationships between the points of truck contact, guideway contact, and the center of gravity are shown in Fig. 4.24, where the magnitudes of the spring elements, active elements, and air cushions are also given.

When the car runs off a 1/4 in. drop in the guideway at 300 miles per hour, the displacements of the front and rear trucks are shown in Fig. 4.25. It is evident that

FIGURE 4.27
Effect of guideway flexibility—shorter span guideway

run down flexible guideway, 300 miles per hour, 100 ft span,
$EI = 2 \cdot 3 \times 10^{10}$ lb ft^2, guideway mass 103·2 lb sec^2/ft^2,
car weight 76 500 lb,
live load deflection 0·069 ft = (span/1460)

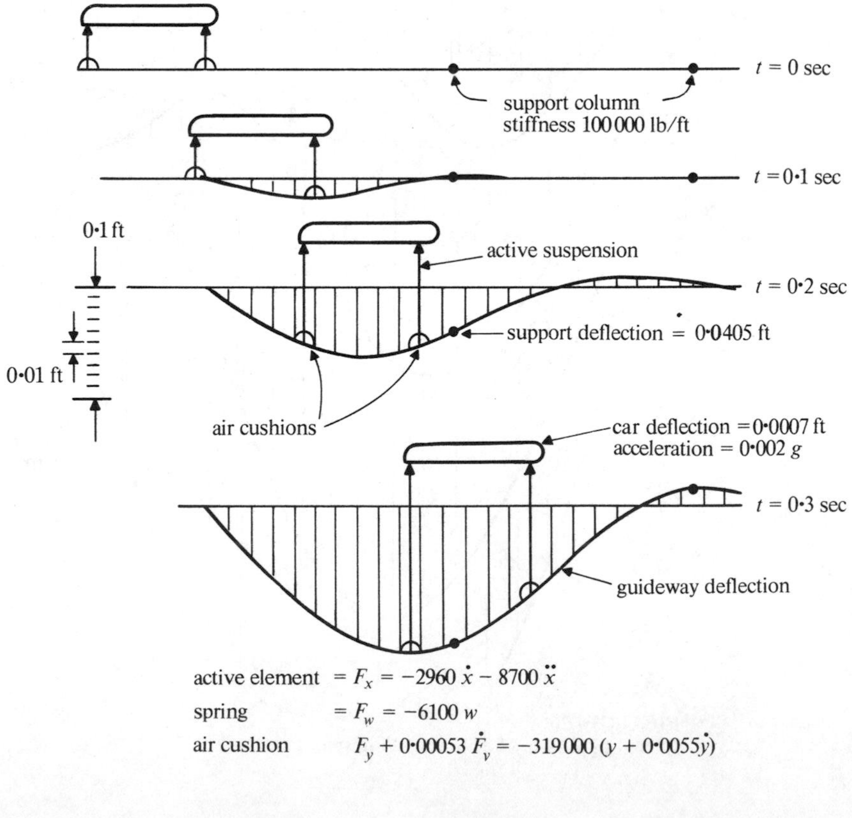

$t = 0$ sec

support column
stiffness 100 000 lb/ft

$t = 0 \cdot 1$ sec

0·1 ft

active suspension

$t = 0 \cdot 2$ sec

support deflection $\dot{=}$ 0·0405 ft

0·01 ft

air cushions

car deflection = 0·0007 ft
acceleration = 0·002 g

$t = 0 \cdot 3$ sec

guideway deflection

$$\text{active element} = F_x = -2960\,\dot{x} - 8700\,\ddot{x}$$
$$\text{spring} = F_w = -6100\,w$$
$$\text{air cushion} \quad F_y + 0 \cdot 00053\,\dot{F}_y = -319\,000\,(y + 0 \cdot 0055\dot{y})$$

the air cushion follows rather quickly in about 0·1 sec, while the car requires several seconds to adjust. The slow car response is largely a result of the large coefficient for the acceleration term in the active element control equation (4.19).

The effect of guideway stiffness is shown in Fig. 4.26. A striking result is shown, namely, that at 300 miles per hour the vehicle with widely spaced supports tends to stay ahead of the guideway deflections. The liveload static deflection of this guideway with the car at mid-span can be computed as 4·4 ft. The maximum guideway deflection at the car was only 0·29 ft, or about a factor of 15 less. The guideway deflection behind the car was much more. The car deflection itself was only about 0·01 ft. In Fig. 4.27 is shown the effect of guideway stiffness with more closely spaced supports. In this case, the deflections are roughly twice the liveload deflection. Note, however, that the suspension coefficients are somewhat changed for this example.

4.7 SUMMARY

We have seen how a variety of problems concerning vehicles and their dynamic behavior can be studied using the ideas and computer programs developed in previous sections. The vehicle suspensions were characterized by spring–damper elements, friction elements, stop elements, creep elements, air cushion elements, and some active elements whose restoring force depends on the velocity or acceleration of the element point of attachment. For the most part, the vehicles were modeled as rigid bodies, which had inertia and mass properties, but whose vibratory motion was not of great significance. In some cases, the guideway along which the vehicle was traveling was flexible. In that case, the idea of time-varying coupling ratios was introduced. The main purpose of the discussions in this chapter was to suggest how dynamical systems of this sort could be modeled, and to demonstrate some of the results attainable by means of the computer programs. The reader will not fail to identify other areas, such as packaging and machinery design, where such methods can also be applied.

4.8 REFERENCES

1. G. R. Potts and H. S. Walker, 'Nonlinear truck ride analysis,' *J. Eng. for Industry, Trans. ASME,* **96,** Ser. B, 597–602, 1974.
2. B. S. Cain, *Vibration of Rail and Road Vehicles,* Pitman Publishing Corp., New York, 1940.
3. A. H. Wickens, 'The dynamic stability of a simplified four-wheeled railway vehicle having profiled wheels,' *Int. J. Solids Structures,* **1,** 385–406, 1965.
4. A. H. Wickens, 'General aspects of the lateral dynamics of railway vehicles,' *J. Eng. for Industry, Trans. ASME,* **91,** Ser. B, 869–878, 1969.
5. T. Matsudaira, N. Matsui, S. Arai, and K. Yokose, 'Problems on hunting of railway vehicle on test stand,' *J. Eng. for Industry, Trans. ASME,* **91,** Ser. B, 879–890, 1969.
6. D. E. Newland, 'Steering a flexible railway truck on curved track,' *J. Eng. for Industry, Trans. ASME,* **91,** Ser. B, 908–918, 1969.
7. H. C. Meacham and D. R. Ahlbeck, 'A computer study of dynamic loads caused by vehicle-track interaction,' *J. Eng. for Industry, Trans. ASME,* **91,** Ser. B, 808–816, 1969.
8. H. H. Richardson and D. W. Wormley, 'Transportation vehicle/beam-elevated guideway dynamic interactions: a state-of-the-art review,' *J. Dynamic Systems, Measurement, and Control, Trans. ASME,* **96,** Ser. G, 169–179, 1974.

5

CONTINUOUS SYSTEMS—THE FINITE ELEMENT METHOD

5.1 INTRODUCTION

When an inquiry is made into the vibratory motion of a structural component, it is now often a straightforward matter to identify its modes of free vibration. The situations we have discussed so far in this book have been concerned with the vibrations of uniform beams. These beams were used as illustrative examples because of the simplicity of their free vibrational characteristics. The modes of free vibrations were used as generalized coordinates. For more complicated structural components, the same concepts apply: their generalized coordinates can be defined as modal deformations. The main difficulty in the general case is to determine these modal deformations. By what means should the modal deformations be sought? We believe that the finite element method provides a powerful numerical approach that can be applied to literally any structure to provide just this required information. It is, therefore, the aim of this chapter to provide a background to the finite element method—in particular, as to its application to the static deformation of elastic structures under loads and the development of the stiffness matrices describing the structure. The stiffness matrices will be used in chapter 6 to determine the vibratory modes of the structure.

The book by Zienkiewicz[1] is an excellent text for the detailed study of the finite element method as applied to engineering science—in particular, as it is applied to elastic or inelastic structures. This chapter presents a self-contained development of the method only in so far as it is directly applicable to the objective of this book—the determination of dynamic responses of structures.

In order to introduce the basic concepts of the finite element method, the first sections of this chapter show how the stiffness properties of a framework of springs can be developed in a logical way. The stiffness properties govern the relationship between forces and deflections of a structure. Next, the same concepts are applied to a two-dimensional elastic continuum. Here, the continuum is broken up into a set of discrete quadrilateral elements, known as finite elements, whose behavior is characterized by the forces and deflections at the nodes of the elements. It is shown how the stiffness properties of these elements are found and how the stiffness properties of an assemblage of such elements can be obtained by direct superposition of the stiffnesses of the individual elements. The concept of isoparametric elements is introduced. These elements have certain properties which make it easy to calculate their properties in a general computational scheme. Finally, three-dimensional bodies are discussed from the same viewpoints. They are represented by hexahedronal elements, whose stiffness properties can be determined in a similar way.

5.2   PRELIMINARY CONCEPTS—A STRETCHED SPRING

Imagine a spring of length L stretched between the points (x_1, y_1) and (x_2, y_2) as shown in Fig. 5.1. If the displacements of the ends in the x- and y-directions are denoted by u_1, u_2, and v_1, v_2, then the stretch is equal to

$$(u_1 - u_2) \cos \alpha + (v_1 - v_2) \sin \alpha \tag{5.1}$$

FIGURE 5.1
A stretched spring

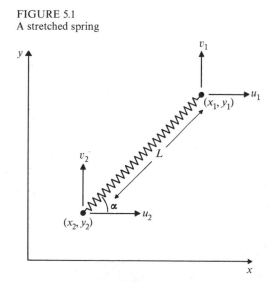

The strain ε in the spring is the stretch divided by the original length L. Written in terms of the coordinates of the ends, the strain is

$$\varepsilon = (u_1 - u_2)\left(\frac{x_1 - x_2}{L^2}\right) + (v_1 - v_2)\left(\frac{y_1 - y_2}{L^2}\right) \tag{5.2}$$

In matrix notation, we can write

$$\varepsilon = [B]\{\delta\} \tag{5.3}$$

where $\{\delta\}$ is a vector of displacements,

$$\{\delta\} = \begin{Bmatrix} u_1 \\ v_1 \\ u_2 \\ v_2 \end{Bmatrix} \tag{5.4}$$

and $[B]$ is given by

$$[B] = \left[\frac{x_1 - x_2}{L^2}, \frac{y_1 - y_2}{L^2}, -\left(\frac{x_1 - x_2}{L^2}\right), -\left(\frac{y_1 - y_2}{L^2}\right)\right] \tag{5.5}$$

Let κ be the force per unit spring strain. Then the internal force F in the spring is

$$F = \kappa\varepsilon = D\varepsilon \tag{5.6}$$

Here, we introduce the notation D as controlling the relationship between the internal spring force (analogous to stress in a continuum) and the internal spring strain. Thus, also, F is related to the displacements by

$$F = D[B]\{\delta\} \tag{5.7}$$

FIGURE 5.2
Internal and external forces on the spring

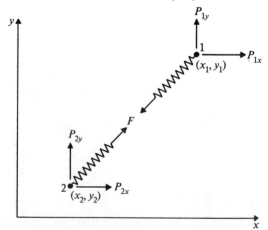

The external forces, $\{P\}^T = [P_{1x}, P_{1y}, P_{2x}, P_{2y}]$, act on the spring as shown in Fig. 5.2. They are related to the internal forces F by

$$
\begin{aligned}
P_{1x} &= F(x_1 - x_2)/L \\
P_{1y} &= F(y_1 - y_2)/L \\
P_{2x} &= -F(x_1 - x_2)/L \\
P_{2y} &= -F(y_1 - y_2)/L
\end{aligned} \right\}
\tag{5.8}
$$

These equations can be consolidated in matrix form as

$$
\left\{
\begin{array}{c}
P_{1x} \\[4pt]
P_{1y} \\[4pt]
P_{2x} \\[4pt]
P_{2y}
\end{array}
\right\}
= L
\begin{bmatrix}
\dfrac{x_1 - x_2}{L^2} \\[8pt]
\dfrac{y_1 - y_2}{L^2} \\[8pt]
\dfrac{-(x_1 - x_2)}{L^2} \\[8pt]
\dfrac{-(y_1 - y_2)}{L^2}
\end{bmatrix}
\{F\}
$$

$$
= L[B]^T \{F\} = L[B]^T D[B] \{\delta\}
\tag{5.9}
$$

We note that the strain energy stored in the spring is given by

$$
U = 1/2\{\delta\}^T \{P\} = 1/2L\{\delta\}^T [B]^T D[B] \{\delta\}
\tag{5.10}
$$

Thus, the strain energy stored per unit length of the spring is

$$
U/L = 1/2\{\delta\}^T [B]^T D[B] \{\delta\}
$$

(In more complicated elements, the strain energy in the element is always obtained by integrating the unit strain energy over the element. In this simple case, multiplying by L accomplishes the integration.)

If we define $[K]$ as the stiffness matrix of the spring, we mean that it relates the displacements $\{\delta\}$ to the external forces $\{P\}$:

$$
\{P\} = [K]\{\delta\}
\tag{5.11}
$$

The work done by the external forces is then also

$$
U = 1/2\{\delta\}^T \{P\} = 1/2\{\delta\}^T [K]\{\delta\}
$$

Comparing the two expressions for U, we have a definition of the stiffness matrix $[K]$ in terms of the geometrical and material properties of the spring:

$$
[K] = L[B]^T D[B]
\tag{5.12}
$$

The displacements in the spring at any point (x, y) between (x_1, y_1) and (x_2, y_2) can be computed in terms of (u_1, v_1) and (u_2, v_2) if we introduce a displacement function, or shape function, that describes how the displacements vary between the two endpoints. For the case of linearly varying displacements (which is reasonable for a spring), we let

$$
\left.
\begin{aligned}
u &= \frac{(x_1 - x)}{x_1 - x_2} u_2 + \frac{(x - x_2)}{x_1 - x_2} u_1 \\[2mm]
v &= \frac{(y_1 - y)}{y_1 - y_2} v_2 + \frac{(y - y_2)}{y_1 - y_2} v_1
\end{aligned}
\right\}
\tag{5.13}
$$

We can write this expression as

$$
\{f\} = \left\{ \begin{matrix} u \\ v \end{matrix} \right\} =
\begin{bmatrix}
\dfrac{x - x_2}{x_1 - x_2} & 0 & \dfrac{x - x_1}{x_2 - x_1} & 0 \\[3mm]
0 & \dfrac{y - y_2}{y_1 - y_2} & 0 & \dfrac{y - y_1}{y_2 - y_1}
\end{bmatrix}
\begin{Bmatrix} u_1 \\ v_1 \\ u_2 \\ v_2 \end{Bmatrix}
\tag{5.14}
$$

But since geometrically

$$
\left.
\begin{aligned}
\frac{x - x_2}{x_1 - x_2} &= \frac{y - y_2}{y_1 - y_2} = N_1 \\[2mm]
\frac{x - x_1}{x_2 - x_1} &= \frac{y - y_1}{y_2 - y_1} = N_2
\end{aligned}
\right\}
\tag{5.15}
$$

we have

$$
\begin{aligned}
\left\{ \begin{matrix} u \\ v \end{matrix} \right\} &=
\begin{bmatrix}
N_1 & 0 & N_2 & 0 \\
0 & N_1 & 0 & N_2
\end{bmatrix}
\{\delta\} \\
&= [\, N_1[I],\, N_2[I]\,]\{\delta\}
\end{aligned}
\tag{5.16}
$$

Here N_1, N_2 are the shape functions, whose particular importance will become apparent as we deal with isoparametric finite elements, and $[I]$ is the identity matrix.

We have now shown how the relationships between external forces, internal forces, and strains are formed by geometric and material properties of an element— in this case, a single spring. Now consider how an assembly of such springs can also be studied by using the information derived from this single basic spring element.

5.3   ASSEMBLY OF A STRUCTURE

The structure shown in Fig. 5.3 consists of four springs assembled to form a quadrilateral. To obtain the total strain energy stored in the four assembled springs, we sum that for each spring taken separately. Denote the nodal displacements of each spring by

$$
\{\delta_1\} = \left\{ \begin{matrix} u_1 \\ v_1 \end{matrix} \right\}, \qquad
\{\delta_2\} = \left\{ \begin{matrix} u_2 \\ v_2 \end{matrix} \right\}, \quad \text{etc.}
$$

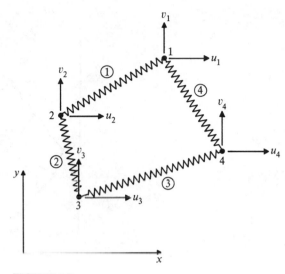

FIGURE 5.3
An assembly of springs

We then have for the assembly:

$$2U = \left\{ \begin{matrix} \delta_1 \\ \delta_2 \end{matrix} \right\}^T [K_1] \left\{ \begin{matrix} \delta_1 \\ \delta_2 \end{matrix} \right\} + \left\{ \begin{matrix} \delta_2 \\ \delta_3 \end{matrix} \right\}^T [K_2] \left\{ \begin{matrix} \delta_2 \\ \delta_3 \end{matrix} \right\}$$

$$+ \left\{ \begin{matrix} \delta_3 \\ \delta_4 \end{matrix} \right\}^T \lfloor K_3 \rfloor \left\{ \begin{matrix} \delta_3 \\ \delta_4 \end{matrix} \right\} + \left\{ \begin{matrix} \delta_4 \\ \delta_1 \end{matrix} \right\}^T \lfloor K_4 \rfloor \left\{ \begin{matrix} \delta_4 \\ \delta_1 \end{matrix} \right\} \qquad (5.17)$$

where $[K_1]$, $[K_2]$, etc., are the stiffness matrices for the individual springs. When we write this expression in the form:

$$2U = \left\{ \begin{matrix} \delta_1 \\ \delta_2 \\ \delta_3 \\ \delta_4 \end{matrix} \right\}^T [K] \left\{ \begin{matrix} \delta_1 \\ \delta_2 \\ \delta_3 \\ \delta_4 \end{matrix} \right\} \qquad (5.18)$$

the matrix $[K]$ is the stiffness of the assembly in the sense that

$$\left\{ \begin{matrix} P_1 \\ P_2 \\ P_3 \\ P_4 \end{matrix} \right\} = [K] \left\{ \begin{matrix} \delta_1 \\ \delta_2 \\ \delta_3 \\ \delta_4 \end{matrix} \right\} \qquad (5.19)$$

where the nodal forces are

$$P_1 = \left\{ \begin{matrix} P_{1x} \\ P_{1y} \end{matrix} \right\}, \qquad P_2 = \left\{ \begin{matrix} P_{2x} \\ P_{2y} \end{matrix} \right\}, \qquad \text{etc.}$$

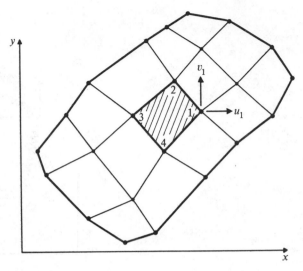

FIGURE 5.4
A two-dimensional region divided into quadrilateral elements

5.4 A TWO-DIMENSIONAL ELASTIC CONTINUUM

Figure 5.4 depicts a two-dimensional elastic structure that has been cut in the xy-plane. It is divided into quadrilateral elements as shown. Consider one of these quadrilaterals, whose corner nodes are denoted by 1, 2, 3, 4. Let us express the displacements (u, v) within the element in terms of those at the nodes. To do this, we introduce a shape function N_1 which has a value of 1 at node 1, while it is zero at the other three nodes. It will be convenient to let N_1 be a polynomial in x and y, so that

$$N_1 = a + bx + cy + dxy \qquad (5.20)$$

The constants a, b, c, and d can be determined by the requirement that

$$\left.\begin{array}{l} N_1 = 1 \text{ at } (x_1, y_1) \\ N_1 = 0 \text{ at } (x_i, y_i) \qquad (i = 2, 3, 4) \end{array}\right\} \qquad (5.21)$$

The appearance of the function is sketched in Fig. 5.5. Thus, a set of simultaneous equations can be written as:

$$\begin{Bmatrix} 1 \\ 0 \\ 0 \\ 0 \end{Bmatrix} = \begin{bmatrix} 1 & x_1 & y_1 & x_1 y_1 \\ 1 & x_2 & y_2 & x_2 y_2 \\ 1 & x_3 & y_3 & x_3 y_3 \\ 1 & x_4 & y_4 & x_4 y_4 \end{bmatrix} \begin{Bmatrix} a \\ b \\ c \\ d \end{Bmatrix} \qquad (5.22)$$

and the constants determined by its solution. Also, we can calculate the partial derivatives at any point (x, y):

$$\left.\begin{aligned}\frac{\partial N_1}{\partial x} &= b + dy \\[2mm] \frac{\partial N_1}{\partial y} &= c + dx\end{aligned}\right\} \tag{5.23}$$

Similarly, now, we introduce three other functions N_2, N_3, and N_4, defined by

$$N_i = a_i + b_i x + c_i y + d_i xy \tag{5.24}$$

whose values are

$$\left.\begin{aligned}N_i &= 1 \text{ at } (x_i, y_i) \\ N_i &= 0 \text{ at } (x_j, y_j) \quad (i \neq j)\end{aligned}\right\} \tag{5.25}$$

The displacements within the element are then written in terms of N_i as

$$\left.\begin{aligned}u &= N_1 u_1 + N_2 u_2 + N_3 u_3 + N_4 u_4 \\ v &= N_1 v_1 + N_2 v_2 + N_3 v_3 + N_4 v_4\end{aligned}\right\} \tag{5.26}$$

or otherwise as

$$\begin{aligned}\{f\} = \begin{Bmatrix} u \\ v \end{Bmatrix} &= [N]\{\delta\} \\ &= [N_1[I], N_2[I], N_3[I], N_4[I]] \end{aligned} \quad \begin{Bmatrix} u_1 \\ v_1 \\ u_2 \\ \vdots \\ v_4 \end{Bmatrix} \tag{5.27}$$

FIGURE 5.5
A typical shape function for a two-dimensional element—the function N_1

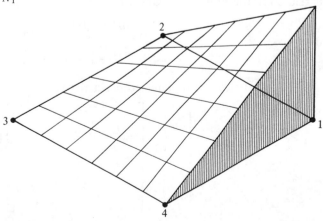

The strain in the element is related to the displacements $\{\delta\}$ through the matrix $[B]$:

$$\{\varepsilon\} = [B]\{\delta\} \tag{5.28}$$

To find $[B]$, we observe that according to elasticity theory, in a two-dimensional continuum:

$$\{\varepsilon\} = \begin{Bmatrix} \varepsilon_x \\ \varepsilon_y \\ \gamma_{xy} \end{Bmatrix} = \begin{Bmatrix} \dfrac{\partial u}{\partial x} \\ \dfrac{\partial v}{\partial y} \\ \dfrac{\partial u}{\partial y} + \dfrac{\partial v}{\partial x} \end{Bmatrix} \tag{5.29}$$

and since, in terms of the polynomials, these derivatives take the form:

$$\frac{\partial u}{\partial x} = \frac{\partial N_1}{\partial x} u_1 + \frac{\partial N_2}{\partial x} u_2 + \frac{\partial N_3}{\partial x} u_3 + \frac{\partial N_4}{\partial x} u_4 \tag{5.30}$$

we get

$$\{\varepsilon\} = \begin{bmatrix} \dfrac{\partial N_1}{\partial x} & 0 & \dfrac{\partial N_2}{\partial x} & 0 & \dfrac{\partial N_3}{\partial x} & 0 & \dfrac{\partial N_4}{\partial x} & 0 \\ 0 & \dfrac{\partial N_1}{\partial y} & 0 & \cdots & & & \cdots & \\ \dfrac{\partial N_1}{\partial y} & \dfrac{\partial N_1}{\partial x} & \dfrac{\partial N_2}{\partial y} & \cdots & & & \cdots & \dfrac{\partial N_4}{\partial x} \end{bmatrix} \begin{Bmatrix} u_1 \\ v_1 \\ u_2 \\ \vdots \\ v_4 \end{Bmatrix} \tag{5.31}$$

which defines the matrix $[B]$.

The stress

$$\{\sigma\}^T = [\sigma_x, \sigma_y, \tau_{xy}]$$

is related to the strain through the expression:

$$\{\sigma\} = [D]\{\varepsilon\}$$

or

$$\{\varepsilon\} = [D]^{-1}\{\sigma\} \tag{5.32}$$

In two-dimensional plane stress elasticity,

$$\begin{Bmatrix} \varepsilon_x \\ \varepsilon_y \\ \gamma_{xy} \end{Bmatrix} = \begin{bmatrix} \dfrac{1}{E} & -\dfrac{v}{E} & 0 \\ -\dfrac{v}{E} & \dfrac{1}{E} & 0 \\ 0 & 0 & \dfrac{2(1+v)}{E} \end{bmatrix} \begin{Bmatrix} \sigma_x \\ \sigma_y \\ \sigma_{xy} \end{Bmatrix} \tag{5.33}$$

Here, E is the elastic modulus, and v is Poisson's ratio. Thus, in plane stress:

$$[D] = \frac{E}{1 - v^2} \begin{bmatrix} 1 & v & 0 \\ v & 1 & 0 \\ 0 & 0 & \dfrac{1-v}{2} \end{bmatrix} \qquad (5.34)$$

In plane strain, it can be shown[2] that

$$[D] = \frac{E(1-v)}{(1+v)(1-2v)} \begin{bmatrix} 1 & \dfrac{v}{1-v} & 0 \\ \dfrac{v}{1-v} & 1 & 0 \\ 0 & 0 & \dfrac{1-2v}{2(1 \quad v)} \end{bmatrix} \qquad (5.35)$$

In general, then, the stress is related to the displacement through

$$\{\sigma\} = [D][B]\{\delta\} \qquad (5.36)$$

The strain energy in a unit area of a unit thickness of the continuum is half the product of stress and strain, $1/2\{\sigma\}^T\{\varepsilon\}$. We obtain the total strain energy U of the element by integrating that expression over the appropriate area. Thus, for the element,

$$U = 1/2 \int_{\text{area}} \{\sigma\}^T \{\varepsilon\} \, dA$$

$$= 1/2 \int_{\text{area}} \{\delta\}^T [B]^T [D][B] \{\delta\} \, dA \qquad (5.37)$$

Since the displacements $\{\delta\}$ are independent of x and y,

$$U = 1/2 \{\delta\}^T \left[\int_{\text{area}} [B]^T [D][B] \, dA \right] \{\delta\} \qquad (5.38)$$

But we also know that

$$U = 1/2 \{\delta\}^T \{P\} = 1/2 \{\delta\}^T [K]\{\delta\} \qquad (5.39)$$

where $[K]$ is the element stiffness matrix. Thus, the stiffness matrix for the element is defined as

$$[K] = \int_{\text{area}} [B]^T [D][B] \, dA \qquad (5.40)$$

The relationship between the displacements and the external loads is

$$\{\delta\} = [K]^{-1}\{P\} \qquad (5.41)$$

Hence, if the external loads are known, the displacements are found through the stiffness matrix. The stiffness matrix is thus a key to the deformation state. To find it, it is necessary to perform the integrations indicated by eq. (5.40). These integrations are usually performed numerically. For this reason, we will pause here to review a particular method of numerical integration.

5.5 NUMERICAL INTEGRATION

We will briefly discuss here a method of numerical integration that is adequate for the purpose of computing the stiffness matrix. Many other schemes are in existence and a large literature exists on the subject. References 1 and 3 should be consulted for further detail.

To start with, we discuss the numerical integration of a third-order polynomial:

$$u = a + b\xi + c\xi^2 + d\xi^3 \tag{5.42}$$

Integrating between the limits of $-1 < \xi < 1$, we get

$$\int_{-1}^{1} u\, d\xi = 2\left(a + \frac{c}{3}\right) \tag{5.43}$$

Now we inquire how the result could also be obtained by selecting two points ξ_1 and ξ_2, and summing the values of the function at those points, when they are multiplied by appropriate weights h_1 and h_2, as in Fig. 5.6(a). Thus,

$$\int_{-1}^{1} u\, d\xi = h_1 u(\xi_1) + h_2 u(\xi_2) \tag{5.44}$$

For these two expressions to be equal, we require that

$$h_1(a + b\xi_1 + c\xi_1^2 + d\xi_1^3) + h_2(a + b\xi_2 + c\xi_2^2 + d\xi_2^3) = 2\left(a + \frac{c}{3}\right)$$

Consequently, by equating the coefficients of $a, b, c,$ and d, we have

$$h_1 + h_2 = 2$$
$$h_1\xi_1 + h_2\xi_2 = 0$$
$$h_1\xi_1^2 + h_2\xi_2^2 = 2/3$$
$$h_1\xi_1^3 + h_2\xi_2^3 = 0$$

As a result of solving these equations for h_1, h_2, ξ_1, ξ_2,

$$h_1 = h_2 = 1$$
$$\xi_1 = -\xi_2 = 1/\sqrt{3} = 0\cdot57735$$

Thus,

$$\int_{-1}^{1} u\, d\xi = u(+0\cdot57735) + u(-0\cdot57735) \tag{5.45}$$

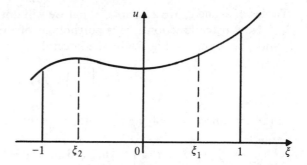

(a) Integration in a single dimension

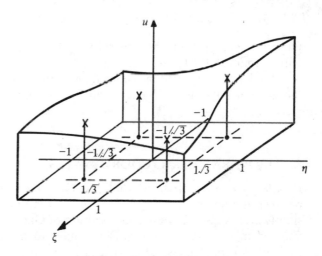

(b) Integration in two dimensions

FIGURE 5.6
Numerical integration

This formula was developed by Gauss, and the process used here is known as Gaussian integration. The points $\xi = \pm 0 \cdot 57735$ are known as the Gauss points. For higher order polynomials, similar formulas can be derived. For example, for polynomials of order $2n - 1$, where

$$u = \sum_0^{2n-1} a_i \xi^i \tag{5.46}$$

the integration is given by

$$\int_{-1}^{1} u \, d\xi = \sum_1^n h_i u(\xi_i) \tag{5.47}$$

Tables of h_i and ξ_i are available,[1,3] but we will not delve into further details here.

Now if the function u is a polynomial of third order in the two dimensions ξ and η, as shown in Fig. 5.6(b), the integral

$$I = \int_{-1}^{1} \int_{-1}^{1} u(\xi, \eta) \, d\xi \, d\eta \qquad (5.48)$$

can be obtained by applying eq. (5.45) twice. First, in the ξ direction:

$$I = \int_{-1}^{1} \left[u(1/\sqrt{3}, \eta) + u(-1/\sqrt{3}, \eta) \right] d\eta$$

Then, in the η direction,

$$I = u(1/\sqrt{3}, 1/\sqrt{3}) + u(1/\sqrt{3}, -1/\sqrt{3}) + u(-1/\sqrt{3}, 1/\sqrt{3}) + u(-1/\sqrt{3}, -1/\sqrt{3})$$
$$(5.49)$$

Observe that, in effect, the integral is simply the sum of the function at the four Gauss points in the (ξ, η) plane.

Similarly, in three dimensions (ξ, η, ζ) there will exist eight Gauss points, and the integral is the sum of the function at these eight points.

5.6   ISOPARAMETRIC COORDINATES

It will have been observed that the Gaussian integrations were all performed between the limits of $-1 < \xi < 1$. The integration formulas that resulted are very simple in this case. However, the position of Gauss points in an element arbitrarily situated in Cartesian space (x, y) will have to be determined for each individual element in turn. This is not a great task, but one which can be avoided by making the transformation shown in Fig. 5.7 from Cartesian coordinates to isoparametric coordinates. In isoparametric space, the elements are all geometrically identical and square, and, as a result, integrations necessary in the stiffness and other matrices are easily, rapidly, and logically performed.

We again introduce the shape functions N_i in isoparametric coordinates, asking that they satisfy the condition

$$\left. \begin{array}{l} N_i = 1 \text{ at } (\xi_i, \eta_i) \\ N_i = 0 \text{ at } (\xi_j, \eta_j) \qquad (i \neq j) \end{array} \right\} \qquad (5.50)$$

A suitable representation is

$$\left. \begin{array}{l} N_1 = 1/4(1 + \xi)(1 + \eta) \\ N_2 = 1/4(1 - \xi)(1 + \eta) \\ N_3 = 1/4(1 - \xi)(1 - \eta) \\ N_4 = 1/4(1 + \xi)(1 - \eta) \end{array} \right\} \qquad (5.51)$$

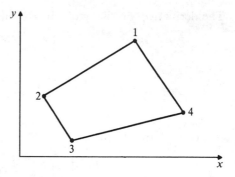

(a) Physical coordinates

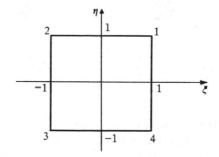

(b) Isoparametric coordinates

FIGURE 5.7
The transformation from physical to isoparametric coordinates

Now the Cartesian coordinates are represented as

$$
\left.
\begin{array}{l}
x = N_1 x_1 + N_2 x_2 + N_3 x_3 + N_4 x_4 \\
y = N_1 y_1 + N_2 y_2 + N_3 y_3 + N_4 y_4
\end{array}
\right\}
\tag{5.52}
$$

Since N_i are functions of ξ, η, the derivatives of x and y with respect to ξ and η are

$$
\left.
\begin{array}{l}
\dfrac{\partial x}{\partial \xi} = \dfrac{\partial N_1}{\partial \xi} x_1 + \dfrac{\partial N_2}{\partial \xi} x_2 + \cdots \\[2ex]
\dfrac{\partial y}{\partial \xi} = \dfrac{\partial N_1}{\partial \xi} y_1 + \dfrac{\partial N_2}{\partial \xi} y_2 + \cdots
\end{array}
\right\}
\tag{5.53}
$$

so we can write

$$
\begin{bmatrix}
\dfrac{\partial x}{\partial \xi} & \dfrac{\partial y}{\partial \xi} \\[2ex]
\dfrac{\partial x}{\partial \eta} & \dfrac{\partial y}{\partial \eta}
\end{bmatrix}
=
\begin{bmatrix}
\dfrac{\partial N_1}{\partial \xi} & \dfrac{\partial N_2}{\partial \xi} & \dfrac{\partial N_3}{\partial \xi} & \dfrac{\partial N_4}{\partial \xi} \\[2ex]
\dfrac{\partial N_1}{\partial \eta} & \dfrac{\partial N_2}{\partial \eta} & \dfrac{\partial N_3}{\partial \eta} & \dfrac{\partial N_4}{\partial \eta}
\end{bmatrix}
\begin{bmatrix}
x_1 & y_1 \\
x_2 & y_2 \\
x_3 & y_3 \\
x_4 & y_4
\end{bmatrix}
\tag{5.54}
$$

The derivatives $\partial N_i/\partial \xi$, etc., can also be written by the chain rule as

$$
\left.
\begin{aligned}
\frac{\partial N_1}{\partial \xi} &= \frac{\partial N_1}{\partial x}\frac{\partial x}{\partial \xi} + \frac{\partial N_1}{\partial y}\frac{\partial y}{\partial \xi} \\[2mm]
\frac{\partial N_1}{\partial \eta} &= \frac{\partial N_1}{\partial x}\frac{\partial x}{\partial \eta} + \frac{\partial N_1}{\partial y}\frac{\partial y}{\partial \eta}
\end{aligned}
\right\}
\tag{5.55}
$$

etc., or in matrix form as

$$
\begin{Bmatrix}
\dfrac{\partial N_i}{\partial \xi} \\[3mm]
\dfrac{\partial N_i}{\partial \eta}
\end{Bmatrix}
=
\begin{bmatrix}
\dfrac{\partial x}{\partial \xi} & \dfrac{\partial y}{\partial \xi} \\[3mm]
\dfrac{\partial x}{\partial \eta} & \dfrac{\partial y}{\partial \eta}
\end{bmatrix}
\begin{Bmatrix}
\dfrac{\partial N_i}{\partial x} \\[3mm]
\dfrac{\partial N_i}{\partial y}
\end{Bmatrix}
\qquad (i = 1, 2, 3, 4)
$$

$$
= [J]
\begin{Bmatrix}
\dfrac{\partial N_i}{\partial x} \\[3mm]
\dfrac{\partial N_i}{\partial y}
\end{Bmatrix}
\tag{5.56}
$$

The matrix $[J]$ is known as the Jacobian. It defines the transformation between the isoparametric and Cartesian coordinates. Observe that the value of $[J]$ can be computed at any point (ξ, η) since its elements are known from eq. (5.54). Thus, it is now possible also to calculate the derivatives of the shape functions N_i in the Cartesian space, simply by inverting eq. (5.56):

$$
\begin{Bmatrix}
\dfrac{\partial N_i}{\partial x} \\[3mm]
\dfrac{\partial N_i}{\partial y}
\end{Bmatrix}
= [J]^{-1}
\begin{Bmatrix}
\dfrac{\partial N_i}{\partial \xi} \\[3mm]
\dfrac{\partial N_i}{\partial \eta}
\end{Bmatrix}
\tag{5.57}
$$

Consider again the displacements u, v, expressed in terms of the shape functions:

$$
\left.
\begin{aligned}
u &= N_1 u_1 + N_2 u_2 + N_3 u_3 + N_4 u_4 \\
v &= N_1 v_1 + N_2 v_2 + N_3 v_3 + N_4 v_4
\end{aligned}
\right\}
\tag{5.58}
$$

The strains are again expressed as

$$
\begin{Bmatrix}
\varepsilon_x \\[3mm]
\varepsilon_y \\[3mm]
\gamma_{xy}
\end{Bmatrix}
=
\begin{Bmatrix}
\dfrac{\partial u}{\partial x} \\[3mm]
\dfrac{\partial v}{\partial y} \\[3mm]
\dfrac{\partial u}{\partial y} + \dfrac{\partial v}{\partial x}
\end{Bmatrix}
\tag{5.59}
$$

These derivatives appear as

$$\frac{\partial u}{\partial x} = \frac{\partial N_i}{\partial x}u_1 + \frac{\partial N_2}{\partial x}u_2 + \frac{\partial N_3}{\partial x}u_3 + \frac{\partial N_4}{\partial x}u_4 \tag{5.60}$$

etc. All the derivatives contain $\partial N_i/\partial x$ and $\partial N_i/\partial y$, which are now known at any point (ξ, η) through the transformation of eq. (5.57). Thus, we can find the strains at any point (ξ, η), too, and in particular we can find the strains at the Gauss points.

We now return to the problem of calculating the stiffness $[K]$. Recall that

$$[K] = \int_{\text{area}} [B]^T [D][B]\,dA \tag{5.61}$$

By using the shape functions of the isoparametric space, and applying Gaussian integration, it is now a simple matter to obtain $[K]$ for each individual element. It is important to note that the volume in the x, y coordinate system associated with each Gauss point is the determinant of the Jacobian at that point.

5.7 ASSEMBLY—THE CONCEPT OF BANDWIDTH

Having computed the stiffness matrix of each individual element, it is possible to find the stiffness of the aggregate structure by direct addition, as was the case with the

FIGURE 5.8
Spring assemblies

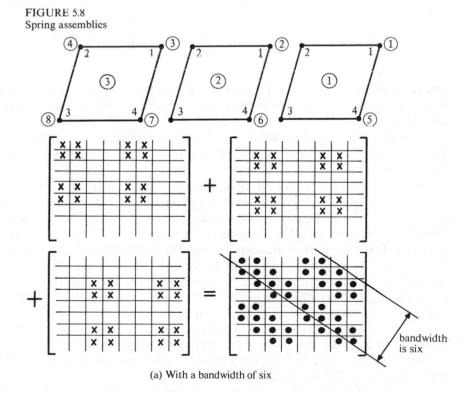

(a) With a bandwidth of six

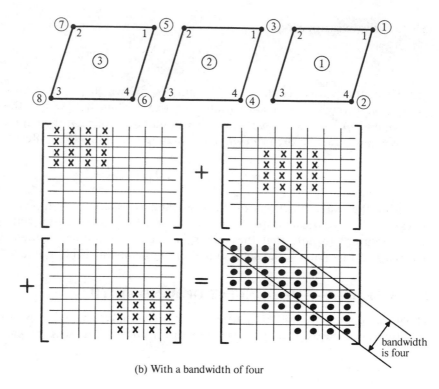

(b) With a bandwidth of four

FIGURE 5.8 —*continued*

spring assembly. However, there are some practical considerations that enter into constructing the most efficient stiffness matrix from the point of view of numerical calculations.

Observe the two demonstrations of Fig. 5.8, wherein the stiffness of three quadrilaterals is added to form the stiffness of the whole. The three quadrilateral elements have their nodes numbered internally, as well as having global numbers externally, which are shown in circles in the figure. Thus, in Fig. 5.8(a), the quadrilateral nodes are globally numbered horizontally. It will be observed that when the stiffness of the final structure is obtained, it has a certain number of off-diagonal terms which are not zero. The elements of the matrix tend to congregate about the diagonal. The half-width of the nonzero congregation is known as the bandwidth, Fig. 5.8(a). The bandwidth plays an important role in numerical calculations, particularly regarding the computation time involved in inverting the matrix, or in extracting eigenvalues from such matrices. The smaller the bandwidth, the faster and more efficient the calculations. In this example, the bandwidth is six.

Figure 5.8(b) shows an alternative approach to choosing the node global numbers. In this case, the global numbering proceeds vertically along the thinnest section of the structure. A similar addition of stiffness terms yields a final stiffness matrix with a smaller bandwidth of four.

It is always important to take great care in numbering the nodes of a structure so that the resulting bandwidths are as small as possible.

5.8   STATIC LOAD ANALYSIS

If the loads P are known at each node, then the global displacements are found by solving the equation

$$\{\delta\} = [K]^{-1}\{P\} \tag{5.62}$$

The stress field in each element is then obtained from the element displacements δ_e:

$$\{\sigma\} = [D][B]\{\delta_e\} \tag{5.63}$$

This stress can be found efficiently if the element product matrices $[D][B]$ have been stored during the process of computing the stiffness matrix.

When distributed force components, such as pressure forces, denoted by X and Y per unit area, act on an element, these loads must be replaced by equivalent nodal loads $\{P_{eq}\}$ We denote the distributed loads by $\{p\} = \lceil X, Y\rceil^T$. The following procedure is then used. An arbitrary virtual displacement is imposed at the nodes. The work done by the distributed load $\{p\}$ is equated to the work done by the equivalent nodal loads $\{P_{eq}\}$. Thus, let $\{\delta_e\}$ be the virtual displacements at the element nodes. Then the work done by the equivalent external nodal forces is $\{\delta_e\}^T\{P_{eq}\}$, and that done by the distributed loads is $\int_A \{\delta_e\}^T [N]^T \{p\}\, dA$. Equating these quantities gives the desired relationship:

$$\{P_{eq}\} = \int_A \lfloor N\rfloor^I \{p\}\, dA \tag{5.64}$$

The summation of the element equivalent loads $\{P_{eq}\}$ is added to the nodal loads to give $\{P\}$ in eq. (5.62).

In practice, the stiffness matrix $[K]$ is not inverted to find the displacements $\{\delta\}$, as indicated in eq. (5.62). Rather, the system $[K]\{\delta\} = \{P\}$ is solved by Gaussian elimination, or an alternative equation-solving technique such as the Gauss–Seidel method (see, for example, Zienkiewicz,[1] pp. 450–457).

A simple illustration of the Gaussian elimination process may be obtained by considering the system of algebraic equations:

$$\left.\begin{array}{l} a_{11}x_1 + a_{12}x_2 + a_{13}x_3 = b_1 \\ a_{21}x_1 + a_{22}x_2 + a_{23}x_3 = b_2 \\ a_{31}x_1 + a_{32}x_2 + a_{33}x_3 = b_3 \end{array}\right\} \tag{5.65}$$

We first eliminate x_1 from the second equation. Multiplying the first equation by a_{21}/a_{11} and subtracting the result from the second equation, we obtain

$$a'_{22}x_2 + a'_{23}x_3 = b'_2 \tag{5.66}$$

where

$$\left.\begin{array}{l} a'_{22} = a_{22} - a_{21}a_{12}/a_{11} \\ a'_{23} = a_{23} - a_{21}a_{13}/a_{11} \\ b'_2 = b_2 - a_{21}b_1/a_{11} \end{array}\right\} \qquad (5.67)$$

Similarly, we eliminate x_1 from the third equation by multiplying the first equation by a_{31}/a_{11} and subtracting the result from the third equation, to obtain

$$a'_{32}x_2 + a'_{33}x_3 = b'_3 \qquad (5.68)$$

where

$$\begin{array}{l} a'_{32} = a_{32} - a_{31}a_{12}/a_{11} \\ a'_{33} = a_{33} - a_{31}a_{13}/a_{11} \\ b'_3 = b_3 - a_{31}b_1/a_{11} \end{array}$$

The equations are now reduced to the form

$$\left.\begin{array}{l} a_{11}x_1 + a_{12}x_2 + a_{13}x_3 = b_1 \\ a'_{22}x_2 + a'_{23}x_3 = b'_2 \\ a'_{32}x_2 + a'_{33}x_3 = b'_3 \end{array}\right\} \qquad (5.69)$$

We now eliminate x_2 from the third equation by multiplying the second equation by a'_{32}/a'_{22} and subtracting the result from the third equation, to obtain

$$a''_{33}x_3 = b''_3 \qquad (5.70)$$

where

$$\begin{array}{l} a''_{33} = a'_{33} - a'_{32}a'_{23}/a'_{22} \\ b''_3 = b'_3 - a'_{32}b'_2/a'_{22} \end{array}$$

The equation now takes on a triangularized form:

$$\left.\begin{array}{l} a_{11}x_1 + a_{12}x_2 + a_{13}x_3 = b_1 \\ a'_{22}x_2 + a'_{23}x_3 = b'_2 \\ a''_{33}x_3 = b''_3 \end{array}\right\} \qquad (5.71)$$

The values of x may now be found by sequentially solving from the lowest equation:

$$\left.\begin{array}{l} x_3 = b''_3/a''_{33} \\ x_2 = (b'_2 - a'_{23}x_3)/a'_{22} \\ x_1 = (b_1 - a_{12}x_2 - a_{13}x_3)/a_{11} \end{array}\right\} \qquad (5.72)$$

This process of initially triangularizing the matrix and subsequent back-substitution is known as Gaussian elimination. It can be applied to any set of nonsingular equations. Sometimes, it happens that one of the coefficients that will form the divisor is zero. In that case, two of the rows are interchanged, which does not change the solution.

The Gaussian elimination technique is well adapted to the solution of the stiffness matrices. Advantage can be taken of the banding of the stiffness matrix in the following way. Consider the banded system:

$$\begin{bmatrix} K_{11} & K_{12} & 0 \\ K_{12}^T & K_{22} & K_{23} \\ 0 & K_{23}^T & K_{33} \end{bmatrix} \begin{Bmatrix} \delta_1 \\ \delta_2 \\ \delta_3 \end{Bmatrix} = \begin{Bmatrix} P_1 \\ P_2 \\ P_3 \end{Bmatrix} \tag{5.73}$$

Here K_{ij} are submatrices that form the banded structure of the stiffness matrix, and δ_i and P_i are submatrices of displacement and load. Usually, in a realistic problem, their size will be less than a tenth the size of the full matrix.

We seek to eliminate δ_1 from the second equation by the same method as before, obtaining for the second equation:

$$K_{22}'\delta_2 + K_{23}'\delta_3 = P_2' \tag{5.74}$$

where

$$K_{22}' = K_{22} - K_{12}^T K_{11}^{-1} K_{12}$$
$$P_2' = P_2 - K_{12}^T K_{11}^{-1} P_1$$

The basic matrix operation is a multiplication of a triple product $K_{12}^T K_{11}^{-1} K_{12}$.

Similarly, we eliminate δ_2 from the third equation by a like series of operations. We thus obtain a triangularized matrix of the form:

$$\begin{bmatrix} K_{11} & K_{12} & 0 \\ & K_{22}' & K_{23}' \\ & & K_{33}'' \end{bmatrix} \begin{Bmatrix} \delta_1 \\ \delta_2 \\ \delta_3 \end{Bmatrix} = \begin{Bmatrix} P_1 \\ P_2' \\ P_3'' \end{Bmatrix} \tag{5.75}$$

Back-substitution then allows the solution for δ_3:

$$\left. \begin{aligned} \delta_3 &= (K_{33}'')^{-1} P_3'' \\ \delta_2 &= (K_{22}')^{-1}(P_2' - K_{23}'\delta_3) \\ \delta_1 &= K_{11}^{-1}(P_1 - K_{12}\delta_2) \end{aligned} \right\} \tag{5.76}$$

The elements of the vector δ_i can be solved for by Gaussian elimination. In this case, the submatrices are much smaller than the original matrix, allowing a rapid solution.

It will be observed that the bandedness of the original stiffness matrix played a significant role, in that some submatrix elements were already zero. The banded structure of a global matrix and its symmetry also assists in computer storage economies, since only the nonzero submatrices above the diagonal need be stored.

Consider a realistically sized banded matrix:

$$\begin{bmatrix} K_{11} & K_{12} & & & & & \\ K_{12}^T & K_{22} & K_{23} & & & 0 & \\ & K_{23}^T & K_{33} & K_{34} & & & \\ & & K_{34}^T & K_{44} & K_{45} & & \\ & 0 & & \ldots\ldots\ldots & & & \\ & & & & \ldots\ldots\ldots & & \\ & & & & & K_{(n-1)n}^T & K_{nn} \end{bmatrix} \tag{5.77}$$

It will be observed that no matter what size the stiffness matrix $[K]$ has, the basic matrix operation consists of the triple multiplications:

$$K_{n(n+1)}^T \, K_{nn}^{-1} \, K_{n(n+1)} \tag{5.78}$$

Thus, the limit on the size of problem that can be solved is not dependent on the size of the stiffness matrix but, rather, on its bandwidth. Thus, bandwidth determines the size of the computer storage required to perform the triple product operation, which generally must be done in the machine core. The remaining submatrices can, if necessary, be stored on core-external memories such as computer tapes and discs, and retrieved as needed.

5.9 BOUNDARY CONDITIONS

If the displacements are prescribed at the nodes which form the boundary of the structure, there are a number of possible approaches to account for them.

The relationship between load and displacement is, in general (here K_{11}, K_{12}, etc., are individual terms in the global stiffness matrix $[K]$):

$$\begin{bmatrix} K_{11} & K_{12} & K_{13} & \ldots \\ K_{21} & K_{22} & K_{23} & \ldots \\ K_{31} & K_{32} & K_{33} & \ldots \\ \vdots & & & \end{bmatrix} \begin{Bmatrix} \delta_1 \\ \delta_2 \\ \delta_3 \\ \vdots \end{Bmatrix} = \begin{Bmatrix} P_1 \\ P_2 \\ P_3 \\ \vdots \end{Bmatrix} \tag{5.79}$$

If $\delta_j = \delta_j^0$, for example, then two numerically distinct methods are available. In the first, the loads P_i are modified by defining a new load vector $P_i^* = P_i - K_{ij}\delta_j^0$, and P_j is set to equal δ_j^0. In addition, the jth row and jth column of the stiffness matrix is set to zero, except for the diagonal term which is set equal to unity. Inspection of the new equations will verify that the boundary conditions are now correctly specified. This method works especially well when the displacement δ_j^0 is specified to be zero. In that case, the loads are unchanged except for $P_j = 0$, however the jth row and column of $[K]$ are modified.

An alternative approach is to multiply the jth diagonal term by a large number, perhaps 10^8, and replace the jth load coefficient by $10^8 K_{jj}\delta_j^0$. In this case, the following modifications would be made:

$$\left.\begin{array}{l} K_{jj}{}^* = 10^8 \, K_{jj} \\ P_j{}^* = 10^8 \, K_{jj} \delta_j^0 \end{array}\right\} \tag{5.80}$$

One caution must be observed. If the large number is too large, the limits on number largeness of the particular computing machine may be exceeded during the solution. In such a case, a smaller number is used. This method gives a solution in which δ_j is very nearly equal to δ_j^0. It works well when δ_j^0 is not zero and saves the cost of modifying the jth row and column of $[K]$.

Occasional numerical difficulties can arise when the magnitude of the stiffness elements are themselves large. In this case, it can happen that while solving the matrix equation, two large numbers are subtracted, leaving a small number of questionable accuracy. A possible cure for this situation is to make the weight factor (10^8 here) be a smaller number.

Boundary conditions involving specified forces are handled as external loads, and present no difficulties.

5.10 THREE-DIMENSIONAL CONTINUA

Three-dimensional elastic bodies are treated in a manner identical to the two-dimensional continuum. The body in this case is divided into a set of hexahedrons, one of which is shown in Fig. 5.9. Once again, we introduce isoparametric coordinates (ξ, η, ζ). The shape functions N are now defined in terms of these isoparametric coordinates as

$$\begin{array}{l} N_1 = 1/8(1 - \xi)(1 - \eta)(1 - \zeta) \\ N_2 = 1/8(1 - \xi)(1 + \eta)(1 - \zeta) \\ \quad \vdots \\ N_8 = 1/8(1 + \xi)(1 - \eta)(1 + \zeta) \end{array} \tag{5.81}$$

and they also satisfy the requirement that

$$\left.\begin{array}{l} N_i = 1 \text{ at } (\xi_i, \eta_i, \zeta_i) \\ N_i = 0 \text{ at } (\xi_j, \eta_j, \zeta_j) \qquad (i \neq j) \end{array}\right\} \tag{5.82}$$

The parametric representation of the Cartesian coordinates is then

$$\left.\begin{array}{l} x = \displaystyle\sum_{i=1}^{8} N_i x_i \\[2mm] y = \displaystyle\sum_{i=1}^{8} N_i y_i \\[2mm] z = \displaystyle\sum_{i=1}^{8} N_i z_i \end{array}\right\} \tag{5.83}$$

in which x_n, y_n, z_n are the coordinates of the nth node, there being eight nodes.

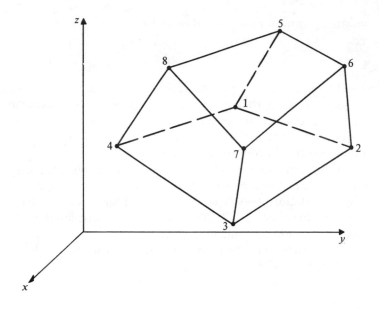

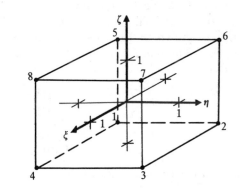

FIGURE 5.9
An eight-noded element and its isoparametric representation

To provide for the transformation of strain derivatives from the isoparametric to the Cartesian coordinates, it is necessary to introduce the Jacobian matrix $[J]$:

$$[J] = \begin{bmatrix} \dfrac{\partial x}{\partial \xi} & \dfrac{\partial y}{\partial \xi} & \dfrac{\partial z}{\partial \xi} \\[2mm] \dfrac{\partial x}{\partial \eta} & \dfrac{\partial y}{\partial \eta} & \dfrac{\partial z}{\partial \eta} \\[2mm] \dfrac{\partial x}{\partial \zeta} & \dfrac{\partial y}{\partial \zeta} & \dfrac{\partial z}{\partial \zeta} \end{bmatrix} \qquad (5.84)$$

Substitution of the assumed displacement form in eq. (5.83) gives

$$[J] = \begin{bmatrix} \dfrac{\partial N_1}{\partial \xi} & \dfrac{\partial N_2}{\partial \xi} & \cdots & \dfrac{\partial N_8}{\partial \xi} \\ \dfrac{\partial N_1}{\partial \eta} & \dfrac{\partial N_2}{\partial \eta} & \cdots & \dfrac{\partial N_8}{\partial \eta} \\ \dfrac{\partial N_1}{\partial \zeta} & \dfrac{\partial N_2}{\partial \zeta} & \cdots & \dfrac{\partial N_8}{\partial \zeta} \end{bmatrix} \begin{bmatrix} x_1 & y_1 & z_1 \\ x_2 & y_2 & z_2 \\ \vdots & \vdots & \vdots \\ x_8 & y_8 & z_8 \end{bmatrix} \tag{5.85}$$

which permits the evaluation of $[J]$ at any given position once the nodal coordinates are specified.

Since Gaussian integration will be used in summing the strain energy in the element, it is important to note that an element volume in the x, y, z coordinate system is related to that in the ξ, η, ζ system by

$$dV_{xyz} = |J| \, dV_{\xi\eta\zeta} \tag{5.86}$$

The derivatives of N_i in the x, y, z coordinate system are related to those in the isoparametric system by an expression that is much the same as that derived in the two-dimensional case, namely:

$$\begin{Bmatrix} \dfrac{\partial N_i}{\partial x} \\ \dfrac{\partial N_i}{\partial y} \\ \dfrac{\partial N_i}{\partial z} \end{Bmatrix} = [J]^{-1} \begin{Bmatrix} \dfrac{\partial N_i}{\partial \xi} \\ \dfrac{\partial N_i}{\partial \eta} \\ \dfrac{\partial N_i}{\partial \zeta} \end{Bmatrix} \tag{5.87}$$

At this point, we introduce three additional shape functions

$$\left. \begin{aligned} N_9 &= 1 - \xi^2 \\ N_{10} &= 1 - \eta^2 \\ N_{11} &= 1 - \zeta^2 \end{aligned} \right\} \tag{5.88}$$

whose values are zero at all eight nodes. These additional functions will enable the addition of more degrees of freedom in the description of the internal displacements.

We specify that the displacements in the x-, y-, z-directions are

$$\left. \begin{aligned} u &= N_1 u_1 + N_2 u_2 + \cdots + N_{11} u_{11} \\ v &= N_1 v_1 + N_2 v_2 + \cdots + N_{11} v_{11} \\ w &= N_1 w_1 + N_2 w_2 + \cdots + N_{11} w_{11} \end{aligned} \right\} \tag{5.89}$$

Here, u_n, v_n, w_n ($n = 1, \ldots, 8$) are the displacements of the eight nodes at the corners of the hexahedron. The additional three sets of displacements ($n = 9, 10, 11$) are internal degrees of freedom. They need not be present in the analysis. They are added here to suggest that there are numerous other ways in which element displacements can be described. Actually, they are used in practice because it is found that their use eliminates errors due to shear within the element. They allow an additional freedom to the deformation of the element. Their magnitudes are chosen, as we shall demonstrate, such that the strain energy in the element is minimized.

These added shape functions now cause the shape of the deformed element to be incompatible with the shape of its neighbor, except at the corner nodes. For this reason, the elements containing shape functions of this type are said to have incompatible modes.

According to the three-dimensional theory of elasticity, the strains are related to displacements by

$$
\left.\begin{aligned}
\varepsilon_x &= \frac{\partial u}{\partial x} \\[2mm]
\varepsilon_y &= \frac{\partial v}{\partial y} \\[2mm]
\varepsilon_z &= \frac{\partial w}{\partial z} \\[2mm]
\gamma_{xy} &= \frac{\partial u}{\partial y} + \frac{\partial v}{\partial x} \\[2mm]
\gamma_{yz} &= \frac{\partial v}{\partial z} + \frac{\partial w}{\partial y} \\[2mm]
\gamma_{zx} &= \frac{\partial w}{\partial x} + \frac{\partial u}{\partial z}
\end{aligned}\right\} \tag{5.90}
$$

Performing the indicated differentiations, we can finally write

$$
\{\varepsilon\} = [B]\{\delta\} \tag{5.91}
$$

where

$$
\{\varepsilon\}^T = \{\varepsilon_x, \varepsilon_y, \varepsilon_z, \gamma_{xy}, \gamma_{yz}, \gamma_{zx}\}
$$

$$
\{\delta\}^T = \{u_1, v_1, w_1 \ldots u_{11}, v_{11}, w_{11}\}
$$

and

$$
[B] =
\begin{bmatrix}
\dfrac{\partial N_1}{\partial x} & 0 & 0 & \dfrac{\partial N_2}{\partial x} & 0 & 0 & \cdots & \dfrac{\partial N_{11}}{\partial x} & 0 & 0 \\[3mm]
0 & \dfrac{\partial N_1}{\partial y} & 0 & 0 & \dfrac{\partial N_2}{\partial y} & 0 & \cdots & 0 & \dfrac{\partial N_{11}}{\partial y} & 0 \\[3mm]
0 & 0 & \dfrac{\partial N_1}{\partial z} & 0 & 0 & \dfrac{\partial N_2}{\partial z} & \cdots & 0 & 0 & \dfrac{\partial N_{11}}{\partial z} \\[3mm]
\dfrac{\partial N_1}{\partial y} & \dfrac{\partial N_1}{\partial x} & 0 & \dfrac{\partial N_2}{\partial y} & \dfrac{\partial N_2}{\partial x} & 0 & \cdots & \dfrac{\partial N_{11}}{\partial y} & \dfrac{\partial N_{11}}{\partial x} & 0 \\[3mm]
0 & \dfrac{\partial N_1}{\partial z} & \dfrac{\partial N_1}{\partial y} & 0 & \dfrac{\partial N_2}{\partial z} & \dfrac{\partial N_2}{\partial y} & \cdots & 0 & \dfrac{\partial N_{11}}{\partial z} & \dfrac{\partial N_{11}}{\partial y} \\[3mm]
\dfrac{\partial N_1}{\partial z} & 0 & \dfrac{\partial N_1}{\partial x} & \dfrac{\partial N_2}{\partial z} & 0 & \dfrac{\partial N_2}{\partial x} & \cdots & \dfrac{\partial N_{11}}{\partial z} & 0 & \dfrac{\partial N_{11}}{\partial x}
\end{bmatrix}
$$

Defining the stress vector by

$$\{\sigma\}^T = \{\sigma_x, \sigma_y, \sigma_z, \tau_{xy}, \tau_{yz}, \tau_{zx}\} \tag{5.92}$$

we can also write

$$\{\sigma\} = [D]\{\varepsilon\} = [D][B]\{\delta\} \tag{5.93}$$

Here, the property matrix for an isotropic material of modulus E and Poisson's ratio v is

$$[D] = \frac{E}{(1+v)(1-2v)}\begin{bmatrix} (1-v) & v & v & 0 & 0 & 0 \\ v & (1-v) & v & 0 & 0 & 0 \\ v & v & (1-v) & 0 & 0 & 0 \\ 0 & 0 & 0 & (1/2-v) & 0 & 0 \\ 0 & 0 & 0 & 0 & (1/2-v) & 0 \\ 0 & 0 & 0 & 0 & 0 & (1/2-v) \end{bmatrix}$$

$$\tag{5.94}$$

We are now able to write down an expression for twice the strain energy U of the element:

$$2U = \int \{\varepsilon\}^T \{\sigma\} \, dV$$
$$= \{\delta\}^T \left[\int_V [B]^T [D][B] \, dV \right] \{\delta\}$$
$$= \{\delta\}^T [K] \{\delta\} \tag{5.95}$$

This equation serves to define the stiffness matrix $[K]$. Now the additional degrees of freedom u_i, v_i, w_i $(i = 9, 10, 11)$ must be eliminated by using the condition that the load is zero for those degrees of freedom. This operation gives

$$\begin{bmatrix} K_{11} & K_{12} \\ K_{21} & K_{22} \end{bmatrix} \begin{Bmatrix} \delta_1^0 \\ \delta_2^0 \end{Bmatrix} = \begin{Bmatrix} P \\ 0 \end{Bmatrix} \tag{5.96}$$

Here, δ_1^0 is the vector of nodal displacements, while δ_2^0 contains the displacements to be eliminated. The matrix $[K]$ is partitioned in a corresponding manner, as shown. The load vector shows that only the nodes are loaded. (In some cases, the degrees of freedom to be eliminated are associated with load terms, as when gravity, centrifugal, or thermal loads are imposed. In such situations, eq. (5.96) is correspondingly modified.) Upon solving eq. (5.96) for δ_2^0, we get

$$\{\delta_2^0\} = -[K_{22}]^{-1}[K_{21}]\{\delta_1^0\} \tag{5.97}$$

Finally, upon eliminating δ_2^0, the load–deflection relationship becomes

$$\{P\} = [K_{11} - K_{12}K_{22}^{-1}K_{21}]\{\delta_1^0\} \tag{5.98}$$

and the deflections can be determined if the external loads are known.

Before closing, it should be noted that we have discussed a three-dimensional eight-noded element having 24 external degrees of freedom and nine internally eliminated degrees of freedom. Zienkiewicz[1] gives extensive discussion on other elements, notably a 20-noded curved isoparametric element (shown in Fig. 5.10), which has 60 degrees of freedom. This element has found extensive use in a variety of investigations.

5.11   ANISOTROPIC MEDIA

Certain materials are inherently anisotropic, i.e., their properties vary with direction. Composite materials are an especially important class of such materials. They are generally fabricated of a number of fibers of very high strength, embedded in a matrix. Examples of these materials include[4,5] graphite fibers in epoxy resin, fiberglass, boron fibers in an aluminum matrix, and eutectic composites such as NiTaC, which are crystallized from a melt. The elastic properties of anisotropic media can be described by methods discussed by Lekhnitskii[6] and, in the case of fibrous composites, Ashton, Halpin, and Petit[7] have shown how to model their properties. Anisotropic media also occur in other areas: e.g., a rolled metal sheet will often exhibit elastic properties which differ in the rolling and thickness directions; a pressed ceramic may exhibit different properties in the direction of pressing and perpendicular to that direction.

Whatever the cause of anisotropy, the elastic properties which relate the stress to the strain can usually be written in a form that is compatible with the development given here. It is necessary only to define a suitable matrix of properties, $[D]$.

Consider the fairly general case where the material is orthotropic. Let the principal directions of the material be denoted by the subscripts 1, 2, 3, while the

FIGURE 5.10
A curved 20-noded element

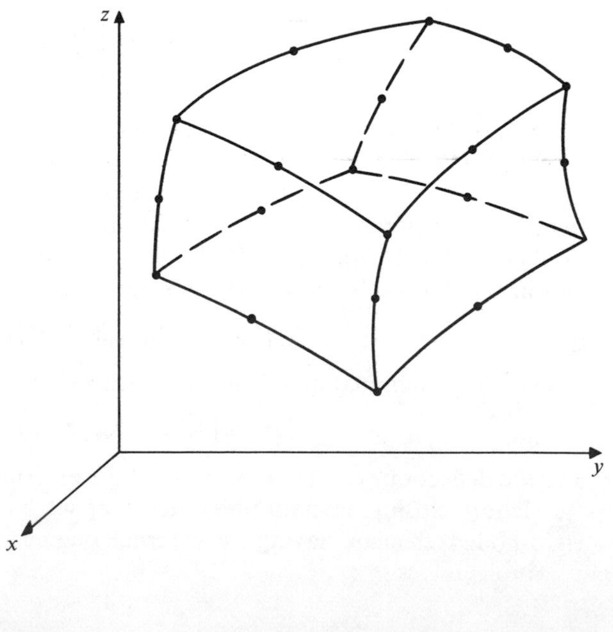

orientation of the $(1, 2, 3)$ axes have direction cosines (l_1, m_1, n_1), (l_2, m_2, n_2), (l_3, m_3, n_3) with respect to the Cartesian (x, y, z) axes. Then the strains in the material coordinates are related to those in the global coordinates by a transformation matrix $[T]$:

$$\{\varepsilon_{1,2,3}\} = [T]^T\{\varepsilon_{x,y,z}\} \qquad (5.99)$$

where

$$[T] = \begin{bmatrix} l_1^2 & l_2^2 & l_3^2 & 2l_1l_2 & 2l_2l_3 & 2l_3l_1 \\ m_1^2 & m_2^2 & m_3^2 & 2m_1m_2 & 2m_2m_3 & 2m_3m_1 \\ n_1^2 & n_2^2 & n_3^2 & 2n_1n_2 & 2n_2n_3 & 2n_3n_1 \\ l_1m_1 & l_2m_2 & l_3m_3 & (l_1m_2 + l_2m_1) & (l_2m_3 + l_3m_2) & (l_3m_1 + l_1m_3) \\ m_1n_1 & m_2n_2 & m_3n_3 & (m_1n_2 + m_2n_1) & (m_2n_3 + m_3n_2) & (m_3n_1 + m_1n_3) \\ n_1l_1 & n_2l_2 & n_3l_3 & (n_1l_2 + n_2l_1) & (n_2l_3 + n_3l_2) & (n_3l_1 + n_1l_3) \end{bmatrix}$$

The stress is given by

$$\{\sigma_{1,2,3}\} = [D]\{\varepsilon_{1,2,3}\} \qquad (5.100)$$

in the material coordinates, or by

$$\{\sigma_{x,y,z}\} = [T][D][T]^T\{\varepsilon_{x,y,z}\} \qquad (5.101)$$

in the global coordinates. Here, the property matrix $[D]^{-1}$ is

$$[D]^{-1} = \begin{bmatrix} \dfrac{1}{E_1} & \dfrac{-v_{12}}{E_2} & \dfrac{-v_{13}}{E_3} & 0 & 0 & 0 \\[2ex] \dfrac{-v_{12}}{E_2} & \dfrac{1}{E_2} & \dfrac{-v_{23}}{E_3} & 0 & 0 & 0 \\[2ex] \dfrac{-v_{13}}{E_3} & \dfrac{-v_{23}}{E_3} & \dfrac{1}{E_3} & 0 & 0 & 0 \\[2ex] 0 & 0 & 0 & \dfrac{1}{G_{12}} & 0 & 0 \\[2ex] 0 & 0 & 0 & 0 & \dfrac{1}{G_{23}} & 0 \\[2ex] 0 & 0 & 0 & 0 & 0 & \dfrac{1}{G_{31}} \end{bmatrix} \qquad (5.102)$$

where E_i are the elastic moduli, v_{ij} are the Poisson's ratios, and G_{ij} are the shear moduli.

5.12 SUMMARY

This chapter has set forth some concepts needed in determining the stiffness matrix $[K]$ of a continuous media for the study of the vibratory behavior of continuous bodies, whether they be modeled in two or in three dimensions. The basic approach is to divide the body into quadrilateral or hexahedronal elements (as the case may be). The stiffness properties of the elements are superimposed in a logical manner to obtain the stiffness of the structure as a whole. This stiffness matrix will be used as a necessary link to the calculation of the free modes of vibration of the structure, as shown in the following chapter.

5.13 REFERENCES

1. O. C. Zienkiewicz, *The Finite Element Method in Engineering Science,* McGraw-Hill Book Company (UK) Ltd, London, 1971.
2. S. Timoshenko and J. N. Goodier, *Theory of Elasticity,* 2nd edn., ch. 2, McGraw-Hill Book Company, Inc., New York, 1951.
3. N. Macon, *Numerical Analysis,* pp. 109–112, John Wiley and Sons, Inc., New York, 1963.
4. A. Kelly, *Strong Solids,* Clarendon Press, Oxford, 1966.
5. J. R. Vinson and T. W. Chou, *Composite Materials and Their Use in Structures,* John Wiley and Sons, Inc., New York, 1975.
6. S. G. Lekhnitskii, *Theory of Elasticity of an Anisotropic Elastic Body,* (Trans. P. Fern, J. J. Brandstaller (Ed.)), Holden-Day, Inc., San Francisco, 1963.
7. J. E. Ashton, J. C. Halpin, and P. H. Petit, *Primer on Composite Materials: Analysis,* Technomic Publishing Company, Inc., Stamford, Connecticut, 1969.

6

VIBRATIONS OF CONTINUOUS SYSTEMS

6.1 INTRODUCTION

The purpose of this chapter is to show how the natural frequencies and mode shapes of continuous systems can be computed by using the finite element method as a basis. The structural component under investigation is broken up into a number of finite elements defined by nodes, as described in chapter 5. The stiffness and mass matrices appropriate to this model are then assembled. The natural frequencies and mode shapes are determined by solving an eigenvalue problem based on these stiffness and mass matrices. A method of condensation is discussed, which serves to reduce the size of the eigenvalue problem. Examples of the use of the condensation method are given, drawn chiefly from the area of turbine design, where the structural components analyzed are turbine buckets.

6.2 MASS MATRICES

When attempting to write the equations of motion for a structure, it is first necessary to somehow distribute the mass of the structure in an appropriate way. The total mass is given by

$$m = \int \rho \, dV \tag{6.1}$$

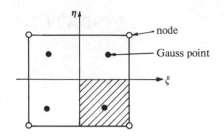

(a) Isoparametric space

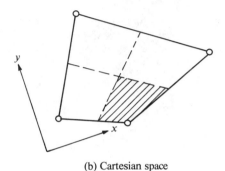

(b) Cartesian space

FIGURE 6.1
Lumping the mass at a nodal point. The shaded area
shows the mass that is lumped at the adjacent node

If this integration is performed by Gaussian integration, it is customary to multiply the volume for each Gauss point of an element, as shown in Fig. 6.1, by the density, and to assign the mass so computed to the neighboring node:

$$m_i = \sum_e \rho \, |J_i| \, dV_{\xi\eta\zeta} = \sum_e \rho \, |J_i| \tag{6.2}$$

where the determinant of the Jacobian gives the volume associated with the Gauss point (see eq. 5.86), and the summation includes all elements having node i as a node point.

Under this scheme, the mass matrix, which represents the contribution of each node, would be a diagonal matrix. The reason why this method is so widely used is that the diagonal mass matrix results in a considerable economy in computational effort compared to other methods where a mass matrix with a larger bandwidth can result.

An alternative and more accurate way to compute a mass matrix is as follows. The kinetic energy T of the element is related to the nodal velocities by

$$2T = \int [\{\dot{u}\}^T \rho \{\dot{u}\} + \{\dot{v}\}^T \rho \{\dot{v}\} + \{\dot{w}\}^T \rho \{\dot{w}\}] \, dV \tag{6.3}$$

Recalling that the velocities $\dot{u}$ can be written in terms of the nodal velocities through the shape functions of eq. (5.81), we have

$$\begin{Bmatrix} \dot{u} \\ \dot{v} \\ \dot{w} \end{Bmatrix} = \begin{bmatrix} N_1 & 0 & 0 & N_2 & 0 & 0 & N_3 & 0 & 0 & \ldots \\ 0 & N_1 & 0 & 0 & N_2 & 0 & 0 & N_3 & 0 & \ldots \\ 0 & 0 & N_1 & 0 & 0 & N_2 & 0 & 0 & N_3 & \ldots \end{bmatrix} \begin{Bmatrix} \dot{u}_1 \\ \dot{v}_1 \\ \dot{w}_1 \\ \vdots \end{Bmatrix} \qquad (6.4)$$

Substitution of these velocities into the energy integral gives

$$2T = \{\dot{\delta}\}^T \left[\int [N]^T \rho [N] \, dV \right] \{\dot{\delta}\} \qquad (6.5)$$

and, as a result, we recognize that the mass matrix can be written as

$$[m] = \int [N]^T \rho [N] \, dV \qquad (6.6)$$

This formulation is recommended by Zienkiewicz.[1] It is sometimes known as a consistent mass matrix. It is a matrix which has the same bandwidth as the stiffness matrix. The consistent mass matrix takes more accurate account of rotary inertia effects than that in eq. (6.2). Where the rotary inertia of the individual element is important, eq. (6.6) should be used rather than eq. (6.2). When a large number of elements are involved, however, the rotary inertia of a single element is usually of negligible importance, and then eq. (6.2) should be used rather than eq. (6.6) for reasons of economy.

6.3   EQUATIONS OF MOTION

Applying Newton's law to the assembled structure will result in a matrix equation that governs the dynamic behavior:

$$\lfloor K \rfloor \{\delta\} + \lfloor C \rfloor \frac{\partial}{\partial t} \{\delta\} + [m] \frac{\partial^2}{\partial t^2} \{\delta\} - \{F(t)\} = 0 \qquad (6.7)$$

Here, we recognize $[K]$ as the stiffness matrix of the structure, $[C]$ as the damping matrix, and $[m]$ as the mass matrix. The external forces acting on the nodes are given by the vector $\{F\}$.

This equation can, in general, be solved by applying the ideas developed in preceding chapters. In essence, we propose to find first the natural frequencies and mode shapes of the structure. Then, with the use of the modal amplitudes as generalized coordinates, we will apply the component element method as described in section 3.12 to compute the time-dependent solutions. Because the determination of mode shapes and natural frequencies plays a central role, this chapter concentrates on a discussion of the free vibrations of continuous structures, and discusses methods whereby the eigenvalues and eigenvectors can be obtained.

6.4   EIGENVALUES AND EIGENVECTORS

When no damping exists and the external forces are zero, the structure can vibrate freely in a manner governed by the equation

$$[K] \{\delta_0\} + [m] \frac{\partial^2}{\partial t^2} \{\delta_0\} = \{0\} \qquad (6.8)$$

Assuming a solution of the form

$$\{\delta_0\} = \{\delta\} \sin \omega t \tag{6.9}$$

it can be shown by direct substitution that

$$([K] - \omega^2[m])\{\delta\} = \{0\} \tag{6.10}$$

A solution exists if the determinant $|[K] - \omega^2[m]|$ is zero. The determinant is of order n, n being the number of degrees of freedom in the structure. Therefore there are n eigenvalues ω_n at which the structure can vibrate, according to this mathematical model. At each frequency ω_n, there is a corresponding vector $\{\delta_n\}$ whose elements are in a specified ratio, but whose absolute magnitude can be arbitrary. This vector is the eigenvector and defines how the structure vibrates in that mode.

The starting point for the determination of the eigenvalues and eigenvectors is to write eq. (6.10) in the form

$$[K]\{\delta_n\} = \omega_n^2[m]\{\delta_n\} \tag{6.11}$$

This equation is valid for the nth mode of vibration. If we now define a modal matrix

$$[\Delta] = [\delta_1 \quad \delta_2 \quad \dots \quad \delta_n] \tag{6.12}$$

then eq. (6.11) can be written as

$$[K][\Delta] = [m][\Delta][D(\omega^2)] \tag{6.13}$$

where $D(\omega^2)$ is a diagonal matrix

$$[D(\omega^2)] = \begin{bmatrix} \omega_1^2 & & & 0 \\ & \omega_2^2 & & \\ & & \ddots & \\ 0 & & & \omega_n^2 \end{bmatrix} \tag{6.14}$$

Introduce now the new matrices $\bar{\Delta}$ and $\bar{K}$:

$$\left.\begin{aligned} \bar{\Delta} &= [m]^{1/2}[\Delta] \\ \bar{K} &= [m]^{-1/2}[K][m]^{-1/2} \end{aligned}\right\} \tag{6.15}$$

where $\bar{K}$ is a symmetric matrix. (If the mass matrix is a diagonal matrix, then its square root is a diagonal matrix containing the square roots of the original elements. However, if $[m]$ is a full matrix, then its square root must be calculated by using the diagonalization technique described by Crandall.[2] This diagonalization is very much akin to the Jacobi method for extracting eigenvalues and more will be said of it in the next section.) Then we can rewrite eq. (6.13) in the form

$$[\bar{K}][\bar{\Delta}] = [\bar{\Delta}][D(\omega^2)] \tag{6.16}$$

The introduction of the new matrices $[\bar{K}]$ and $[\bar{\Delta}]$ is necessary to bring the eigenvalue problem into a form amenable to solution by Jacobi's method. This form is known as the special eigenvalue problem. The eigenvalues and eigenvectors can be determined from this form by the use of the Jacobi method, which involves the diagonalization of the matrices by a technique of successive rotation as described by Crandall.[3]

It should be noted that numerous other methods of determining eigenvalues and eigenvectors exist. Many, like the Jacobi method, give all the eigenvalues of the system at the same time. However, for some structural problems where only a few of the lowest modes of vibration are of interest, a method that yields these modes and no others would be economical. Bathe and Wilson[4] discuss and compare a number of methods, including the Jacobi method, the HQRI (Householder–QR–Inverse Iteration) method, a determinant search technique, and a subspace iteration method. They show how the choice of technique is dependent not only on the number of eigenvalues desired but also upon the bandwidth and the size of the matrices.

6.5 JACOBI'S METHOD FOR CALCULATING EIGENVALUES

It is desired to compute the eigenvectors and eigenvalues of the system of equations in the form of the special eigenvalue problem

$$K\Delta = \Delta D(\omega^2) \qquad (6.17)$$

(In this section, we shall, for clarity, drop the brackets which have up until now been used to identify matrices. We also drop the superscript bars of eqs. (6.15) and (6.16).)

We first require that the eigenvectors δ_n are normalized so that the modal mass is unity. Thus,

$$\delta_n^T \delta_n = 1 \qquad (6.18)$$

In addition, the orthogonality condition in eq. (3.27) applies, so that,

$$\delta_i^T \delta_j = 0 \qquad (i \neq j) \qquad (6.19)$$

With this understanding, it will be seen that

$$\Delta^T \Delta = \begin{bmatrix} \delta_1 \\ \delta_2 \\ \vdots \\ \delta_n \end{bmatrix} [\delta_1 \delta_2 \dots \delta_n] = \begin{bmatrix} 1 & & & 0 \\ & 1 & & \\ & & 1 & \\ 0 & & & 1 \end{bmatrix} = I \qquad (6.20)$$

Thus, the modal matrix Δ has the unique property that its transpose is also its inverse. Returning now to eq. (6.17), and premultiplying by Δ^T, we obtain

$$\Delta^T K \Delta = D(\omega^2) \qquad (6.21)$$

The Jacobi method seeks to find a matrix Δ which, by the operations in the equation above, will diagonalize the matrix K. This diagonal matrix will contain the eigenvalues and is, in fact, the matrix $D(\omega^2)$. The manner in which this is done depends on a number of observations which we set forth below.

First, starting with the matrix K, we operate on it with a series of square matrices T_r, as follows:

$$\left.\begin{aligned}
K &= K_0 \\
T_1^T K_0 T_1 &= K_1 \\
T_2^T K_1 T_2 &= K_2 \\
&\;\;\vdots \\
T_r^T K_{r-1} T_r &= K_r
\end{aligned}\right\} \tag{6.22}$$

As a result, we finally have the sequence of operations

$$T_r^T \dots T_2^T T_1^T K T_1 T_2 \dots T_r = K_r \tag{6.23}$$

We wish to select the matrices T_r so that this equation converges toward the form

$$\Delta^T K \Delta = D(\omega^2) \tag{6.24}$$

We can identify here that

$$\left.\begin{aligned}
T_1 T_2 \dots T_r &\to \Delta \\
K_r &\to D(\omega^2)
\end{aligned}\right\} \tag{6.25}$$

Since one property of Δ is that its transpose is also its inverse (i.e., $\Delta^T \Delta = I$), we also will require that each matrix T_r satisfies this condition (i.e., $T_r^T T_r = I$). In doing so, the condition that

$$(T_1 T_2 \dots T_r)^T (T_1 T_2 \dots T_r) = I \tag{6.26}$$

will also apply.

In order to produce the diagonalization of K, the matrices T_r employed in Jacobi's method have the special property that each T_r causes a particular off-diagonal term in K_{r-1} to vanish as K_r is created. Let the elements of K_{r-1} be a_{jk} and those of K_r be b_{jk}. If a_{pq} is not equal to zero, we define T_r such that

$$
T_r = \quad
\begin{array}{c}
\\ \\ \\ p \\ \\ \\ q \\ \\ \\
\end{array}
\begin{bmatrix}
1 & 0 & \cdots & 0 & 0 & 0 & 0 & \cdots & 0 \\
0 & 1 & \cdots & 0 & 0 & 0 & 0 & \cdots & 0 \\
\cdots & \cdots & \cdots & \cdots & \cdots & \cdots & \cdots & \cdots & \cdots \\
0 & 0 & \cdots & C & 0 & 0 & -S & \cdots & 0 \\
0 & 0 & \cdots & 0 & 1 & 0 & 0 & \cdots & 0 \\
0 & 0 & \cdots & 0 & 0 & 1 & 0 & \cdots & 0 \\
0 & 0 & \cdots & S & 0 & 0 & C & \cdots & 0 \\
\cdots & \cdots & \cdots & \cdots & \cdots & \cdots & \cdots & \cdots & \cdots \\
0 & 0 & \cdots & 0 & 0 & 0 & 0 & \cdots & 1
\end{bmatrix}
\tag{6.27}
$$

where $C = \cos \theta$ and $S = \sin \theta$ and the value of θ will be subsequently chosen.

T_r represents a rotation through an angle θ in the plane (pq). Its transpose is also its inverse, satisfying the requirement of eq. (6.26). The multiplication

$$T_r^T K_{r-1} T_r = K_r \tag{6.28}$$

will show that all elements in K_{r-1} are the same as in K_r, except those in the pth and qth rows and columns, which become

$$
\left.
\begin{aligned}
b_{pp} &= a_{pp} \cos^2 \theta + 2a_{pq} \sin \theta \cos \theta + a_{qq} \sin^2 \theta \\
b_{pq} &= a_{pq}(\cos^2 \theta - \sin^2 \theta) - (a_{pp} - a_{qq}) \sin \theta \cos \theta \\
b_{qq} &= a_{pp} \sin^2 \theta - 2a_{pq} \sin \theta \cos \theta + a_{qq} \cos^2 \theta \\
b_{pj} &= a_{pj} \cos \theta + a_{qj} \sin \theta \quad \Big\} \\
b_{qj} &= -a_{pj} \sin \theta + a_{qj} \cos \theta \Big\} \quad (j \neq p, q)
\end{aligned}
\right\} \tag{6.29}
$$

The element b_{pq} vanishes if we take the angle θ such that

$$\tan 2\theta = \frac{2a_{pq}}{a_{pp} - a_{qq}} \tag{6.30}$$

The vanishing of b_{pq} is obtained at the expense of changing the other off-diagonal terms b_{pj} and b_{qj}. But we note that the sum of the squares of the remaining diagonal terms is unchanged no matter what the value of θ, for it can be verified that

$$b_{pj}^2 + b_{qj}^2 = a_{pj}^2 + a_{qj}^2 \qquad (j \neq p, q) \tag{6.31}$$

Thus, the transformation has performed a definite step toward the diagonalization of K. Jacobi's method is based on imposing a sequence of rotations, each of which is chosen to make an off-diagonal term equal to zero. In practice, the largest terms in K_r are sometimes made to vanish, but to avoid the necessity of searching for the largest term at each rotation, a systematic row-by-row sequence is often employed. This sequential operation is found to converge very rapidly after a small number of complete sweeps through the matrix.

It should be noted here that quite a number of subtle variations are introduced in practice, each designed to reduce the computing time. For example, in a method known as the threshold Jacobi method, the rotation is omitted if $|a_{pq}|$ is already less than a certain threshold value. At each sweep, this threshold value is reduced to a smaller number.

The square root of a matrix

In view of the discussion concerning the properties of the modal matrix Δ, we are now in a position to explain just how the general eigenvalue problem of eq. (6.13), namely

$$K\Delta = m\Delta D(\omega^2) \tag{6.32}$$

is reduced to the form of the special eigenvalue problem

$$\bar{K}\bar{\Delta} = \bar{\Delta} D(\omega^2) \tag{6.33}$$

through the transformation

$$\begin{aligned}\overline{\Delta} &= m^{1/2}\Delta \\ \overline{K} &= m^{-1/2}Km^{-1/2}\end{aligned}\Bigg\}$$ (6.34)

It is the determination of $m^{1/2}$ that is the key to the process. The square root of a matrix m can be found as follows.

Take any square symmetric matrix A. Since it is symmetric, Jacobi's method is applicable. Calculate the eigenvalues λ and the normalized eigenvectors Δ_A of A, so that

$$A\Delta_A = \Delta_A D(\lambda)$$ (6.35)

Then by postmultiplying by Δ_A^T,

$$A = \Delta_A D(\lambda)\Delta_A^T$$ (6.36)

Now, it can be shown[3] that

$$A^a = \Delta_A D(\lambda^a)\Delta_A^T$$ (6.37)

where a is any exponent. Thus, we can write

$$\begin{aligned}A^{1/2} &= \Delta_A D(\lambda^{1/2})\Delta_A^T \\ A^{-1/2} &= \Delta_A D(\lambda^{-1/2})\Delta_A^T\end{aligned}\Bigg\}$$ (6.38)

The square root of the mass matrix m is therefore found by first diagonalizing m as described for the Jacobi method. The diagonal matrix and the associated eigenvector matrix can then be used to construct $m^{1/2}$ and $m^{-1/2}$ as required for the transformations (6.34).

6.6   EIGENVALUE ECONOMIZER METHODS

In most complex structures, there may be up to 200 or more elements and perhaps 600 to 1000 nodes. This means that there may be some 2000 or more degrees of freedom, resulting in a like number of predictable natural frequencies. It is both economically unfeasible to compute such a large number of eigenvalues, and also quite unnecessary. Most structural responses can be adequately computed using 20 to 50 eigenvalues and eigenvectors. Consequently, methods have been devised by which the size of the matrices involved can be reduced to the size of those latter numbers. Such methods are known as eigenvalue economizer methods and in practice must be used to secure good economy in computation without significantly sacrificing accuracy.

The idea of reducing the size of the problem was introduced originally by Irons[5,6] and Guyan.[7] Let the degrees of freedom $\{\delta\}$ of the structure be divided into two parts. The first set is specified as master displacements $\{\delta_m\}$, while the second set are slaves $\{\delta_s\}$. The slave displacements are governed uniquely by the master displacements. The total displacement vector is therefore related to the master displacements by the transformation

$$\{\delta\} = [L]\{\delta_m\}$$ (6.39)

If $[L]$ is known, then reduced stiffness and mass matrices can be obtained by requiring that the kinetic and potential energies of the structure remain the same. Thus,

$$2T = \{\dot{\delta}\}^T [m] \{\dot{\delta}\} \quad \left.\right\}$$
$$2U = \{\delta\}^T [K] \{\delta\} \quad \left.\right\} \tag{6.40}$$

Substituting the transformation into master displacements, eq. (6.39) gives

$$2T = \{\dot{\delta}_m\}^T [L]^T [m] [L] \{\dot{\delta}_m\} \quad \left.\right\}$$
$$2U = \{\delta_m\}^T [L]^T [K] [L] \{\delta_m\} \quad \left.\right\} \tag{6.41}$$

Thus the mass and stiffness matrices of the reduced system, which is defined completely by the master degrees of freedom, are

$$[m'] = [L]^T [m] [L] \quad \left.\right\}$$
$$[K'] = [L]^T [K] [L] \quad \left.\right\} \tag{6.42}$$

Hence, the equations of motion become

$$[K'] \{\delta_m\} - \omega^2 [m'] \{\delta_m\} = \{0\} \tag{6.43}$$

The eigenvalues and eigenvectors of the reduced system are calculated by means of Jacobi's method, as before. This reduction process is also known as condensation.

Obviously, the implementation of this procedure depends on two things. The first is the choice of the master displacements. The second is the determination of the transformation matrix $[L]$. Assuming that the master displacements have been chosen (their choice will be discussed subsequently), a conceptual way to illustrate the determination of $[L]$ is as follows.

Imagine that the imposition of the master displacements on the structure will so load it that its deformations are essentially equivalent to the behavior of the actual structure. The external forces $\{P\}$ at the master nodes that are necessary to accomplish this deformation are not known. We also assume that no forces act on the slave nodes. Then we can write

$$\left\{\begin{array}{c} P \\ 0 \end{array}\right\} = [K] \{\delta\} = \begin{bmatrix} K_{11} & K_{12} \\ K_{12}^T & K_{22} \end{bmatrix} \left\{\begin{array}{c} \delta_m \\ \delta_s \end{array}\right\} \tag{6.44}$$

Solving now for $\{\delta_s\}$ in terms of $\{\delta_m\}$, we find that

$$\{\delta_s\} = -[K_{22}^{-1} \quad K_{12}^T] \{\delta_m\} \tag{6.45}$$

The desired transformation is therefore

$$[L] = \begin{bmatrix} I \\ -K_{22}^{-1} & K_{12}^T \end{bmatrix} \tag{6.46}$$

This method, although strictly correct, is of conceptual value only, for the following reason. Assume that the reduction of degrees of freedom was to have been from 1000 to 50. Then the size of K_{22} is 950 × 950. This is still a very large matrix, and it turns out that the amount of computational effort required to invert it is excessive.

In addition, the use of this method requires the sorting of the stiffness matrix into elements associated with the master or the slave displacements, an operation which can inhibit the logical flow of numerical computations.

A more economical approach to obtaining $[L]$ derives from the observation that a column of $[L]$ gives the nodal deflections for a unit value of one of the master deflections when all others are zero. Thus, $[L]$ can be determined by solving the equation

$$[K'][L] = [F] \tag{6.47}$$

where $[K']$ is a modified stiffness matrix (following section 5.9, first method), with the rows and columns corresponding to each master displacement set to zero, except that each master displacement diagonal term is set to unity. The $[F]$ matrix (following section 5.9) has a column for each master displacement. Its elements for each master displacement column are generally equal to the negative of the corresponding stiffness terms removed from $[K]$, except for a 1 at the master displacement row. An example will clarify these modifications. The equation

$$\begin{bmatrix} K_{11} & 0 & K_{13} & 0 & K_{15} \\ 0 & 1 & 0 & 0 & 0 \\ K_{31} & 0 & K_{33} & 0 & K_{35} \\ 0 & 0 & K_{43} & 1 & 0 \\ K_{51} & 0 & K_{53} & 0 & K_{55} \end{bmatrix} \begin{bmatrix} L_{11} & L_{12} \\ 1 & 0 \\ L_{31} & L_{32} \\ 0 & 1 \\ L_{51} & L_{52} \end{bmatrix} = \begin{bmatrix} -K_{12} & -K_{14} \\ 1 & 0 \\ -K_{32} & -K_{34} \\ 0 & 1 \\ -K_{52} & -K_{54} \end{bmatrix} \tag{6.48}$$

represents the modifications due to the existence of master displacements δ_2 and δ_4. An equation like eq. (6.48) can be solved by Gaussian elimination using submatrices (as described for eq. 5.75), with a reasonable economy of effort even if the original system is fairly large.

It now becomes clear that there is a close analogy between the relationship of master and slave displacements, and the relationship of the nodal and element displacements described in chapter 5. The columns of the matrix $[L]$ can be viewed as generalized shape functions.

6.7   CHOOSING THE MASTER DISPLACEMENTS

The choice of the master displacements now needs to be discussed. There are no general rules for choosing these displacements. In the reduction, or condensation, of the mass and stiffness matrices, practically the only requirement that was imposed was that of the preservation of the kinetic and strain energies. To make most effective use of the master deflections, therefore, it is desirable to choose those deflections that may dominate in the energetic exchanges during vibrational motion. However, for a complex structure, it is rather difficult to know beforehand which deflections will be most effective.

For simpler systems, such as cantilevered structures, Irons[5] has suggested that certain nodal deflections do not contribute greatly. They can be chosen as slaves. Some of these slave deflections are the lengthwise movements that are associated with high frequency modes; movements near the built-in end of the structure, which contribute little to the kinetic energy; and the slopes of the structure, as opposed to displacements themselves. He points out that the choosing of every remaining third or fifth displacement node should be adequate. One should note that the errors in the choice of master deflections will not be reflected severely in the eventual frequencies that are computed through this method. This observation is analogous to Rayleigh's principle concerning the determination of frequencies from an assumed mode shape: a first-order error in mode shape, except in unusual circumstances, is reflected as a second-order error in frequency.

Levy[8] gives two analogous guidelines. The first is the selection of displacements associated with the largest entries in the mass matrix. If there is truly an outstanding mass, then one displacement might indeed be associated with it, since its deflection would be somewhat uncoupled from the remainder of the structure. For complex structures, however, it is rare to find such large differences in mass terms. The second is the selection of deflections that have the largest movements in the modes of interest. This selection requires some *a priori* knowledge of the structural behavior. However, the engineer usually has some intuitive feel for the vibrational modes and can often make reasonable guesses as to their behavior. In this regard, the analogy between master displacements in the structure and nodal displacements in the elements should be recalled. Experience and intuition has also played a role in the choice of the shape functions of the element.

To illustrate the effect of the choice of master displacements, consider the turbine blade shown in Fig. 6.2. It is fixed at its base and is modeled with 40 elements of the type described in section 5.10. There are 90 nodes. The lowest five natural frequencies were computed using the economizer method for a number of different choices of the master displacements. The master displacements, shown by arrows, and the corresponding frequencies, are illustrated in Fig. 6.3. In Fig. 6.3(a), 10 lateral master displacements were used, resulting in the highest set of frequencies. In Fig. 6.3(b), 20 masters were used, again in the lateral direction. The lowest frequency is not significantly different, although the remainder are considerably lower than before. In Fig. 6.3(c), 20 masters were used, some of which are radial and some lateral. Now the frequencies are all lower yet. Finally, in Fig. 6.3(d), 20 somewhat differently distributed masters (some chordwise) were employed, which yielded somewhat lower frequencies yet. The lowest frequency computed by using the complete original matrices was 8200 Hz.

This exercise shows, first, that reasonable accuracy can be expected for the lowest modes, but that the accuracy drops off for higher modes.

Second, the number of master displacements employed is of importance. Various investigators have commented on this aspect. The experience of Anderson, Irons, and Zienkiewicz[9] is that at least 20 master deflections be retained. Ramsden and Stoker[10] concluded, in a specific case where 1800 degrees of freedom appeared, that 70 master displacements were sufficient for obtaining 20 vibration modes with reasonable accuracy. They also pointed out that it may be quite reasonable to use a diagonal mass matrix rather than a more accurate consistent mass matrix, since condensation from 1800 to 70 degrees of freedom masks the differences between the

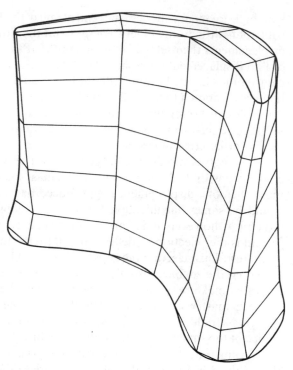

FIGURE 6.2
A turbine blade model

two formulations, Jennings[11] has developed further examples showing the effect of the number of master displacements. He proposes that, in general, for the mass condensation approach, the ratio of masters to the required number of natural frequencies be 3·5 or less.

Third, a change in the directions as well as position of the master displacements is of importance in determining the frequencies.

Finally, the resulting frequencies are always higher if condensation is employed than if the original matrices are used. The reason is that the condensation process introduces additional constraints into the original description of the structure, thereby causing the frequencies to be higher. Anderson, Irons, and Zienkiewicz[9] have shown that within the range of the first K frequencies of the original system, there cannot be more than $(K - 1)$ frequencies of the reduced system. Thus, the frequencies of the reduced and original systems appear to be interlaced.

6.8   OTHER ECONOMIZER CONCEPTS—ITERATIVE TECHNIQUES

Because of the possible errors associated with the somewhat loose choice of master deflections, several investigators[8,10,11,12] have evolved methods whereby improvements in the frequency calculations can be made. These schemes are generally based

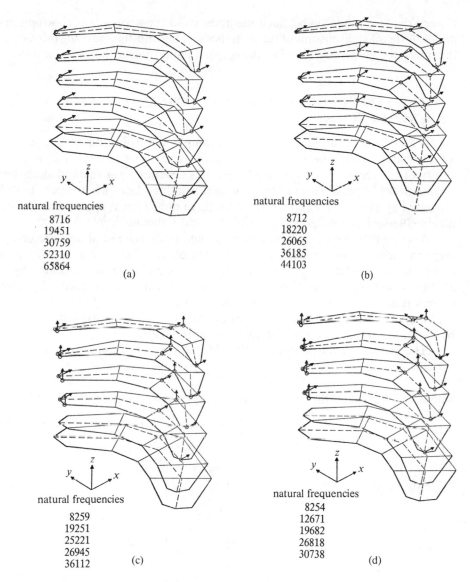

natural frequencies
8716
19451
30759
52310
65864

(a)

natural frequencies
8712
18220
26065
36185
44103

(b)

natural frequencies
8259
19251
25221
26945
36112

(c)

natural frequencies
8254
12671
19682
26818
30738

(d)

FIGURE 6.3
Master displacements, shown by arrows, and corresponding natural frequencies (in Hz)

on iterative techniques and appear to lead to better accuracy than the basic mass condensation scheme described here. Bathe, Wilson, and Peterson[12] use an iterative method whereby an initial guess is máde as to the shape functions referred to previously in connection with the transformation matrix $[L]$. They initially use a very rough estimate, perhaps a full matrix of unit elements. An iteration rapidly gives a revised estimate for the mode shape, which is in turn re-used. Convergence is rapid toward the final mode shape. Such methods avoid the necessity of choosing master nodes. A detailed discussion of these methods is beyond the scope of this book and the interested reader is referred to the original papers for further information.

6.9   THE EFFECTS OF CENTRIFUGAL FORCES ON NATURAL FREQUENCIES

When a body is rotating, as does, for example, a turbine blade, the centrifugal forces tend to stiffen the blade and serve to increase its natural frequencies. The effects of the centrifugal forces are taken into account by a method described by Zienkiewicz.[13] Special cases in the case of a rotating plate-like structure have been discussed by Zienkiewicz and Cheung.[14] A three-dimensional application to rotating turbine blades has been given by Trompette and Lalanne.[15]

Assume that the centrifugal forces at each node are radial and are computed through the expression $r_i m_i \Omega^2$, where Ω is the angular velocity in rad/sec, r_i is the radius of the ith node from the center of rotation, and m_i is the accumulated mass at the ith node due to contributions from the adjacent Gauss points. By applying these forces as external loads, we can compute the initial centrifugal stresses $\{\sigma^0\}$ in the body. The basic assumption for this analysis is that this stress state is unaffected by the small lateral deflections of the blade during vibrations.

The increase in elastic strain in the blade during vibration is given by

$$\{\varepsilon\} = \{\varepsilon^L\} \tag{6.49}$$

where $\{\varepsilon^L\}$ represents the linear strain and is given, as in eq. (5.90), by

$$\{\varepsilon^L\}^T = \begin{bmatrix} \varepsilon_x \ \varepsilon_y \ \varepsilon_z \ \gamma_{xy} \ \gamma_{yz} \ \gamma_{zx} \end{bmatrix}$$

so that

$$\{\varepsilon^L\} = \begin{Bmatrix} \dfrac{\partial u}{\partial x} \\[2mm] \dfrac{\partial v}{\partial y} \\[2mm] \dfrac{\partial w}{\partial z} \\[2mm] \dfrac{\partial u}{\partial y} + \dfrac{\partial v}{\partial x} \\[2mm] \dfrac{\partial v}{\partial z} + \dfrac{\partial w}{\partial y} \\[2mm] \dfrac{\partial w}{\partial x} + \dfrac{\partial u}{\partial z} \end{Bmatrix} \tag{6.50}$$

The unit shortening against the centrifugal stress is defined as $\{\varepsilon^N\}$. When multiplied by the centrifugal stresses, it gives the potential energy stored in the centrifugal field. This unit shortening is

$$\{\varepsilon^N\} = \begin{Bmatrix} 1/2\left[\left(\dfrac{\partial w}{\partial x}\right)^2 + \left(\dfrac{\partial v}{\partial x}\right)^2\right] \\[2ex] 1/2\left[\left(\dfrac{\partial u}{\partial y}\right)^2 + \left(\dfrac{\partial w}{\partial y}\right)^2\right] \\[2ex] 1/2\left[\left(\dfrac{\partial v}{\partial z}\right)^2 + \left(\dfrac{\partial u}{\partial z}\right)^2\right] \\[2ex] \left(\dfrac{\partial w}{\partial x}\right)\left(\dfrac{\partial w}{\partial y}\right) \\[2ex] \left(\dfrac{\partial u}{\partial y}\right)\left(\dfrac{\partial u}{\partial z}\right) \\[2ex] \left(\dfrac{\partial v}{\partial x}\right)\left(\dfrac{\partial v}{\partial z}\right) \end{Bmatrix} \tag{6.51}$$

Quadratic terms involving $\partial u/\partial x$, $\partial v/\partial y$, and $\partial w/\partial z$ have been omitted from eq. (6.51), since they are an order of magnitude smaller for the small deflections used in vibration analysis. The definitions of the strains $\{\varepsilon^L\}$ and $\{\varepsilon^N\}$ are valid for all deformation states where u, v, and w are small in comparison with the geometry.

There is an increment in potential energy due to the nonlinear portion of strain, and for a unit volume it is given by

$$U = \{\sigma^0\}^T \{\varepsilon^N\} \tag{6.52}$$

This potential energy is the work done against the centrifugal field during vibration. In view of the definition of $\{\varepsilon^N\}$, the increment of potential energy per unit volume can also be written as

$$U = \frac{1}{2}\begin{Bmatrix} \dfrac{\partial w}{\partial x} \\[1.5ex] \dfrac{\partial w}{\partial y} \\[1.5ex] \dfrac{\partial u}{\partial y} \\[1.5ex] \dfrac{\partial u}{\partial z} \\[1.5ex] \dfrac{\partial v}{\partial z} \\[1.5ex] \dfrac{\partial v}{\partial x} \end{Bmatrix}^T \begin{bmatrix} \sigma_x^0 & \tau_{xy}^0 & 0 & 0 & 0 & 0 \\ \tau_{xy}^0 & \sigma_y^0 & 0 & 0 & 0 & 0 \\ 0 & 0 & \sigma_y^0 & \tau_{yz}^0 & 0 & 0 \\ 0 & 0 & \tau_{yz}^0 & \sigma_z^0 & 0 & 0 \\ 0 & 0 & 0 & 0 & \sigma_z^0 & \tau_{xz}^0 \\ 0 & 0 & 0 & 0 & \tau_{xz}^0 & \sigma_x^0 \end{bmatrix} \begin{Bmatrix} \dfrac{\partial w}{\partial x} \\[1.5ex] \dfrac{\partial w}{\partial y} \\[1.5ex] \dfrac{\partial u}{\partial y} \\[1.5ex] \dfrac{\partial u}{\partial z} \\[1.5ex] \dfrac{\partial v}{\partial z} \\[1.5ex] \dfrac{\partial v}{\partial x} \end{Bmatrix} \tag{6.53}$$

The central matrix of stress is denoted henceforth by $[\Sigma^0]$. This potential energy per unit volume is integrated over the element volume V_e to obtain the energy increment for the element.

Referring now to eq. (5.89), we can relate the nodal displacements to the displacements through the introduction of shape functions N_i:

$$\begin{Bmatrix} u \\ v \\ w \end{Bmatrix} = \begin{bmatrix} N_1 & 0 & 0 & N_2 & 0 & 0 & N_3 & 0 & 0 & \cdots \\ 0 & N_1 & 0 & 0 & N_2 & 0 & 0 & N_3 & 0 & \cdots \\ 0 & 0 & N_1 & 0 & 0 & N_2 & 0 & 0 & N_3 & \cdots \end{bmatrix} \begin{Bmatrix} u_1 \\ v_1 \\ w_1 \\ u_2 \\ \vdots \end{Bmatrix} \tag{6.54}$$

Differentiation gives

$$\begin{Bmatrix} \dfrac{\partial w}{\partial x} \\[2mm] \dfrac{\partial w}{\partial y} \\[2mm] \dfrac{\partial u}{\partial y} \\[2mm] \dfrac{\partial u}{\partial z} \\[2mm] \dfrac{\partial v}{\partial z} \\[2mm] \dfrac{\partial v}{\partial x} \end{Bmatrix} = \begin{bmatrix} 0 & 0 & \dfrac{\partial N_1}{\partial x} & 0 & 0 & \dfrac{\partial N_2}{\partial x} & 0 & 0 & \dfrac{\partial N_3}{\partial x} & \cdots \\[2mm] 0 & 0 & \dfrac{\partial N_1}{\partial y} & 0 & 0 & \dfrac{\partial N_2}{\partial y} & 0 & 0 & \dfrac{\partial N_3}{\partial y} & \cdots \\[2mm] \dfrac{\partial N_1}{\partial y} & 0 & 0 & \dfrac{\partial N_2}{\partial y} & 0 & 0 & \dfrac{\partial N_3}{\partial y} & 0 & 0 & \cdots \\[2mm] \dfrac{\partial N_1}{\partial z} & 0 & 0 & \dfrac{\partial N_2}{\partial z} & 0 & 0 & \dfrac{\partial N_3}{\partial z} & 0 & 0 & \cdots \\[2mm] 0 & \dfrac{\partial N_1}{\partial z} & 0 & 0 & \dfrac{\partial N_2}{\partial z} & 0 & 0 & \dfrac{\partial N_3}{\partial z} & 0 & \cdots \\[2mm] 0 & \dfrac{\partial N_1}{\partial x} & 0 & 0 & \dfrac{\partial N_2}{\partial x} & 0 & 0 & \dfrac{\partial N_3}{\partial x} & 0 & \cdots \end{bmatrix} \begin{Bmatrix} u_1 \\ v_1 \\ w_1 \\ u_2 \\ v_2 \\ w_2 \\ u_3 \\ v_3 \\ w_3 \\ \vdots \end{Bmatrix}$$

$$= [G]\{\delta\} \tag{6.55}$$

The matrix $[G]$ can therefore be computed through eq. (5.87). The additional potential energy in an element due to centrifugal force can now be written as

$$U_e = 1/2\{\delta\}^T \int_{V_e} [G]^T [\Sigma^0][G]\,dV_e \{\delta\}$$

$$= 1/2\{\delta\}^T [K_R]\{\delta\} \tag{6.56}$$

This expression serves to define the added stiffness $[K_R]$ due to rotation.

To determine the natural frequencies of a body during rotation, it is merely necessary to compute the centrifugal stress field in order to obtain the centrifugal stiffness matrix $[K_R]$. This stiffness increment is then added to the original stiffness matrix $[K]$ in order to form the eigenvalue problem

$$[[K] + [K_R] - \omega^2[m]]\{\delta\} = 0 \tag{6.57}$$

The frequencies and mode shapes obtained from this eigenvalue problem will be those of the body as it is influenced by the centrifugal forces.

6.10 A CASE STUDY—GAS TURBINE BUCKET VIBRATIONS

A number of results will be given now for a bucket in the second stage of a gas turbine. This work was reported by Zirin[16] who used computer programs based on the methods discussed in this book to develop the analysis.

The bucket shown in Fig. 6.4 has a tip shroud and its shank is contoured, with integral cover plates. The bucket is broken into finite elements according to the scheme shown in Fig. 6.5. The geometrical detail of the bucket is well retained except at the tip shroud, where the tip seals are neglected. The bucket is fixed at the widest dovetail hook. The model consists of 293 nodes and 124 elements.

The natural frequencies were first computed for a static bucket with no rotational forces imposed. The resulting lowest four frequencies are given in Table 6.1. There, the calculated frequencies are compared with experimental measurements. Good agreement within two per cent is observed in all modes except

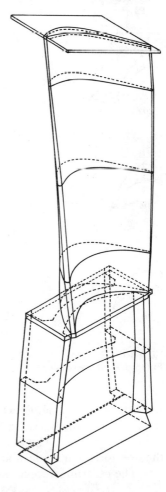

FIGURE 6.4
Second stage bucket of a gas turbine

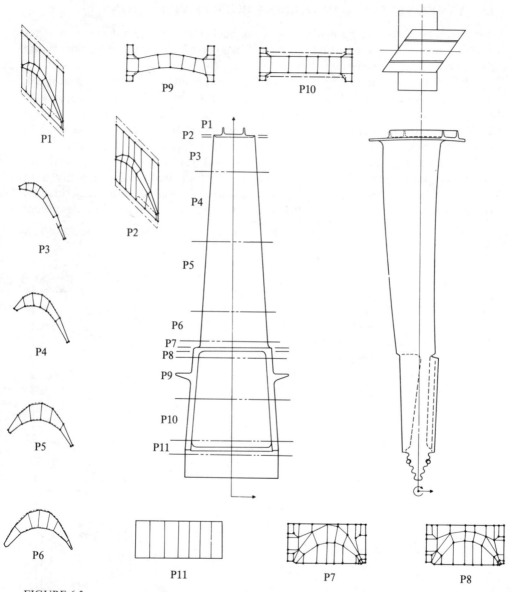

FIGURE 6.5
Finite element breakdown of a turbine bucket

the first torsional mode, where the error is 6 per cent. This error is attributed to the neglect of the tip seals. The mode shapes are shown in Fig. 6.6 relative to the undeformed shape. The first tangential bending mode, the first axial bending mode, the first torsional mode, and a higher complex mode are illustrated.

The centrifugal forces acting on a rotating bucket cause stresses which act to

Table 6.1   SOME NATURAL FREQUENCIES OF A GAS TURBINE BLADE—CALCULATED AND EXPERIMENTAL

	0 rpm			3600 rpm	
Frequency	Vibration test (Hz)	Analysis (Hz)	Percentage discrepancy	Analysis (Hz)	Percentage increase (over 0 rpm)
First tangential	194	190	2	225	16
First axial	465	464	0	486	4·5
First torsional	784	834	6	801	2·2
Complex	910	920	1	928	2

stiffen the bucket for the small increment in stress during vibration. A method to account for this stiffening effect has been discussed in the previous section. The centrifugal forces at the nodal points are $x_i m_i \Omega^2$, 0, $z_i m_i \Omega^2$. Here, x_i and z_i are the distances of the ith node from the center of rotation. It is assumed that the rotation occurs about the y-axis, and Ω is the angular velocity in rad/sec. The mass m_i is the accumulated mass at the ith node due to the contributions from the adjacent Gauss points.

This increase in the magnitude of the appropriate stiffness elements now allows computation of frequencies and mode shapes with the effect of rotation accounted for. The results of a calculation of the turbine blade natural frequencies for rotation of the turbine at 3600 rpm are also given in Table 6.1. It is observed that the effect of the centrifugal force on the natural frequencies is to raise them: the lowest mode of vibration is the most affected, showing a 16 per cent increase, while the highest complex mode shows only a two per cent increase.

6.11   SUMMARY

In this chapter, we have shown how the natural frequencies and mode shapes of a continuous structure can be obtained by using the finite element method. The concept of master and slave displacements was introduced as an aid to size reduction in the computation. The use of master displacements enables the calculation of a manageable number of modes of free vibration, for a structure that is modeled with fine detail (and consequently many degrees of freedom). The choice of the master deflections is a matter that is largely intuitive, although new methods are now being developed to allow a computer to logically make the same calculations by iterative techniques. Illustrations of the methods used, and their results, were made for some turbine blades whose natural frequencies and mode shapes were calculated.

There exist a number of computer programs that are able to make the calculations described in this chapter. We wish to refer to two of them here. Both are available to analysts who wish to use them. The first is known as NASTRAN,[17] and was developed with aerospace structures in mind. The program is not only able to compute the natural frequencies of continuous structures modeled with the sort of two- and three-dimensional finite elements described in chapter 5, but it is also

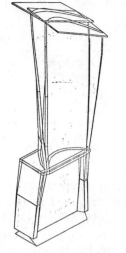

(a) First tangential bending mode

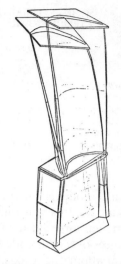

(b) First axial bending mode

(c) First torsional mode

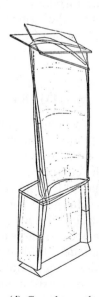

(d) Complex mode

FIGURE 6.6
First four modes of a gas turbine bucket

suitable for calculations involving framed structures and structures made of plates and stringers. Thus, it is a versatile and general purpose program of wide applicability. The second program is known as SAP.[12,18] It is particularly valuable for the study of three-dimensional solid structures, since it allows a wide choice of three-dimensional element types (e.g., 8-noded and 20-noded isoparametric elements). Versions of the SAP program additionally contain elements for plates, beams, shells, pipes and two-dimensional elastic continua, so that structures which can be modeled by combinations of these elements can be analyzed.

6.12 REFERENCES

1. O. C. Zienkiewicz, *The Finite Element Method in Engineering Science,* McGraw-Hill Book Company (UK) Ltd., London, 1971.
2. S. H. Crandall, *Engineering Analysis,* p. 122, McGraw-Hill Book Company, Inc., New York, 1956.
3. S. H. Crandall, *op. cit.,* pp. 118–122.
4. K. J. Bathe and E. L. Wilson, 'Solution methods for eigenvalue problems in structural mechanics,' *Int. J. Num. Methods in Engineering,* **6**, 213–226, 1973.
5. B. Irons, 'Eigenvalue economizers in vibration problems,' *J. Roy. Aero. Soc.,* **67**, 526, 1963.
6. B. Irons, 'Structural eigenvalue problems: elimination of unwanted variables,' *AIAA J.,* **3**, 961–962, 1965.
7. R. J. Guyan, 'Reduction of stiffness and mass matrices,' *AIAA J.,* **3**, 380, 1965.
8. R. Levy, 'Guyan reduction solutions recycled for improved accuracy,' *NASTRAN: User's Experience,* NASA TM X-2378, pp. 201–220, Sept. 1971.
9. R. G. Anderson, B. M. Irons, and O. C. Zienkiewicz, 'Vibration and stability of plates using finite elements,' *Int. J. Solids and Structures,* **4**, 1031–1055 (particularly 1047), 1968.
10. J. N. Ramsden and J. R. Stoker, 'Mass condensation: a semi-automatic method for reducing the size of vibration problems,' *Int. J. Num. Methods in Engineering,* **1**, 333–349, 1969.
11. A. Jennings, 'Mass condensation and simultaneous iteration for vibration problems,' *Int. J. Num. Methods in Engineering,* **6**, 543–552, 1973.
12. K. J. Bathe, E. L. Wilson, and F. E. Peterson, 'SAP IV—A structural analysis program for static and dynamic response of linear systems,' Report No. EERC 73-11, Earthquake Engineering Research Center, College of Engineering, University of California, Berkeley, California, 1973.
13. O. C. Zienkiewicz, *op. cit.,* pp. 427–431.
14. O. C. Zienkiewicz and Y. K. Cheung, *The Finite Element Method in Structural and Continuum Mechanics,* pp. 183–185, McGraw-Hill Book Company (UK) Ltd., London, 1967.
15. P. Trompette and M. Lalanne, 'Vibration analysis of rotating turbine blades,' ASME Paper No. 74-WA/DE-23, 1974.
16. R. Zirin, 'Static and dynamic analysis of a turbine bucket,' personal communication, July 1973.
17. T. G. Butler and D. Michel, 'NASTRAN—A summary of the functions and capabilities of the NASA structural analysis computer system,' NASA SP-260, 1971. (Further bibliographical information on NASTRAN is provided in this reference.)
18. E. L. Wilson, K. J. Bathe, F. E. Peterson, and H. H. Dovey, 'SAP—A structural analysis program for linear systems,' *Nuclear Engineering and Design,* **25**, 257–274, 1973.

7

A CASE STUDY—AIRCRAFT ENGINE FAN BLADES

7.1 INTRODUCTION

In order to amplify some of the concepts discussed in previous chapters, we propose to apply them to a dynamic study of an aircraft engine fan blade.

A modern aircraft engine is shown diagrammatically in Fig. 7.1. It is known as a high bypass turbofan engine. Its principle of operation is as follows. Air is drawn into the inlet by a large fan. This air runs into two distinct areas of the engine. An inner core of air is compressed by the compressor, heated in a combustion chamber, and then passed through the turbine to provide power for the rotation of the fan. The bulk of the air, however, runs through the outer ring of the engine—the bypass—and provides thrust for the aircraft. Engines of this type consequently have rather large fans made of from 30 to 40 blades. Because of the size of these blades (about 2 to 4 ft in length), it has been an engineering goal to fabricate them from fibrous composite materials in order to achieve a high strength to weight ratio.

A design challenge of considerable importance from the point of view of safety is to create fan blades that can withstand the shock of impacts of objects ingested into the inlet during the operation of the engine. The ingestion of such foreign objects is a rare, but possible, event. Hail stones, bolts, sand, gravel, and birds can present a potential hazard, and the blades must survive the impact by such things. To design for such events, an analytical means for predicting the dynamic response

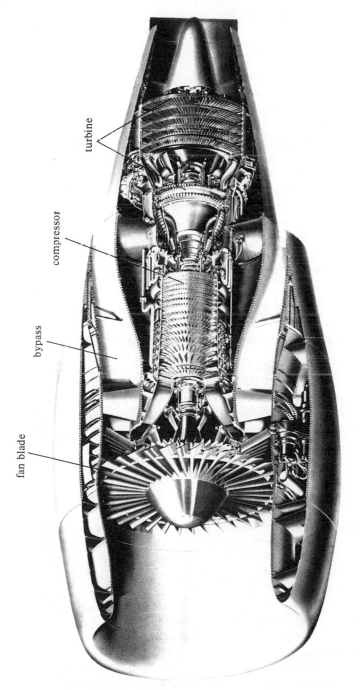

turbine

compressor

bypass

fan blade

FIGURE 7.1
A high bypass turbofan engine, showing the fan blades (courtesy of General Electric Company)

of the blade must be devised. In this chapter, we will develop one possible route that can be followed with respect to the analysis of such impacts. The blade will be modeled by a finite element technique in order to identify generalized coordinates for use in the dynamic analysis. The impact will be dynamically modeled by the component element method, and illustrative results of what can be achieved by this means will be presented.

The analysis of the vibrational modes of turbomachinery blades has a large literature. Investigations of particular interest to the present discussion have been carried out by Ahmad, Anderson and Zienkiewicz,[1,2] by Hofmeister and Evensen,[3] and by others,[4,5] who used a finite element analysis based on shell elements in studying the vibrations of turbine blades. Other fundamental investigations on vanes and blades have been carried out based on twisted cantilever plates, in order to gain insight into the effect of the vane twisting and aspect ratio. Such investigations have been made by Henry and Lalanne,[6] and by MacBain.[7] They used plate-like finite elements for their calculations. Three-dimensional isoparametric elements have been applied to rotating turbine blade vibration studies by Trompette and Lalanne.[8] Valuable experimental insights into the mode shapes of vanes and blades have been obtained through the technique of holographic interferometry,[7,9] and the resulting photographs of the mode shapes are effective in verifying the predictions of the finite element methods.

7.2   VIBRATIONAL ANALYSIS OF FAN BLADES

The appearance of the fan blade is sketched in Fig. 7.2. It is basically a thin plate-like structure which is twisted along its length. Its base is flared in order to provide for a secure grip by the disc, which restrains the blade against centrifugal forces during its rotation. This blade is fabricated of a fibrous composite made of high-strength graphite fibers in a matrix of epoxy resin. The airfoil section of the blade is made up of many layers of the graphite epoxy, each of which is 0·005 in. in thickness. The orientation of the layers are shown in Fig. 7.2, and are at $0°$ and $\pm 22°$ to the radial direction. In the dovetail region, the section is flared by means of short added layers, as also indicated in Fig. 7.2. In this region, the fiber orientations are $\pm 22°$ and $\pm 45°$.

In order to perform the dynamic impact analysis, the modal deformations of the blade are selected as generalized coordinates. To determine the mode shapes and natural frequencies of the blade, the finite element method is used. The blade is modeled by three-dimensional hexahedrons as shown in Fig. 7.3. Note that the finite element model consists of 133 elements and 320 nodes, since the eight-noded box with nine internal degrees of freedom is used here. A single hexahedron is used through the thickness of the blade. There are here 960 degrees of freedom in the finite element model before its reduction, or condensation, due to the introduction of master displacements.

The material behavior is of the anisotropic type, due to the presence of the fiber orientations. The elastic properties are different in the radial, chordwise, and

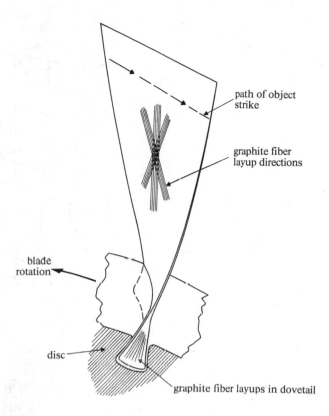

path of object
strike

graphite fiber
layup directions

blade
rotation

disc

graphite fiber layups in dovetail

FIGURE 7.2
Fan blade configuration and material characteristics

thickness directions. Appropriate elastic moduli, Poisson's ratios, and shear moduli were calculated in order to use the property matrix of eq. (5.102).

The master displacements in this example were chosen as shown by arrows in Fig. 7.3. There are 28 of them. According to the criteria discussed in section 6.8, we can expect reasonably accurate modal information for the first seven or eight modes. For this dynamic analysis, the first eight modes were used as generalized coordinates.

The centrifugal forces acting on the rotating fan blades cause stresses which act to stiffen the blade. A method to account for this stiffening effect is given by Zienkiewicz,[10] and was discussed in section 6.10. The centrifugal forces at the nodal points are $x_i m_i \Omega^2$, 0, $z_i m_i \Omega^2$. Here, x_i and z_i are the distances of the ith node from the center of rotation. It is assumed that the rotation occurs about the y-axis. The angular velocity in rad/sec is Ω. The mass m_i is the accumulated mass at the ith node due to the contributions from the adjacent Gauss points. This increase in the magnitude of the appropriate stiffness elements now allows computation of frequencies and mode shapes with the effect of rotation accounted for. For the present

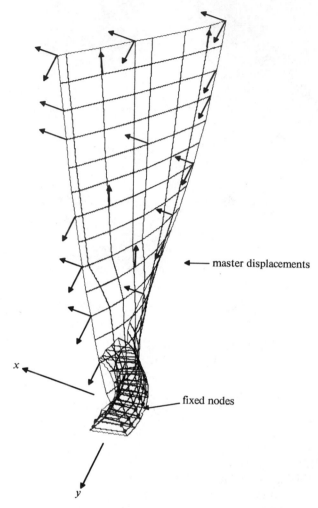

master displacements

x

fixed nodes

y

FIGURE 7.3
Master displacements chosen for the natural frequency calculation

application on the fan blade dynamics, the mode shapes and natural frequencies
were calculated for the case when the engine speed was 3600 rpm.

The blade is placed rather loosely into the dovetail slots in the disc, and seats
tightly during rotation due to the centrifugal forces. For purposes of calculation,
an estimate is made of the contact points between the blade and disc during
rotation. In this case, it was determined that only two of the lowest nodal lines
were in contact. As a result, these nodes were assumed to be fixed, as shown in
Fig. 7.3.

If more accurate detail were required in the specification of the boundary
conditions to account for the elastic compliance of the disc against the blade, then

boundary conditions for the contacting nodes could be modeled by springs in the following manner.

Let (u_A, v_A, w_A) and (u_B, v_B, w_B) be the displacements at two nodes A and B, which are to be connected by a spring k_s. If the direction cosines of the line of action of the spring are (l_s, m_s, n_s), then the relative displacement of the two nodes along that line is

$$\delta_s = (u_A - u_B)l_s + (v_A - v_B)m_s + (w_A - w_B)n_s \tag{7.1}$$

The force developed in the spring is

$$F_s = k_s \delta_s \tag{7.2}$$

and it opposes the displacement. To overcome the force, there are required external forces in the x-, y-, z-directions, whose magnitudes at nodes A and B are

$$
\left.
\begin{aligned}
f_{sxA} &= l_s F_s \\
&= (l_s^2 u_A + l_s m_s v_A + l_s n_s w_A - l_s^2 u_B - l_s m_s v_B - l_s n_s w_B)k_s \\
f_{syA} &= (l_s m_s u_A + m_s^2 v_A + m_s n_s w_A - l_s m_s u_B - m_s^2 v_B - m_s n_s w_B)k_s \\
&\vdots \\
f_{szB} &= (-l_s n_s u_A - m_s n_s v_A - n_s^2 w_A + l_s n_s u_B + m_s n_s v_B + n_s^2 w_B)k_s
\end{aligned}
\right\}
\tag{7.3}
$$

Thus the matrix

$$k_s
\begin{bmatrix}
l_s^2 & l_s m_s & l_s n_s \\
l_s m_s & m_s^2 & m_s n_s \\
l_s n_s & m_s n_s & n_s^2
\end{bmatrix}
\tag{7.4}$$

must be added to the stiffness matrix at node A in the node A equations, and at node B in the node B equations, while it is subtracted from the stiffness matrix at node B in the node A equations, and at node A in the node B equations. Now, to model the situation where the compliance of the disc adjacent to the boundary node A is to be accounted for, it is necessary only to add the matrix at the boundary node. The choice of the stiffness k_s will depend, of course, on the disc–blade compliance, and the direction cosines will be determined from the direction perpendicular to the contact plane. Thus, it is seen that fairly complex boundary conditions can be modeled in a rather simple manner, which is consistent with the general scheme of the finite element method.

The first eight mode shapes of the blade are shown in Fig. 7.4. They were obtained by using the master displacements shown in Fig. 7.3. In the illustrations of the mode shapes, only one face of the blade is actually drawn, to avoid a confusion of lines. The outline of the undeformed shape of the blade is shown, in order to show the deformation in each mode. The modes are all normalized so that the modal mass is unity. We can identify several of the simpler modes: the first bending mode, the first torsional mode, a chordwise bending mode, modes in which the blade corner movements predominate, and, at higher frequencies, plate-like modes showing deflections in both radial and chordwise directions.

undeformed
outline

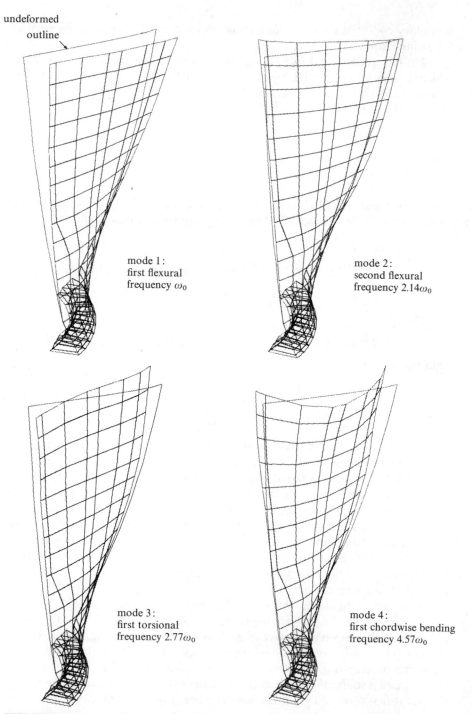

mode 1:
first flexural
frequency ω_0

mode 2:
second flexural
frequency $2.14\omega_0$

mode 3:
first torsional
frequency $2.77\omega_0$

mode 4:
first chordwise bending
frequency $4.57\omega_0$

FIGURE 7.4
Some mode shapes of the fan blade. (*Note:* the order and magnitude of the blade frequencies will vary with blade design.)

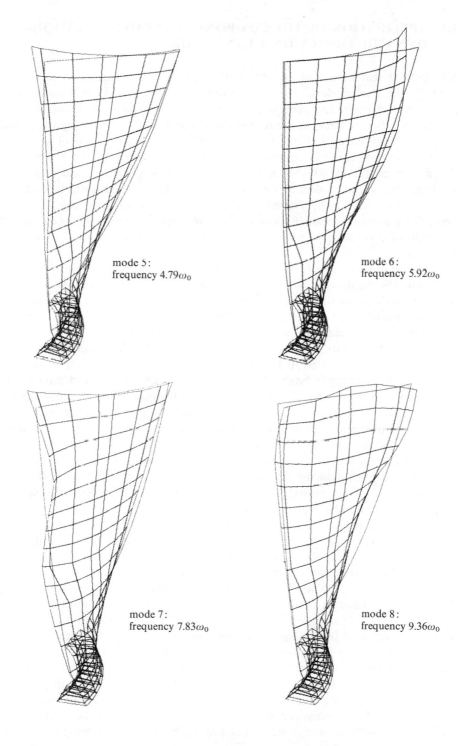

mode 5:
frequency $4.79\omega_0$

mode 6:
frequency $5.92\omega_0$

mode 7:
frequency $7.83\omega_0$

mode 8:
frequency $9.36\omega_0$

7.3 APPLICATION OF THE COMPONENT ELEMENT METHOD— DYNAMIC IMPACT ON A FAN BLADE

In studying the impact of an object on a blade, a determination must first be made of the level of detail that is required in the analysis. In this study, we have set out to determine the general displacement of the blade, with the objective of computing also the stress at the root of the blade, which is caused by the impacting forces. A knowledge of these stresses will allow a determination of what impact conditions could cause complete fracture of the blade at the root region. A more detailed investigation, which is not attempted here, might have asked what stresses were induced in the blade near the impact region. In that case, local impact damage criteria could have been studied in addition. A study of local conditions would necessitate a more detailed modeling of the impact region, a larger number of generalized coordinates and a finer time and space representation of the blade's behavior.

The manner in which the impact itself occurs and is modeled deserves close attention. Physically, the incoming object has a known velocity relative to the blade. Thus, its path across the blade can be determined from the point at which it strikes near the leading edge, to the point at which it leaves the blade at the trailing edge. The trajectory that was used in this study is shown in Fig. 7.2.

The object and its means of interaction with the blade is modeled as a spring–mass system, as illustrated in Fig. 7.5(a). This concept was used in section 3.17 to study a mass striking a beam. The magnitude of the spring and damper are so chosen that the incoming object does not vibrate relative to the blade during the contact phase. In this case, the spring is chosen so as to simulate the elastic compliance of the incoming object, while the damping is chosen so as to induce nearly critical damping in this system.

In the event that it is desired to model a more complex object, such that the object may break up during the impact, the scheme outlined in Fig. 7.5(b) might be used. There, the three masses could be allowed to separate when the displacement between them exceeded a given amount.

The impact response is now easily calculated by applying the component element method as it has been described in section 3.16. In section 3.17, the method was applied to the vibration of a beam struck by a falling mass. In that instance, the coupling ratios are constant in time. In the present instance, however, since the object striking the fan blade sweeps across the blade during the impact, the coupling ratios must be allowed to change in time. Such time-varying coupling ratios were also encountered in section 4.5, where the case of a mass traversing a flexible beam was discussed, and in section 4.6, on the air cushion vehicle dynamics. For this study, therefore, the computer program described in section 3.16 was modified so that the coupling ratios could appropriately change as the impact object swept across the blade. The generalized coordinates used in the calculation were the eight modal deflections of the blade and the component of the impacting object's motion normal to the blade. The generalized forces acting on the blade are computed by using the modal deflections at the traversing impact point as coupling ratios. Note that

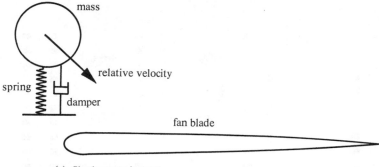

(a) Single mass impact

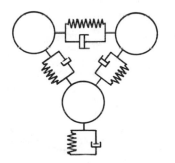

(b) More complex impact model

FIGURE 7.5
Impact models

the only data necessary as far as displacements are concerned are these nodal displacements along the path of impact. The generalized masses are the impacting mass, and those computed from the modal deformations:

$$m_n = \int_v \rho(u_n^2 + v_n^2 + w_n^2)\, \mathrm{d}V \qquad (7.5)$$

The resulting physical deformations of the blade are calculated from a knowledge of the time variation of modal generalized coordinates. Thus, to obtain the deflection of a given node in the blade, in a given direction, one sums the products of the modal displacements at that point in each mode and the generalized coordinate corresponding to each mode.

The deformations of the blade at each point in time can be plotted automatically by means of peripheral computer plotters. A sequence of displacement plots can be joined together in order to create an animated effect. A Stromberg–Carlson 4020 plotter was used to make a movie of the blade response. Part of the sequence is shown in Fig. 7.6, where the initial frames of the moving picture are illustrated. The small dot on the first 12 frames is the impacting object during its contact period. Subsequent frames show the blade in free dynamic oscillation.

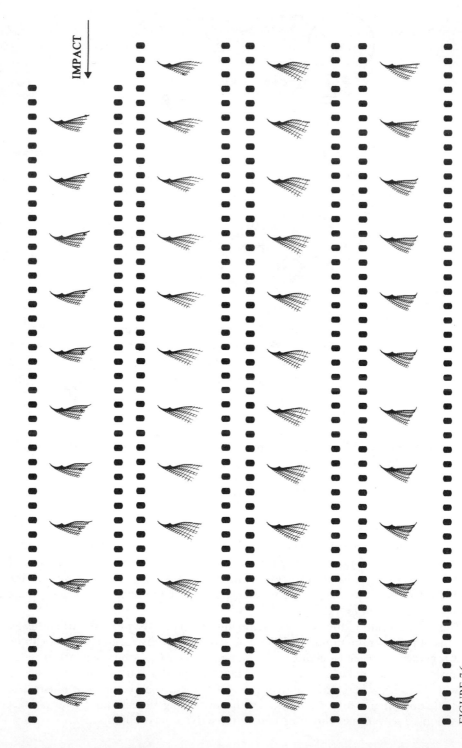

FIGURE 7.6
Movie sequence of an impact by a 4 oz object (time interval between frames 0·001 sec)

The dynamic stresses in the blade can be computed in two ways. The first method is to determine the normalized stress distribution in each mode. These stresses are then summed after appropriate multiplication by the modal generalized coordinate. The second method involves the calculation of the inertia forces during the response, and the subsequent solution of sequential static deformations under these inertia loads. The latter method is much more accurate for stresses, but is also more time-consuming. The total stress in the blade is the sum of the dynamic stress and the steady centrifugal stress.

7.4 SUMMARY

This chapter has presented a description of how a typical structural dynamic analysis is made. We have discussed the use of the finite element method to obtain the natural frequencies and mode shapes of the structure. Master displacements were selected that appeared appropriate to the desired degrees of freedom. The dynamic impact of the fan blade was modeled by using the mode shapes as generalized coordinates. The path along which the impacting object moved across the blade determined the variation of the coupling ratios between the object and the modal displacements, which are necessary for the calculation of the generalized force on the blade during each time step. A modified version of the computer program described in section 3.16 was used to determine the dynamic response. The dynamic response in the form of displacements was plotted in a useful graphical form to help the analyst understand how the blade behaved during the impact. The process described is typical of a dynamic response calculation. In a design situation, of course, an iterative procedure would be followed until a satisfactory design is achieved.

7.5 REFERENCES

1. R. Ahmad, R. Anderson, and O. C. Zienkiewicz, 'Vibration of thick curved shells, with particular reference to turbine blades,' *J. Strain Analysis,* **5**, 200–206, 1970.
2. O. C. Zienkiewicz, *The Finite Element Method in Engineering Science,* pp. 354–359, McGraw-Hill Book Company (UK) Ltd., London, 1971.
3. L. D. Hofmeister and D. A. Evensen, 'Vibration problems using isoparametric shell elements,' *Int. J. Num. Meth. in Engng.,* **5**, 142–145, 1972.
4. A. W. Filstrup, 'Finite element analysis of a gas turbine blade,' ASME Paper No. 74-WA/GT-11, 1974.
5. J. W. Allen and L. B. Erickson, 'NASTRAN analysis of a turbine blade and comparison with test and field data,' ASME Paper No. 75-GT-77, 1975.
6. R. Henry and M. Lalanne, ·'Vibration analysis of rotating compressor blades,' *J. Engng. for Industry, Trans. ASME,* **96**, Ser. B, 1028–1035, 1974.
7. J. C. MacBain, 'Vibratory behavior of twisted cantilever plates.' *J. of Aircraft.* **12**, 343–349, 1974.
8. P. Trompette and M. Lalanne, 'Vibration analysis of rotating turbine blades,' ASME Paper No. 74-WA/DE-23, 1974.
9. J. P. Waters and H. G. Aas, 'Holographic analysis of turbine blades,' ASME Paper No. 71-GT-84, 1971.
10. O. C. Zienkiewicz, *op. cit.,* pp. 427–431.

8

RESPONSE OF BUILDINGS TO EARTHQUAKES

8.1   INTRODUCTION

In this chapter, we inspect some aspects of the response of buildings to earthquakes. Such seismic analyses are an important application of the ideas and tools of structural dynamics. A thorough study of all aspects of seismic analysis is beyond the scope of the chapter: an excellent treatise on this subject has been written by Newmark and Rosenblueth.[1] We shall restrict our attention to the aspects of analysis having to do with the applications of the component element method developed in previous chapters and the finite element methods of structural analysis. From the structural analyst's viewpoint, a building can be modeled as an assemblage of mass and stiffness components. For tall or flexible buildings, a dynamic analysis of the building as it is excited at its foundation by an earthquake is often sufficient to establish its dynamic behavior. However, in areas of high seismic activity, it has been found that stiffer structures are necessary to withstand seismic shocks. In that case, for heavy buildings, it has been found that the motion of the buildings tends to interact with the soil, so that the dynamic model has to account in some way for this interaction. For these reasons, very massive and stiff buildings such as nuclear power plants also would interact with the soil. Thus, the study of soil–structure interaction has received increasing attention in recent years.

In this chapter, we first show how earthquakes are characterized by design response spectra or by time history specifications of the excitation. A method of obtaining upper-bound response estimates directly from the response spectra is discussed. Structural idealizations, mainly drawn from the nuclear reactor engineering field, are discussed, and a number of methods of describing the interaction between the structure and the soil are illustrated.

8.2 EARTHQUAKE CHARACTERISTICS

From the point of view of a building designer, an earthquake is characterized by the movement of the ground at the building site. The acceleration, velocity, and displacement of the ground in the vertical and two perpendicular horizontal directions must be known, as well as the duration of the earthquake. However, this information is different for each earthquake. The duration generally depends not only on the distance of the site from the focus and the magnitude, but also upon the type of soil at the site. For further details, the reader should consult Newmark and Rosenblueth.[1] One record from the well-known earthquake that was recorded in 1940 at El Centro, California, is shown in Fig. 8.1. There, the north–south horizontal component is illustrated. The deduced velocities and displacements associated with the acceleration record are also shown. The duration of this earthquake was of the

FIGURE 8.1
El Centro earthquake of 1940, north–south component (after Newmark and Rosenblueth,[1] reproduced by permission of the Portland Cement Association, Skokie, Illinois)

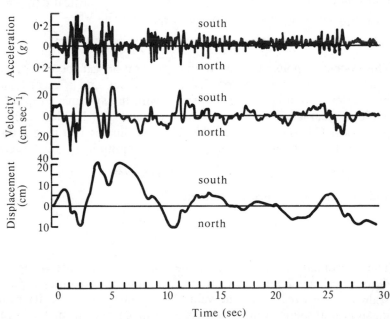

order of 30 sec and the magnitude of the maximum acceleration was about $0.33g$. According to Newmark and Rosenblueth, earthquakes such as this one, that are characterized by rather long, irregular motions, are associated with moderate distances from the earthquake focus and occur on firm ground. The focus of the El Centro earthquake was 30 miles away horizontally and was 15 miles deep. Other earthquakes may consist of a single violent shock. They are associated with firm ground and occur near the epicenter. Motions having several prevailing periodic components are associated with layered mantles of softer soil, which tend to filter the ground motions that propagate from the bedrock.

The wide range of possible ground motions makes it impractical for the designer to use the time history of a single earthquake to make predictions as to what will occur to the building. Rather, he is forced to extract the pertinent information on what is known about earthquakes in general, and to apply that knowledge to obtain a conservative design for the building site under question. A method that yields a conservative design for any site has been developed over a number of years. It is based on the idea of using design response spectra to describe earthquake characteristics in a statistical manner. The response spectrum essentially describes the frequency content of a given time history of excitation.

8.3   GENERATION OF RESPONSE SPECTRA FROM A GIVEN TIME HISTORY

Suppose that we apply a recorded time history of acceleration as a ground motion stimulus to a single mass–spring–damper system. Let the system be described such that it has an undamped natural frequency ω and a critical damping factor ζ. For a mass–spring–damper system we can therefore write

$$\omega = \sqrt{k/m} \qquad \text{and} \qquad \zeta = c/c_c = c/2m\omega$$

The system response to the ground acceleration can be computed (by using the computer program of section 2.4, for example), and a record made of how the maximum absolute displacement varies with the system's natural frequency ω and damping characteristic ζ. The composite result can be plotted on a special graph paper which is shown in Fig. 8.2. The coordinates in Fig. 8.2 take advantage of the fact that in simple harmonic motion of amplitude d_0, the deflection is given by

$$d = d_0 \sin \omega t$$

and the velocity and acceleration are

$$v = \omega d_0 \cos \omega t$$

$$a = -\omega^2 d_0 \sin \omega t$$

The coordinates in Fig. 8.2 plot the maximum absolute computed displacement d_M on the lines sloping upward at 45° against a horizontal logarithmic frequency scale. On the same figure, the value of ωd_M (a pseudo velocity nearly equal to the maximum absolute computed velocity) can be read from the logarithmic vertical

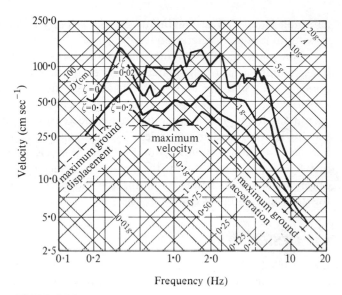

FIGURE 8.2
Response spectra for single mass elastic systems, 1940 El Centro
earthquake, N–S component (after Newmark and Rosenblueth,[1]
reproduced by permission of the Portland Cement Association,
Skokie, Illinois)

scale. The value of $\omega^2 d_M$ (a pseudo acceleration nearly equal to the maximum
absolute computed acceleration) can be read on the lines sloping downward at
$-45°$.

The motion of the elastic system under consideration is described by the
equation

$$m(\ddot{y} + \ddot{x}) + c\dot{y} + ky = 0 \tag{8.1}$$

where x is the ground displacement and y is the relative displacement between the
ground and the mass m. Solving eq. (8.1) for the pseudo acceleration $\omega^2 y$ gives

$$\omega^2 y = \left(\frac{k}{m}\right)y = -(\ddot{y} + \ddot{x}) - \left(\frac{c}{m}\right)\dot{y}$$

The actual acceleration of the mass m is $(\ddot{y} + \ddot{x})$. Thus, when the damping c is small,
the pseudo acceleration is the negative of the actual acceleration.

Equation (8.1) may be rewritten in terms of the natural frequency ω and the
damping parameter ζ as

$$\ddot{y} + 2\zeta\omega\dot{y} + \omega^2 y = -Ag \tag{8.2}$$

where A is the ground acceleration in gs and g is the acceleration of gravity.
Equation (8.2) is solved for a given time history of the ground acceleration by
using, for example, the computer program described in section 2.4. Usually, the
maximum relative displacement y_{max} is plotted for many frequencies ω and a given

damping factor ζ. The maximum pseudo velocity is then given by ωy_{max}, and the maximum pseudo acceleration is $\omega^2 y_{\text{max}}/g$.

Newmark and Rosenblueth[1] show results of carrying out the solutions for the relative motions when the system is excited by the El Centro earthquake N–S component. They plot the maxima for each frequency in a series of response spectra, as shown in Fig. 8.2. We observe that the effect of damping is strong, and that even a small amount of damping greatly decreases the responses. The shape of the spectra is controlled more or less by the maximum ground accelerations at high frequency, by the maximum ground velocity at medium range frequencies in the order of 0·5 Hz to 2 Hz, and by the ground displacement at lower frequencies. This observation follows from comparing the shape of the spectra with the envelopes of the ground motions as shown in Fig. 8.2.

8.4   DESIGN RESPONSE SPECTRA

Each earthquake produces its own characteristic spectra. Many earthquakes have been recorded in some detail and it appears that their spectra are often of the same general shape as that obtained for the El Centro earthquake. A compilation of many earthquake records observed in the United States has been made by Newmark, Blume and Kapur.[2] They normalized these records to all have a $1g$ maximum acceleration. They generated a series of spectra for various values of the damping parameter ζ such that the points of any given earthquake spectrum would fall within the bounds of the design spectra. Their recommended design response spectra for horizontal earthquake components are shown in Fig. 8.3(a). The design response spectra for vertical components is similar and is shown in Fig. 8.3(b) for two per cent critical damping. We repeat that the earthquakes are scaled such that the maximum ground acceleration has a value of $1g$, which is the asymptotic value of all the response spectra at high frequencies.

These spectra are useful because they embrace the spectra of many selected observed earthquakes. Therefore, a designer of a building can safely select these design response spectra as inputs that describe the likely frequency distribution in excitations of the ground at any site in the United States. He can be sure that earthquakes at a specific site will have spectra that fall within these bounds. He must select a maximum g-level for the earthquake by other means.

In order to obtain the design spectra appropriate to the building site, it is only necessary to make an estimate of the maximum acceleration magnitude which the site could experience. In general, this is done through a consideration of the geological features of the site and the seismicity of the region. The resulting maximum acceleration estimate is used to obtain site design response spectra akin to Fig. 8.3 by scaling the spectra of Fig. 8.3 in the appropriate manner.

8.5   UPPER-BOUND RESPONSE OF LINEAR SYSTEMS

An estimate of the maximum response of a linear system to the earthquake design response spectrum can be made by taking advantage of the fact that, for certain

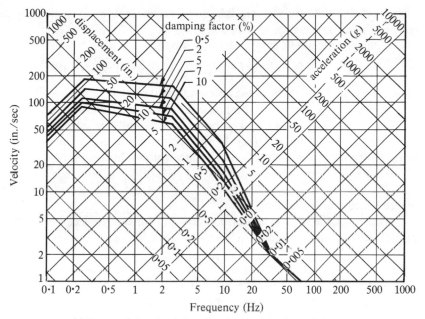

(a) Spectra for various damping factors − horizontal components

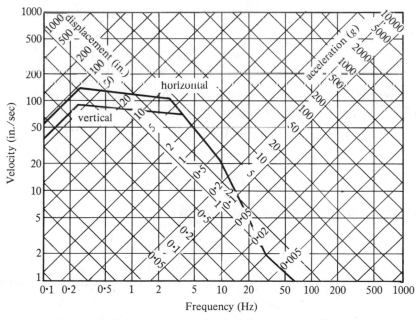

(b) Relation between horizontal and vertical spectra at 2% damping

FIGURE 8.3
Design response spectra (after Newmark, Blume, and Kapur,[2] reproduced from *J. Power Division, Proc. ASCE,* **99,** 1973, American Society of Civil Engineers, with permission)

damping distributions, the modal equations of motion in such a system are un-coupled, as was shown in section 3.4.

Consider the linear system described by the equation

$$[m]\{\ddot{\delta}\} + [c]\{\dot{\delta}\} + [k]\{\delta\} = -[m]\{D\}a(t) \tag{8.3}$$

Here, $a(t)$ represents the ground acceleration as a time history, while $\{D\}$ is a vector whose elements are unity in the positions where a displacement is in the direction of the ground motion, and zero otherwise. We shall consider only the cases where damping is proportional either to the mass matrix or to the stiffness matrix, or both, so that the modal equations are uncoupled. Thus, we set

$$[c] = \alpha[m] + \beta[k] \tag{8.4}$$

where α and β are constants.

The response $\{\delta\}$ is expressed as a sum of the responses in the several normal modes of vibration so that

$$\{\delta\} = [\delta_{nk}]\{A(t)\} \tag{8.5}$$

The matrix $[\delta_{nk}]$ represents the normal modes—the subscript n represents the nth mode while the subscript k represents the kth mass. Substitution into eq. (8.3) yields

$$[m][\delta_{nk}]\{\ddot{A}\} + (\alpha[m] + \beta[k])[\delta_{nk}]\{\dot{A}\} + [k][\delta_{nk}]\{A\} = -[m]\{D\}a(t) \tag{8.6}$$

Now we recall that the orthogonality relations for normal modes (section 3.4) demand that

$$\left. \begin{array}{l} [\delta_{nk}]^T[m][\delta_{nk}] = [M_n] \\ [\delta_{nk}]^T[k][\delta_{nk}] = [M_n\omega_n^2] \end{array} \right\} \tag{8.7}$$

where $[M_n]$ is the modal mass matrix. As a result, after premultiplication by $[\delta_{nk}]^T$, eq. (8.6) becomes uncoupled in every mode:

$$\{\ddot{A}\} + [2\zeta_n\omega_n]\{\dot{A}\} + [\omega_n^2]\{A\} = -[M_n]^{-1}[\delta_{nk}]^T[m]\{D\}a(t) \tag{8.8}$$

The equation can be written out for each nth mode and results in

$$\ddot{A}_n + 2\zeta_n\omega_n\dot{A}_n + \omega_n^2 A_n = P_n a(t) \tag{8.9}$$

Here, ζ_n is the damping constant, given by

$$\zeta_n = \frac{1}{2}\left(\frac{\alpha}{\omega_n} + \beta\omega_n\right) \tag{8.10}$$

and P_n is the nth modal participation factor. The participation factors are given by

$$\{P\} = -[M_n]^{-1}[\delta_{nk}]^T[m]\{D\} \tag{8.11}$$

and they determine to what extent each mode participates in the final response.

The uncoupled equation now represents a single degree-of-freedom system of frequency ω_n and damping factor ζ_n, acted on by the earthquake excitation $P_n a(t)$.

With this understanding, the site design response spectra can be used to obtain the maximum response of this mode. The resulting maximum response is scaled from the site design response spectra by multiplying it by the modal participation factor P_n. Let this maximum response be denoted by S_n. The response of the system at each point is given by

$$[S_{nk}] = [\delta_{nk}][S_n]_D \tag{8.12}$$

where $[S_n]_D$ is a diagonal matrix of maximum responses and $[S_{nk}]$ is the response of the kth mass in the nth mode.

An upper bound of the response can be found in a number of ways, as described, for example, by Stoykovitch.[3] Essentially, one adds the responses of each mode in a given fashion. If the frequencies are closely spaced, the responses are assumed to be in phase, and are added directly. On the other hand, if the modal frequencies differ by more than about 10 per cent, it would be overly conservative to add them directly on the assumption that they are in phase. Often, they are then added as the square root of the sum of the squares of the responses.

For many purposes, the upper-bound responses found by a process such as the one described here are adequate for design purposes. Occasionally, it is desirable to know more precisely what the actual response will be. A closer estimate of the response can be obtained by a time history process, wherein the structure is excited by an acceleration time history which has the design response spectrum as its own spectrum. To derive such a time history, it is necessary to operate on the design response spectrum in a manner described in the following section.

8.6   GENERATION OF ARTIFICIAL TIME HISTORIES FROM GIVEN DESIGN RESPONSE SPECTRA

A time history analysis of the response of a structure can be made by applying the component element method that was evolved in chapter 3. The ground excitation must first be obtained, however. Whereas it is possible to use a particular earthquake record, it is feasible also to construct an artificial earthquake record whose spectrum envelops the site design response spectrum.

A number of methods have been devised for generating the time history of an artificial earthquake whose spectra are replicas of the design response spectra. Tsai[4] and Rizzo, Shaw, and Jarecki[5] have developed such methods. Essentially, they depend on the manipulation of the amplitude and phase of a Fourier representation of an existing accelerogram trace, for example of the El Centro N–S record. Using this record as input, they obtained a trial response spectrum. For frequencies where the trial response spectrum was higher than desired, a 'filtering' action is applied to the initial time history. For frequencies where the trial response was lower than desired, a damped sinusoid of that frequency was added to the initial time history. The modified time history was then used to compute a second response spectrum. The process of modification of time history and computation of new response spectra was repeated until a satisfactory agreement between computed and desired response spectra was achieved. The last time history used was taken as the desired time history.

We wish to describe here another simple method[6] which differs from the above in certain important respects. First, to insure a full coverage of the frequency range, a large number of closely spaced frequencies are chosen such that they cover the frequency range of the desired response spectrum. Adjacent frequencies satisfy the condition

$$\frac{\Delta F}{F} < 2\left(\frac{c}{c_c}\right) \tag{8.13}$$

where ΔF is the frequency separation at frequency F and (c/c_c) is the damping ratio. The choice is such that the 'half-power' points (on a resonance curve like Fig. 1.2, the half-power points are where the amplitude is $1/\sqrt{2}$ of the maximum amplitude) of adjacent frequencies overlap. Since the desired time history is considered to have inputs at each of these frequencies, it is apparent that it will be 'rich' over the full range of frequency, and no region of the desired frequency range will not be covered.

To take into account the time–severity characteristics of actual earthquakes, an envelope curve $F(t)$ is used. $F(t)$, for example, could be the positive envelope to the El Centro N–S earthquake. In that case, $F(t)$ is at all times positive and has a value of about $0.3g$ at 2.5 sec, a value of about $0.1g$ at 4.0 sec, a value of about $0.2g$ at 5 sec, etc., with a final value of about $0.05g$ at 30 sec.

The acceleration time history of the earthquake is expressed by

$$F(t) \sum_{i=1}^{N} (-1)^i A_i \sin(2\pi F_i t) \tag{8.14}$$

where the coefficients A_i are to be determined, the frequencies F_i have already been specified to have a spacing such that successive frequencies have overlapping half-power points, and N is the total number of frequencies required to cover the frequency range. The values of the A_is are assumed to vary linearly in magnitude between frequencies at which the desired response spectrum is specified. The use of the factor $(-1)^i$ in eq. (8.14) improves the solution.

The final step in the procedure is to determine the coefficients A_i. These are found using an iteration process. Initially, the magnitudes of the A_is are taken to be proportional to the corresponding response spectrum g-values at frequencies F_i on the desired response spectrum curve. Using the initial trial time history, a response spectrum curve is computed and compared with the desired response spectrum curve. For the second trial time history, the new magnitudes of the A_is are obtained by multiplying the initial values of the A_is by the ratio of the desired response spectrum at frequency F_i to that computed in the first iteration. This iteration process is continued until the computed response spectrum is close enough to the desired one.

The computation of the response spectrum curve involves the determination of the maximum response of many single degree-of-freedom systems having frequencies covering the desired frequency range. This computation is done using a numerical integration procedure like the one described in section 2.4. In particular, a solution is obtained for the equation of motion of a single degree-of-freedom system:

$$m\frac{d^2(y + x)}{dt^2} + c\frac{dy}{dt} + ky = 0 \qquad (8.15)$$

where x is the ground motion, y is the relative displacement, t is time, c is the damping, m is mass, and k is the stiffness. The maximum absolute value of (ky/gm) is taken as the acceleration on the response spectrum curve corresponding to a frequency $(1/2\pi)\sqrt{k/m}$.

As an example of the results obtainable by this method, we illustrate in Fig. 8.4 the convergence to the design response spectrum of Fig. 8.3(b) for two per cent critical damping. The first three iterations are shown in Fig. 8.4. The time history that is derived is shown in Fig. 8.5. It was obtained using the El Centro N–S envelope described above. Naturally, it is not unique. There are many other time histories (actually an infinite number) whose spectra could approximate the design response spectra. However, by using the envelope curve and the duration of a realistic earthquake, a time history is generated that has characteristics that are in reasonable

FIGURE 8.4
Response spectrum curves for two per cent critical damping

—————— desired spectrum

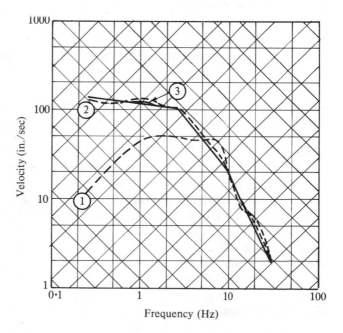

agreement with physical reality. Furthermore, it is a conservative time history in the sense that all frequency ranges are present. The response of a building can now be calculated by applying the earthquake record of Fig. 8.5. The component element method can be used for this purpose.

Very often, it is necessary to attach a piece of equipment to the building, perhaps on one of the upper floors. In that case, it is desirable to specify a design response spectrum which the equipment must survive. The equipment design response spectrum (also known as the floor response spectrum) can be derived in the following manner. Suppose that having made a time history analysis of the building, we record the acceleration time history at the equipment location. This time history is now used to construct a new spectrum, shown in Fig. 8.6(b). This particular example was derived by Stoykovich[3] for a nuclear plant whose dynamic model is illustrated in Fig. 8.6(a). It will be observed that the response spectrum of Fig. 8.6(b) has three peaks at periods of 0·2, 0·4, and 0·6 sec. Because of the possible variability in the properties of the building (e.g., elastic modulus, damping, etc.), it is likely that the spectral estimates can, in reality, show some deviation from the predicted values. As a result, in order to obtain a conservative equipment design, the spectrum is smoothed as shown in Fig. 8.6(c) and the resonance peaks are widened somewhat. In the case illustrated here, Stoykovich found some sensitivity to the way in which the interaction between the soil and the building is modeled. The two peaks at periods of 0·4 and 0·6 sec are a result of a parametric variation of the soil–structure interaction model. The smoothed response spectrum for these peaks therefore embraces both of them, this being a conservative estimate of the position of the peak.

Variations in parameters describing the structural model may, as we have seen, have an effect on the resulting floor response spectra, from which equipment is to be designed. Liu, Child, and Nowotny[7] have suggested methods whereby uncertainties in structural models can be accounted for in the generation of floor response spectra. They specifically considered the effect of statistical distributions of the soil shear wave velocity, moments of inertia, elastic moduli and masses, as well as damping ratios in the soil and in the structural materials. They concluded that for quantities that significantly affect the natural frequencies of the structure such as soil shear wave velocity, elastic moduli, etc., the response spectra are smoothed by statistical distributions, in that the peaks become lower and valleys become higher.

The floor response spectrum that results from the operations described above can be used to derive a suitable time history of excitation by applying the ideas already discussed. These excitations, in turn, can be used to generate the response of a substructure or a piece of equipment mounted at that location. In carrying out this calculation, it is assumed that the mass of the substructure is small compared to the mass of the original primary structure, so that its movements do not substantially affect those of the primary structure. It should also be noted that the results of this calculation will result in movements that are larger than if an interaction were to occur. This is because the possible effect that the smaller mass may have as a dynamic vibration absorber is suppressed when no interaction is assumed to occur.

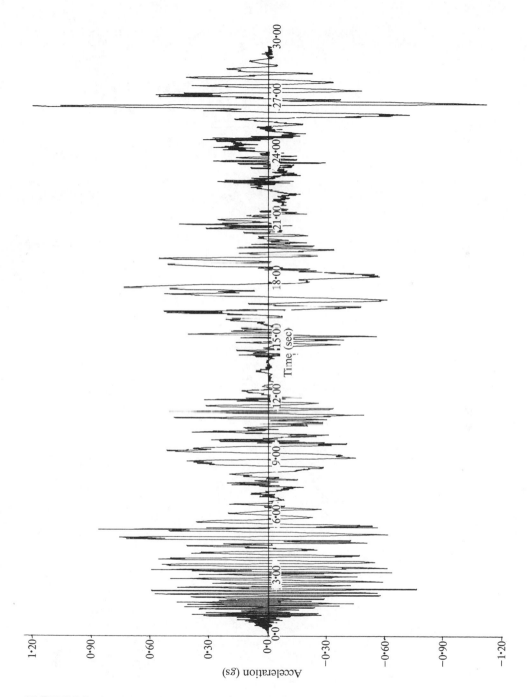

FIGURE 8.5
Time history for design response spectrum—1940 El Centro N–S envelope, at iteration 3

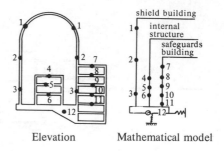

Elevation   Mathematical model

(a) Building model

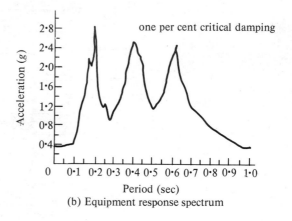

(b) Equipment response spectrum

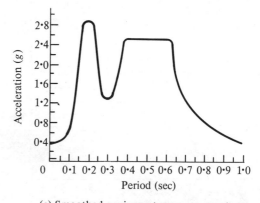

(c) Smoothed equipment response spectrum

FIGURE 8.6
Development of equipment response spectrum (after
Stoykovich,[3] reproduced from *Structural Design of
Nuclear Plant Facilities,* **1,** 1973, American Society of
Civil Engineers, with permission)

8.7   SOIL–STRUCTURE INTERACTION

The manner in which the structural foundation interacts with the soil on which it rests plays an important role in the structural response to earthquakes. In many analyses, the foundation is imagined to be attached to a rigid ground through a number of springs and dampers, as illustrated in a schematic fashion in Fig. 8.7. There, a rectangular foundation is restrained in a vertical and two horizontal directions by translational springs and dampers. In addition, because the foundation can also rock about its horizontal axes, and twist about the vertical axis, rotational springs and dampers are also provided. The appropriate spring and damper constants have been estimated through the work of many investigators. In general, estimates of the spring stiffnesses are made by calculating the behavior of a rigid circular or rectangular slab resting on an elastic soil of semi-infinite extent, as illustrated in Fig. 8.8. The damping constants have been estimated in such a manner as to obtain reasonable dynamic magnification factors of the foundation system at resonance. Additionally, it has been found that it is desirable to add a virtual mass to the foundation in order to account for the motion of the soil beneath it.

The calculation of the spring and damping constants rests, for the most part, on the fundamental work of Lamb,[8] Reissner,[9,10] and Bycroft,[11] who studied the vertical, torsional, and rocking vibrations of rigid circular discs resting on semi-infinite elastic foundations. From these basic solutions a number of investigators have developed spring and damping constants that are suitable over ranges of frequencies of interest in seismic analyses. Such results have been summarized by

FIGURE 8.7
Rocking, torsional, and translational springs and dampers acting on a foundation slab to simulate soil–structure interaction effects

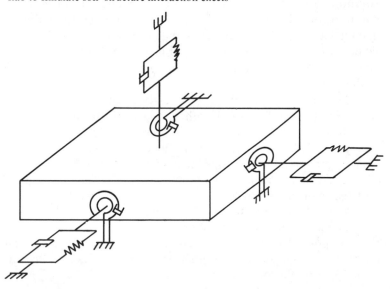

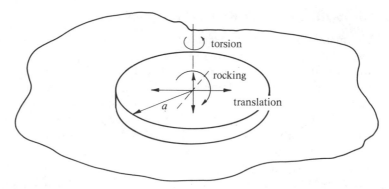

FIGURE 8.8
Rocking, torsional, and translational motions of a circular foundation on a three-dimensional semi-infinite elastic continuum

Richart, Hall, and Woods,[12] by Whitman,[13] and by Newmark and Rosenblueth.[14] We reproduce the data quoted by Newmark and Rosenblueth as Table 8.1.

Table 8.1 describes the appropriate springs and dampers for both circular slabs of radius a and rectangular slabs of side dimensions B and L. The slabs rest on an elastic soil of shear modulus μ, elastic modulus E, Poisson's ratio v, and mass density ρ. The rectangular slab has a moment of inertia I about the horizontal axis at the soil–slab interface, and a moment of inertia J about its vertical axis. The area of the slab is A. Some of the constants in the rectangular slab stiffness vary with aspect ratio B/L, and are shown in the charts below the table.

When a structural foundation is embedded in the soil, as is often the case, modifications to the spring and damper constants may be necessary. Whitman[13] describes a number of approaches to these modifications. Tajimi[15] has obtained solutions for a rigid circular cylinder embedded in an elastic half-space, and his solution could be used to construct similar estimates for spring constants. Similarly, Tajimi has constructed solutions that account for the finite thickness of the soil layer, so that situations where a soil layer rests on bedrock may be studied. Other effects studied recently include the slipping of the foundation and hysteretic damping of the soil.[16]

The numerous details such as embedment, soil layering, and damping have prompted the development of other models of soil–structure interaction. The finite element method has been used to analyze the effect of the soil around a structure. Examples of this approach are provided by Clough,[17] who discusses how a finite element model can be developed to account for the difference in soil layer properties. That particular model is axially symmetric. The layout of the finite element grid and the manner in which the grid is spaced to model the soil layering is shown in Fig. 8.9. Yamada[18] has studied a two-dimensional soil model for the case of a pier embedded in a soil. His model is shown in Fig. 8.10. The extent of the soil is an important question for models of this type, since the finite element model requires that boundary conditions be specified at points that are not infinitely far removed from the foundation.

Table 8.1 **STIFFNESSES, DAMPING CONSTANTS, AND VIRTUAL MASSES FOR SOIL–STRUCTURE INTERACTION** (after Newmark and Rosenblueth,[14] reproduced from *Fundamentals of Earthquake Engineering*,© 1971, Prentice-Hall Inc., Englewood Cliffs, New Jersey, 1971, by permission of the publisher)

| Degree of freedom | Stiffness | | Damping constant* | Height of soil prism for virtual mass (h) |
	Circular base	Rectangular base		
Vertical	$4\mu a/(1 - v)$	$Ec_S A^{1/2}/(1 - v^2)$	$5\cdot42(k\rho h^3)^{1/2}$	$0\cdot27 A^{1/2}$
Horizontal	$5\cdot8\pi\mu a(1 - v^2)(2 - D)^2$	$Ek_T A^{1/2}/(1 - v^2)$	$41\cdot1(k\rho h^3)^{1/2}$	$0\cdot05 A^{1/2}$
Rocking	$2\cdot7\mu a^3$ $(v = 0)$	$EIk_\phi/A^{1/2}(1 - v^2)$	$0\cdot97(k\rho h^5)^{1/2}$	$0\cdot35 A^{1/2}$
Torsion	$16\mu a^3/3$	$1\cdot5\,EJk_T/A^{1/2}(1 - v^2)$	$3\cdot76(k\rho h^5)^{1/2}$	$0\cdot25 A^{1/2}$

* $k = 4\mu a/(1 - v)$.

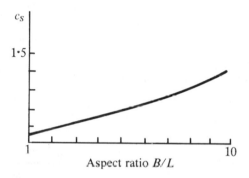

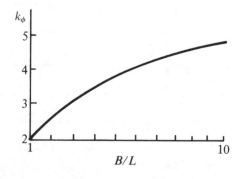

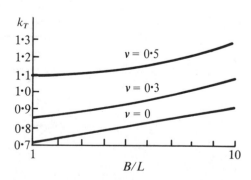

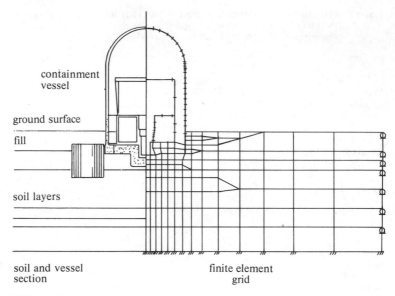

containment
vessel

ground surface

fill

soil layers

soil and vessel
section

finite element
grid

FIGURE 8.9
Soil model in axisymmetric finite elements (after Clough,[17] reproduced from *Recent Advances in Matrix Methods of Structural Analysis and Design,* 1970, © 1971 by the University of Alabama Press, with permission)

FIGURE 8.10
Soil–structure interaction model by finite elements (after Yamada,[18] reproduced from *Recent Advances in Matrix Methods of Structural Analysis and Design,* 1970 © 1971 by the University of Alabama Press, with permission)

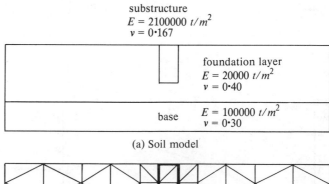

substructure
$E = 2100000 \ t/m^2$
$v = 0 \cdot 167$

foundation layer
$E = 20000 \ t/m^2$
$v = 0 \cdot 40$

base $E = 100000 \ t/m^2$
$v = 0 \cdot 30$

(a) Soil model

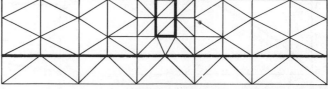

(b) Finite element layout

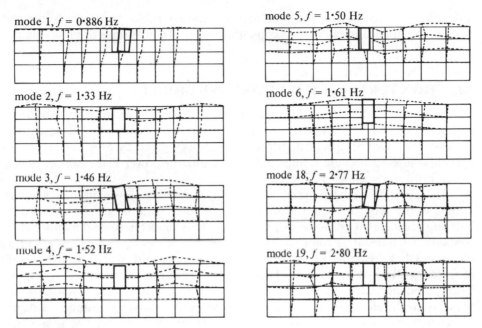

FIGURE 8.11
Vibrational modes in a soil containing an embedded structure (after Yamada,[18] reproduced from
Recent Advances in Matrix Methods of Structural Analysis and Design, 1970, © 1971 by the
University of Alabama Press, with permission)

Yamada[18] finds that, for this particular case, the placement of the boundary
at a distance of about three to five times the width of the pier appears to be
appropriate. Similarly, work by Lysmer and his associates[19] has shown that a
reasonable horizontal distance is $2.5D$ and a reasonable depth is $1.5D$, where D is
the diameter of the foundation. Some free vibration modes of a block of soil with
the embedded pier are shown in Fig. 8.11. There, it is seen that numerous modes
exist within a rather narrow range of frequencies.

Finite element soil models have been studied in many geometric configurations.
Two-dimensional models in plane strain (like that in Fig. 8.10) have been used by
Isenberg and Adham,[20] Chu, Agrawal, and Singh,[21] and Lysmer and his associates,[19]
who have also studied the axisymmetric geometry. Approaches to three-dimensional
models have been discussed by Lysmer, Seed, and their associates.[23]

Finally, it has been found[19] that the nonlinearity of the material properties
of the soil is of considerable importance. The shear modulus and the damping
both depend to a considerable extent on the strain experienced by the soil.[22] The
finite element method appears to be able to model nonlinearities, and Lysmer and
his associates[19,23] have developed a means whereby, through an iterative process
involving the solution of a series of linear problems, the nonlinearity in soil properties
can be partially accounted for.

From this short summary, it will be concluded that the state of knowledge in soil–structure interaction is in a state of flux, although rapid strides are being made through the application of models consisting either of soil springs or finite elements.

8.8 STRUCTURAL IDEALIZATIONS AND MODELS

The computer simulation of buildings requires that the structure be idealized in such a manner that its stiffness and mass distribution is adequately represented. A discussion of such an idealization process has been given by Stoykovich.[3] It is of particular interest because it illustrates how a nuclear power plant can be modeled structurally and what level of detail is feasible to use.

Stoykovich's model is developed in two stages, as shown in Fig. 8.12. The outline of the plant building is shown in Fig. 8.12(a). A finite element model of the building is generated as shown in Fig. 8.12(b). This model is three-dimensional and contains much detail. A count of the nodal points shows that there are about 330 nodes and that a dynamic system based on this model would have some 1000 degrees of freedom. Now, this is a fairly large number, and even with today's computers it would be difficult to carry out a time history response analysis with the model. Stoykovich therefore uses a condensation scheme to reduce the number of degrees of freedom in the model to a more manageable number. The scheme is similar to the one described in section 6.4 on eigenvalue economizer methods. The reduction in the number of degrees of freedom is made while preserving the distribution in stiffness of the original structure. The final dynamic model is shown in Fig. 8.12(c) and consists of 12 masses. In other analyses, it would be entirely feasible, if necessary, to use up to 50 or more masses in such a model.

The model in Fig. 8.12(c) is a commonly used one. The beam structures represent various portions of the plant. Since the shield building, safeguard building, and internal structures are essentially freestanding on the foundation, no coupling springs are put between the beam models. If the case arises where there is coupling, it is a simple matter to insert such coupling elements. The basic analysis does not change in that case. Such a situation has been considered by Muto and his associates,[24,25] whose model is shown in Fig. 8.13. They have made detailed calculations and experimental comparisons of the response of buildings to earthquake and vibrational excitations.

The stiffness and mass properties of the models are obtainable directly by treating the buildings as beams. The mass of the building is then distributed at the chosen nodal points in an approximate manner by lumping at the node the mass of the building adjacent to the node. The stiffnesses of the elements are then determined from beam stiffness formulas, such as those derived in sections 3.18 and 3.19, that contain the moments of inertia of the building cross-section. This simplified approach is adequate for a preliminary dynamic calculation.

Takemori and Hama[26] have studied the differences in the predictions of a number of models, which are illustrated in Fig. 8.14. In model 1, each portion of the building—the interior concrete structure, the steel containment vessel, and the annulus shield wall—is considered to be separately fixed to a rigid foundation,

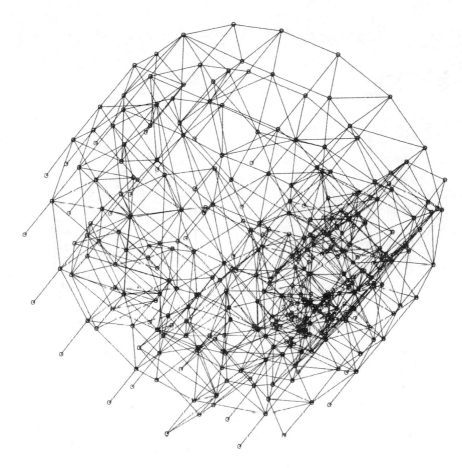

(b) Finite element model

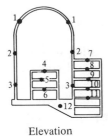

Elevation

(a) Reactor building

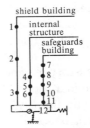

Mathematical model

(c) Condensed dynamic model

FIGURE 8.12
Structural idealizations of a nuclear power plant (after Stoykovich,[3] reproduced from *Structural Design of Nuclear Plant Facilities*, **1,** 1973, American Society of Civil Engineers, with permission)

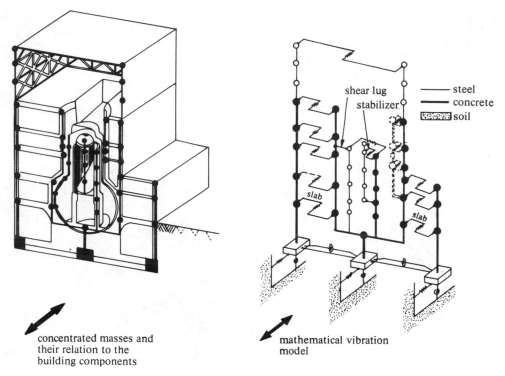

concentrated masses and
their relation to the
building components

mathematical vibration
model

FIGURE 8.13
Structural model of a nuclear plant, with coupling between structural components (after Muto and his
associates,[24] reproduced from *Proc. Second Int'l. Conf. on Structural Mechs. in Reactor Technology,* 1973,
with permission of the Commission of the European Communities)

where the ground excitation is assumed to act. In model 2, the buildings are set
on a rigid platform, which is coupled to the ground by a flexible structure. In
this case, the building components are coupled through the platform. In model 3,
soil–structure interaction is accounted for by the introduction of translational and
rotating springs of the type discussed in the previous section. Finally, in model 4,
a three-dimensional representation is used in which coupling is permitted between
the x- and y-directions of motion.

Dynamic analyses (which used the modal superposition method rather than the
time history methods discussed in this book) were made by Takemori and Hama[26]
on the four models by applying a number of earthquake records as time histories,
including the 1940 El Centro N–S record. They compare the natural frequencies
and responses of the models. Model 1 has the highest natural frequencies; then
come model 2, model 4, and model 3, in order of decreasing natural frequencies.
The progressive decrease in natural frequencies can be attributed to the decreasing
stiffness of the foundation support. In terms of structural response, they conclude
that model 3 is more conservative than the other models in that, in general, larger
stresses and responses are predicted by this model.

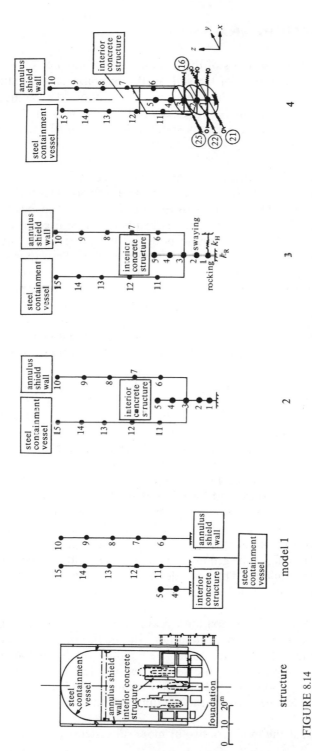

FIGURE 8.14
Various structural idealizations used in dynamic analysis of nuclear plant (after Takemori and Hama,[26] reproduced from *Proc. Second Int'l. Conf. on Structural Mechs. in Reactor Technology,* 1973, with permission of the Commission of the European Communities)

8.9   COMPONENT MODE ANALYSES

We have seen in previous sections of this chapter that a number of methods are available for modeling the effect of the soil on the response of a building. It has been found that ground spring idealizations based on solutions derived from a semi-infinite elastic half-space are suitable when the soil is very deep. If the building is situated on bedrock, then an analysis is suitable which assumes an excitation at the base of the building, with the base fixed. It is for the situations between these extremes, when the soil is only moderately deep, that finite element methods, or other means of accounting for this interaction, have to be developed.

In previous chapters, we have discussed a number of ways in which a dynamic system may be modeled. One method called for the use of the beam elements described in section 3.18. Such component elements could be used in conjunction with the beam models of the previous section. On the other hand, the building could also be modeled by using its modes of free vibration. This approach would be compatible with building models based on either the beam models, or on the more detailed finite element model. In either modeling approach, the response can be computed by applying the component element method. The behavior of the soil can be modeled by using the soil modes of vibration as component elements. In the case where we wish to study the dynamics of the combined building and soil system, the question arises as to which modes of vibration should be used as the dynamic components.

A number of observations have led to a rather specific choice of the type of modes that are to be used. First, the combined dynamic system of the soil and building often possesses a complex collection of numerous modes. Because of this circumstance, it is desirable to separate the building (or buildings) and the soil into two distinct and separate dynamical systems. Second, because the design of the building may be an iterative process, the building modes may change during the design evolution. Thus, for a specific site, the soil modes in the absence of the building can be calculated once and for all, and used to describe the soil system regardless of the building modes. Similarly, it may be desirable to use a given building design at different sites. In that case, the building modes can be determined once and for all to describe the building dynamics, while the soil modes change from site to site. Finally, the damping of the ground is often different, and may be substantially larger, than that of the building. Thus, one desires to assign different damping parameters (for example, α and β in eq. 8.4) to the building modes and to the soil modes.

Having concluded that two separate dynamical subsystems are desirable, the question remains as to what boundary conditions should be applied to these systems. A logical and convenient choice is to use, for the building, modes that are calculated from a fixed-base assumption. For the soil, a reasonable choice is to use those modes derived from a model like the one shown in Fig. 8.13, but in the absence of the building (the interface between the building and the soil being assumed free of loads).

Analyses which use the modes of two or more subsystems are known under the generic name of *component mode analyses*. Component mode analyses have been developed by Hurty[27,28] and others. The basic idea is that the full structure can be separated into a main structure (here, the soil) and components, or branch structures through a series of transformations that form the basis of the component mode method. Chu, Agrawal, and Singh[21] have applied a component mode method to the study of soil–structure interaction. They have used a method described by Benfield and Hruda[29] as a basis. Their analysis of how a building is seismically excited involves three basic steps. First, an artificial time history is generated which is compatible with the given site design response spectra. Second, an estimate is made of the motion of the bedrock that would be required to produce this time history at the soil surface. Finally, the bedrock motions are imposed as accelerations on the soil structure combination, to give a dynamic response in the building by means of a component mode method.

We wish to describe here a derivation of a somewhat different component mode method that is based on the use of the fixed-base modes of one or more buildings, and the modes of the soil subsystem.

The method of analysis is presented here in the context of a system consisting of one or more buildings on the ground. However, it is also equally valid for other systems which can be subdivided into one or more subsystems. Thus, the method can be used also to study the dynamics of packages, or of cabinets containing pieces of equipment subject to shock excitation. Consequently, in what follows, the word 'ground' refers to the primary system to which the excitation is applied, and the word 'building' refers to the secondary system which is excited through the connection at its base to the 'ground.'

The building subsystem

The component mode analysis to follow makes use of the fixed-base modes of the building and any auxiliary structures. The structural modal equations are uncoupled. Equations (8.16) to (8.23) which follow are given for a particular structure. Similar equations apply to additional structures. The geometric coordinates are x in the horizontal direction and z in the vertical direction.

We use the equation

$$M_1\ddot{y}_1 + 2\zeta_1\omega_1 M_1\dot{y}_1 + M_1\omega_1^2 y_1 = Q_1 \qquad (8.16a)$$

as the first modal equation of the structure, where

$$M_1 = \int \rho[(\xi^{(1)})^2 + (\gamma^{(1)})^2]\, d(\text{vol}), \text{ the modal mass}$$
$$\rho = \text{mass per unit volume}$$
$$\xi^{(1)} = \text{mode shape function in the first mode (fixed-base) in the}$$
$$x\text{-direction}$$
$$\gamma^{(1)} = \text{the same in the } z\text{-direction}$$
$$y_1 = \text{modal amplitude in the first mode}$$
$$2\zeta_1\omega_1 M_1 = \text{modal damping in mode 1}$$
$$\omega_1 = \text{natural frequency of mode 1 in rad/sec}$$
$$Q_1 = \text{generalized force in mode 1}$$

For the second structural mode, we use the equation

$$M_2\ddot{y}_2 + 2\zeta_2\omega_2 M_2\dot{y}_2 + M_2\omega_2^2 y_2 = Q_2 \tag{8.16b}$$

Similar equations can be written for additional modes as necessary. The process to be described converges rapidly as the number of modes increases.

For a structure excited only by motions of its base, the generalized force is given by

$$Q_1 = -\ddot{w}V_1 - \ddot{u}P_1 - \ddot{\theta}J_1 \tag{8.17a}$$

where

> w = displacement of the base in the z-direction
> u = displacement of the base in the x-direction
> θ = rotation of the base (positive from z to x)
> $V_1 = \int \rho\gamma^{(1)}\,d(\text{vol})$, the structural modal participation in vertical translation, of the first mode
> $P_1 = \int \rho\xi^{(1)}\,d(\text{vol})$, the structural modal participation in horizontal translation, of the first mode
> $J_1 = \int \rho(z\xi^{(1)} - x\gamma^{(1)})\,d(\text{vol})$, the structural modal participation in rotation for the first mode

Similarly for the other structural modes, we have

$$Q_2 = -\ddot{w}V_2 - \ddot{u}P_2 - \ddot{\theta}J_2, \text{ etc.} \tag{8.17b}$$

The horizontal x-displacement δ_x and the vertical z-displacement δ_z are given at any point in the subsystem by

$$\left.\begin{aligned}\delta_x &= y_1\xi^{(1)} + y_2\xi^{(2)} + \cdots + u + \theta z \\ \delta_z &= y_1\gamma^{(1)} + y_2\gamma^{(2)} + \cdots + w - \theta x\end{aligned}\right\} \tag{8.18}$$

The vertical reaction of the structure against the base is

$$Q_z = -\int \rho\ddot{\delta}_z\,d(\text{vol})$$

$$= -\ddot{y}_1 \int \rho\gamma^{(1)}\,d(\text{vol}) - \ddot{y}_2 \int \rho\gamma^{(2)}\,d(\text{vol}) - \cdots - \ddot{w} \int \rho\,d(\text{vol}) + \ddot{\theta}\int \rho x\,d(\text{vol})$$

$$= -\ddot{y}_1 V_1 - \ddot{y}_2 V_2 - \cdots - \ddot{w}M_0 + \ddot{\theta}JX_0 \tag{8.19a}$$

and, similarly, the horizontal reaction is

$$Q_x = -\ddot{y}_1 P_1 - \ddot{y}_2 P_2 - \cdots - \ddot{u}M_0 - \ddot{\theta}JZ_0 \tag{8.19b}$$

where

$$\begin{aligned}M_0 &= \int \rho\,d(\text{vol}), & \text{the structural mass}\\ JX_0 &= \int \rho x\,d(\text{vol}), & \text{the } x\text{-moment of mass}\\ JZ_0 &= \int \rho z\,d(\text{vol}), & \text{the } z\text{-moment of mass}\end{aligned}$$

The moment reaction of the structure against the base is

$$Q_\theta = +\int \rho\ddot{\delta}_z x\,d(\text{vol}) - \int \rho\ddot{\delta}_x z\,d(\text{vol})$$

$$= -\ddot{y}_1 J_1 - \ddot{y}_2 J_2 + \cdots + \ddot{w}JX_0 - \ddot{u}JZ_0 - \ddot{\theta}I_0 \tag{8.19c}$$

where

$I_0 = \int \rho(x^2 + z^2)\, d(\text{vol})$, the moment of inertia about the (x, z) origin

Equations (8.16) can be solved for the acceleration $\ddot{y}$, giving

$$
\left.
\begin{aligned}
\ddot{y}_1 &= -2\zeta_1\omega_1\dot{y}_1 - \omega_1^2 y_1 + \frac{Q_1}{M_1} \\[2mm]
\ddot{y}_2 &= -2\zeta_2\omega_2\dot{y}_2 - \omega_2^2 y_2 + \frac{Q_2}{M_2},\ \text{etc.}
\end{aligned}
\right\}
\tag{8.20}
$$

Then, using eqs. (8.17) for the generalized forces, we have

$$
\left.
\begin{aligned}
\ddot{y}_1 &= -2\zeta_1\omega_1\dot{y}_1 - \omega_1^2 y_1 + \ddot{w}\left(\frac{V_1}{M_1}\right) - \ddot{u}\left(\frac{P_1}{M_1}\right) - \ddot{\theta}\left(\frac{J_1}{M_1}\right) \\[2mm]
\ddot{y}_2 &= -2\zeta_2\omega_2\dot{y}_2 - \omega_2^2 y_2 - \ddot{w}\left(\frac{V_2}{M_2}\right) - \ddot{u}\left(\frac{P_2}{M_2}\right) - \ddot{\theta}\left(\frac{J_2}{M_2}\right),\ \text{etc.}
\end{aligned}
\right\}
\tag{8.21}
$$

Substituting eqs. (8.21) into (8.19) gives, for the base reactions,

$$
\begin{aligned}
Q_x = \ &P_1\omega_1^2 y_1 + P_2\omega_2^2 y_2 + \cdots \\
&+ 2\zeta_1\omega_1 P_1\dot{y}_1 + 2\zeta_2\omega_2 P_2\dot{y}_2 + \cdots \\
&+ \ddot{w}(P_1 V_1/M_1 + P_2 V_2/M_2 + \cdots) \\
&+ \ddot{u}(P_1^2/M_1 + P_2^2/M_2 + \cdots - M_0) \\
&+ \ddot{\theta}(P_1 J_1/M_1 + P_2 J_2/M_2 + \cdots - JZ_0)
\end{aligned}
\tag{8.22a}
$$

$$
\begin{aligned}
Q_z = \ &V_1\omega_1^2 y_1 + V_2\omega_2^2 y_2 + \cdots \\
&+ 2\zeta_1\omega_1 V_1\dot{y}_1 + 2\zeta_2\omega_2 V_2\dot{y}_2 + \cdots \\
&+ \ddot{w}(V_1^2/M_1 + V_2^2/M_2 + \cdots - M_0) \\
&+ \ddot{u}(P_1 V_1/M_1 + P_2 V_2/M_2 + \cdots) \\
&+ \ddot{\theta}(V_1 J_1/M_1 + V_2 J_2/M_2 + \cdots + JX_0)
\end{aligned}
\tag{8.22b}
$$

$$
\begin{aligned}
Q_\theta = \ &J_1\omega_1^2 y_1 + J_2\omega_2^2 y_2 + \cdots \\
&+ 2\zeta_1\omega_1 J_1\dot{y}_1 + 2\zeta_2\omega_2 J_2\dot{y}_2 + \cdots \\
&+ \ddot{w}(J_1 V_1/M_1 + J_2 V_2/M_2 + \cdots + JX_0) \\
&+ \ddot{u}(J_1 P_1/M_1 + J_2 P_2/M_2 + \cdots - JZ_0) \\
&+ \ddot{\theta}(J_1^2/M_1 + J_2^2/M_2 + \cdots - I_0)
\end{aligned}
\tag{8.22c}
$$

It is shown, subsequently, that all the parenthetical expressions in eqs. (8.22) are zero. As a result, eqs. (8.22) can be simplified to

$$
\begin{aligned}
Q_x = \ &P_1\omega_1^2 y_1 + P_2\omega_2^2 y_2 + \cdots \\
&+ 2\zeta_1\omega_1 P_1\dot{y}_1 + 2\zeta_2\omega_2 P_2\dot{y}_2 + \cdots
\end{aligned}
\tag{8.23a}
$$

$$
\begin{aligned}
Q_z = \ &V_1\omega_1^2 y_1 + V_2\omega_2^2 y_2 + \cdots \\
&+ 2\zeta_1\omega_1 V_1\dot{y}_1 + 2\zeta_2\omega_2 V_2\dot{y}_2 + \cdots
\end{aligned}
\tag{8.23b}
$$

$$Q_\theta = J_1\omega_1^2 y_1 + J_2\omega_2^2 y_2 + \cdots$$
$$+ 2\zeta_1\omega_1 J_1\dot{y}_1 + 2\zeta_2\omega_2 J_2\dot{y}_2 + \cdots \tag{8.23c}$$

(It should be noted that when all of the modes are not used, the parenthetical expressions will not be exactly zero. In such a case, eqs. (8.22) rather than eqs. (8.23) must be used. When z-direction displacements are expected, the results are improved by including some modes having z-direction motions.)

Demonstration of the relationships required to simplify eqs. (8.22)

A short demonstration of the relationships required to go from eqs. (8.22) to eqs. (8.23) will be given. It concerns the interrelationship of the modal participations and the various moments of the structure; these relationships become clear when we consider the results of applying to the structure first a unit translation, and then a unit rotation. Throughout the demonstration, use is made of the orthogonality of the normal modes of vibration of a linear system (as the building is assumed to be).

Consider that the structure is given a unit uniform displacement in the x-direction and that its motion is zero in the z-direction. This motion can be modeled by a series of m modes with coefficients A to be determined:

$$1 = \sum_{i=1}^{m} A_i \xi^{(i)} \tag{8.24}$$

$$0 = \sum_{i=1}^{m} A_i \gamma^{(i)} \tag{8.25}$$

Multiplying both sides of eq. (8.24) by ρ and integrating over the volume, we have

$$\int \rho \, d(\text{vol}) = \int \rho \sum_{i=1}^{m} A_i \xi^{(i)} \, d(\text{vol}) \tag{8.26}$$

which, according to the definitions of the modal participations P, gives

$$M_0 = A_1 P_1 + A_2 P_2 + \cdots \tag{8.27}$$

The same procedure used on eq. (8.25) yields

$$0 = A_1 V_1 + A_2 V_2 \tag{8.28}$$

Now, multiplying eq. (8.24) by $\rho\xi^{(1)}$ and eq. (8.25) by $\rho\gamma^{(1)}$ and adding, we have

$$\int \rho\xi^{(1)} \, d(\text{vol}) = \int \rho \{ A_1[(\xi^{(1)})^2 + (\gamma^{(1)})^2] + A_2[\xi^{(1)}\xi^{(2)} + \gamma^{(1)}\gamma^{(2)}] \ldots \} \, d(\text{vol}) \tag{8.29}$$

However, the orthogonality of these modes results in the elimination of all quantities except the first on the right-hand side of the above equation. As a result, we find that

$$P_1 = A_1 M_1 \tag{8.30}$$

so that

$$A_1 = P_1/M_1 \tag{8.31}$$

and, similarly,

$$A_2 = P_2/M_2, \quad A_3 = P_3/M_3, \quad \text{etc.} \tag{8.32}$$

Substituting these values of the amplitudes A_i into eqs. (8.27) and (8.28), we get

$$\left.\begin{array}{l} M_0 = P_1^2/M_1 + P_2^2/M_2 + \cdots \\ 0 = P_1 V_1/M_1 + P_2 V_2/M_2 + \cdots \end{array}\right\} \tag{8.33}$$

Similarly, we can show that

$$M_0 = V_1^2/M_1 + V_2^2/M_2 + \cdots \tag{8.34}$$

These three relationships are pertinent to the simplification of eqs. (8.22a) and (8.22b).

We now apply to the structure a unit rotation about the origin, whereupon the z-displacement is denoted by $(-x)$ and the x-displacement is denoted by z. This rotation is replaced by a series expansion in the modes with coefficients B_i to be determined:

$$z = \sum_{i=1}^{m} B_i \xi^{(i)} \tag{8.35}$$

$$-x = \sum_{i=1}^{m} B_i \gamma^{(i)} \tag{8.36}$$

If we multiply both sides of eq. (8.35) by $\rho\xi^{(1)}$ and both sides of eq. (8.36) by $\rho\gamma^{(1)}$, add and integrate over the volume, we have

$$\int \rho(z\xi^{(1)} - x\gamma^{(1)}) \, d(\text{vol}) = \int B_1 \rho[(\xi^{(1)})^2 + (\gamma^{(1)})^2] \, d(\text{vol}) \tag{8.37}$$

(the other modes having been eliminated by orthogonality), giving

$$J_1 - B_1 M_1, \quad B_1 - J_1/M_1 \tag{8.38}$$

Similarly,

$$B_2 = J_2/M_2, \quad B_3 = J_3/M_3, \quad \text{etc.}$$

If we multiply eq. (8.35) by ρz and eq. (8.36) by $(-\rho x)$, add and integrate over the volume, we have

$$\int \rho(z^2 + x^2) \, d(\text{vol}) = \int \rho \left[z \sum_{i=1}^{m} B_i \xi^{(i)} - x \sum_{i=1}^{m} B_i \gamma^{(i)} \right] d(\text{vol})$$

giving

$$I_0 = \sum_{i=1}^{m} B_i J_i \tag{8.39}$$

With eq. (8.38), this becomes

$$I_0 = J_1^2/M_1 + J_2^2/M_2 + \cdots \tag{8.40}$$

which is used to simplify eq. (8.22c). If we multiply both sides of eq. (8.35) by ρ and integrate,

$$\int \rho z \, d(\text{vol}) = \int \rho \sum_{i=1}^{m} B_i \zeta^{(i)} \, d(\text{vol}) \tag{8.41}$$

giving

$$JZ_0 = B_1 P_1 + B_2 P_2 + \cdots \tag{8.42}$$

With eq. (8.38), this becomes

$$JZ_0 = J_1 P_1/M_1 + J_2 P_2/M_2 + \cdots \tag{8.43}$$

If we multiply eq. (8.36) by ρ and integrate over the volume, we similarly get

$$-JX_0 = J_1 V_1/M_1 + J_2 V_2/M_2 \tag{8.44}$$

These relationships are used to simplify eq. (8.22c).

The soil subsystem

Now let us look at the soil modes relative to a position in the soil far enough removed from the structures to be unaffected by their presence, for example at the bedrock. Let

$Y_1, Y_2, \ldots$ be the modal amplitudes in the ground modes

$\phi^{(1)}, \phi^{(2)}, \ldots$ be the ground mode shapes

$\phi_{bxn}^{(1)}, \phi_{bxn}^{(2)}, \ldots$ be the ground modal amplitudes in the x-direction at the base of the nth structure

$\phi_{bzn}^{(1)}, \phi_{bzn}^{(2)}, \ldots$ be the ground modal amplitudes in the z-direction at the base of the nth structure

$\phi_{b\theta n}^{(1)}, \phi_{b\theta n}^{(2)}, \ldots$ be the ground modal rotations at the base of the nth structure

The base motion of the nth structure is then given by:

$$\left.\begin{aligned}
u_n &= U + Y_1 \phi_{bxn}^{(1)} + Y_2 \phi_{bxn}^{(2)} + \cdots \\
w_n &= W + Y_1 \phi_{bzn}^{(1)} + Y_2 \phi_{bzn}^{(2)} + \cdots \\
\theta_n &= Y_1 \phi_{b\theta n}^{(1)} + Y_2 \phi_{b\theta n}^{(2)} + \cdots
\end{aligned}\right\} \tag{8.45}$$

where U and W are the distant ground motions. (The distant ground motion is considered to be free of rotation.)

The ground modal equations are given by

$$\left.\begin{aligned}
M_{G1} \ddot{Y}_1 + 2\zeta_{G1}\omega_{G1} M_{G1} \dot{Y}_1 + M_{G1}\omega_{G1}^2 Y_1 &= Q_{G1} \\
M_{G2} \ddot{Y}_2 + 2\zeta_{G2}\omega_{G2} M_{G2} \dot{Y}_2 + M_{G2}\omega_{G2}^2 Y_2 &= Q_{G2}, \text{ etc.}
\end{aligned}\right\} \tag{8.46}$$

where

$$M_{G1} = \int \rho [\phi^{(1)}]^2 \, d(\text{vol}), \text{ the modal mass of the ground}$$
$$2\zeta_{G1}\omega_{G1} M_{G1} = \text{modal damping in ground mode 1}$$
$$\omega_{G1} = \text{natural frequency of ground mode 1 (rad/sec)}$$
$$Q_{G1} = \text{generalized force for ground mode 1, etc.}$$

Coupling the building and soil subsystems

If the reactions of soil and building are now equated, the desired coupling of both subsystems is achieved. We have, therefore, the relationships between the generalized forces in the ground modes; the base forces Q_{xn}, Q_{zn}, $Q_{\theta n}$ for the buildings; and the soil support motions U and W:

$$Q_{G1} = -P_{G1}\ddot{U} - V_{G1}\ddot{W} + \sum_n (Q_{xn}\phi_{bxn}^{(1)} + Q_{zn}\phi_{bzn}^{(1)} + Q_{\theta n}\phi_{b\theta n}^{(1)}) \qquad (8.47)$$

with similar equations for the other ground modes. In eq. (8.47),

$P_{G1} = \int \rho_G \phi_x^{(1)} \, d(\text{vol})$, where subscript x indicates the x-direction component

$V_{G1} = \int \rho_G \phi_z^{(1)} \, d(\text{vol})$, where subscript z indicates the z-direction component

ρ_G = mass density of ground

$\ddot{U}$ = soil acceleration in the x-direction some distance from the structures (for example, at the bedrock)

$\ddot{W}$ = soil acceleration in the z-direction some distance from the structures (for example, at the bedrock)

Taking derivatives of eqs. (8.45) for the nth structure,

$$\ddot{u}_n = \ddot{U} + \ddot{Y}_1\phi_{hxn}^{(1)} + \ddot{Y}_2\phi_{hxn}^{(2)} + \cdots$$
$$\ddot{w}_n = \ddot{W} + \ddot{Y}_1\phi_{bzn}^{(1)} + \ddot{Y}_2\phi_{bzn}^{(2)} + \cdots \qquad (8.48)$$
$$\ddot{\theta}_n = \ddot{Y}_1\phi_{b\theta n}^{(1)} + \ddot{Y}_2\phi_{b\theta n}^{(2)} + \cdots$$

Substituting these values in eqs. (8.17) gives, for the nth structure,

$$\left. \begin{array}{l} Q_{1n} = -P_{1n}\ddot{u}_n - V_{1n}\ddot{w}_n - J_{1n}\ddot{\theta}_n \\ Q_{2n} = -P_{2n}\ddot{u}_n - V_{2n}\ddot{w}_n - J_{2n}\ddot{\theta}_n, \text{ etc.} \end{array} \right\} \qquad (8.49)$$

Equations (8.46) can be solved for the ground accelerations $\ddot{Y}$ using eq. (8.47) as follows:

$$\ddot{Y}_1 = -2\beta_{G1}\omega_{G1}\dot{Y}_1 - \omega_{G1}^2 Y_1 - (P_{G1}/M_{G1})\ddot{U} - (V_{G1}/M_{G1})\ddot{W}$$
$$+ \sum_n [\phi_{bxn}^{(1)}Q_{xn} + \phi_{bzn}^{(1)}Q_{zn} + \phi_{b\theta n}^{(1)}Q_{\theta n}]/M_{G1} \qquad (8.50a)$$

$$\ddot{Y}_2 = -2\beta_{G2}\omega_{G2}\dot{Y}_2 - \omega_{G2}^2 Y_2 - (P_{G2}/M_{G2})\ddot{U} - (V_{G2}/M_{G2})\ddot{W}$$
$$+ \sum_n [\phi_{bxn}^{(2)}Q_{xn} + \phi_{bzn}^{(2)}Q_{zn} + \phi_{b\theta n}^{(2)}Q_{\theta n}]/M_{G2}, \text{ etc.} \qquad (8.50b)$$

Solution method

With the series of equations we have derived, it is a relatively simple matter to organize them so that a numerical solution is obtained for the modal amplitudes of the building and soil, y and Y. Having obtained these amplitudes, the relative displacements of the building with respect to its foundation, d_x and d_z in the x- and

z-directions, respectively, can be obtained by summing the modes:

$$\left.\begin{array}{l} d_x = \sum_{i=1}^{N} y_i \zeta^{(i)} \\[2mm] d_z = \sum_{i=1}^{N} y_i \gamma^{(i)} \end{array}\right\} \qquad (8.51)$$

The input that is needed for these calculations consists of the following quantities, which may be determined from a free vibration analysis of the soil and buildings using the finite element method (or, in the case of simpler modeling, from a beam model of the building and a lumped spring–damper foundation model):

the total mass of each building, M_0
the first moments about the x- and z-axes, JX_0 and JZ_0
the moment of inertia about the origin, I_0
the N structural modal masses, M, of each building
the N structural frequencies, ω, of each building
the N structural damping constants, ζ
the N structural modal participations in translation, P, of each building
the N structural modal participations in the vertical direction, V, of each building
the N structural modal participations in rotation, J, of each building
similar modal information for the soil modes, namely, M_G, ω_G, ζ_G, P_G, and V_G
the ground modal amplitudes at the base of each building, denoted by ϕ_{bxn}, ϕ_{bzn}, and $\phi_{b\theta n}$
the seismic excitation of the bedrock, $\ddot{U}$ and $\ddot{W}$

Once these basic quantities have been collected, the solution technique is similar to that described in chapter 3 for the component element method. The analysis proceeds as follows:

(a) Use the present and previous values of modal amplitudes y and Y to determine $\dot{y}$ and $\dot{Y}$. (With a variable s and a time step Δ, letting s_0 be the present value, s_{-1} the previous value, and s_{-2} two steps back, we take $\dot{s}_0 = [3s_0 - 4s_{-1} + s_{-2}]/2\Delta$.)
(b) Get the base reaction forces Q_x, Q_z, and Q_θ from eqs. (8.23) for each structure. If eqs. (8.22) must be used, estimate the second derivatives from the backward difference formula

$$\ddot{s}_0 = [2s_0 - 5s_{-1} + 4s_{-2} - s_{-3}]/\Delta^2$$

(c) Get the modal acceleration $\ddot{Y}$ for each ground mode from eqs. (8.50).
(d) Get the base accelerations $\ddot{u}_n$, $\ddot{w}_n$, and $\ddot{\theta}_n$ for each structure from eqs. (8.48).
(e) Get the generalized forces for each structure mode from eqs. (8.49).
(f) Get the structural modal accelerations $\ddot{y}$ from eqs. (8.20).
(g) Determine new values of y and Y from the previous values and the present value of $\ddot{y}$ and $\ddot{Y}$. Use is made of the difference relation

$$s_{+1} = \ddot{s}_0 \Delta^2 + 2s_0 - s_{-1}$$

Repeat steps (a) through (g) for successive time steps.

This analysis may easily be automated in a computer program rather similar to those described in chapter 3. Such a computer program is briefly described in the next paragraphs, and a listing is given. Some elementary examples are also discussed, which show how the program can be used.

A computer program for component mode analysis

In the following pages, a computer program listing for component mode analysis is given, whose logic is based on steps (a) through (g) of the previous paragraph. The required input is shown in Table 8.2. It involves the modal descriptions of each building and the ground. These modal descriptions take the form of the mass and inertia properties of the building and its modal participation factors, and the modal participation factors of the ground, as well as the natural frequency of each mode. The system of units is pound, second, inch; however, this system can be changed by substituting the appropriate value of g (in the program, it is 386 in./sec^2). To use the International System in newton, second, metre units, g should be set to the value 9·807 m/s^2, as shown in the program. The format and order of the input is shown in Table 8.3; the symbols are described in Table 8.2.

Table 8.2   DESCRIPTION OF INPUT VARIABLES

Variables	Description
NSS	The number of structures
NS	Number of structure modes for each structure
NG	Number of ground modes
NFILE	Input–output control:
	0 = data for ground acceleration is on cards
	1 = data for x-acceleration is on file 1
	2 = as for 1; output is on file 3 every TSTP
	3 = as for 1; output is on file 3 every DEL
	4, 5, 6 = as for 1, 2, 3 except on files 1 and 2 for x- and z-input
DEL	Desired integration time step; the program may select a smaller DEL for convergence
TSTP	Desired printing time interval for output
TTL	Total time
SMO	Total mass of structure
JXO	First moment; integral of x times mass (origin of (x, z) coordinate system of each structure is at its point of attachment to the ground)
JZO	First moment; integral of z times mass
SIO	Moment of inertia about the x, z origin
SM	Structure modal masses in sequence. The last mode should be the highest. Up to 75 modes are permitted (units in lb sec^2/in.)
SW	Structure modal frequencies (in Hz)
SB	Ratios of structure modal damping to the critical damping
SP	Structural modal participations in x-translation (units in lb sec^2/in.)
SV	Structural modal participations in z-translation
SJ	Structural modal participations in rotation
GX	Base x-direction translation in first, second, ..., ground modes
GZ	Base z-direction translation for each ground mode
GR	Base rotations for each ground mode

Table 8.2 — *continued*

Variables	Description
GM	Ground modal masses with the highest mode last. Up to 75 modes are permitted
GW	Ground modal frequencies (in Hz)
GB	Ratio of ground modal damping to the critical damping
GP	Ground x-direction modal participations
GV	Ground z-direction modal participations
ATTX	Time of ground x-acceleration
ATX	Ground x-acceleration (acceleration is assumed to vary linearly between points on the ATX versus ATTX curve)
ATTZ	Time of ground z-acceleration
ATZ	Ground z-acceleration

This program can be used to study the response of one or more buildings on the ground, where an acceleration input is applied to the ground. As we mentioned previously, the word 'building' applies also to any secondary subsystem such as an electronic package or machine, which is linked to a primary system, such as a cabinet or external package to which the excitation is applied.

Table 8.3   INPUT DATA SYMBOLS AND FORMAT

Format	Variables
(10I4)	NSS, (NS(K), K = 1, NSS), NG, NFILE
(8F10.5)	DEL, TSTP, TTL

For each structure J, read the following, where NJ is the number of modes in the structure:

Format	Variables
(8F10.5)	SMO(J), JXO(J), JZO(J), SIO(J)
(8F10.5)	(SM (K, J), K = 1, NJ)
(8F10.5)	(SW (K, J), K = 1, NJ)
(8F10.5)	(SB (K, J), K = 1, NJ)
(8F10.5)	(SP (K, J), K = 1, NJ)
(8F10.5)	(SV (K, J), K = 1, NJ)
(8F10.5)	(SJ (K, J), K = 1, NJ)
(8F10.5)	(GX (K, J), K = 1, NJ)
(8F10.5)	(GZ (K, J), K = 1, NJ)
(8F10.5)	(GR (K, J), K = 1, NJ)
	Repeat this set for each structure.

For the ground, read the following, where NG is the number of ground modes:

Format	Variables
(8F10.5)	(GM(K), K = 1, NG)
(8F10.5)	(GW(K), K = 1, NG)
(8F10.5)	(GB(K), K = 1, NG)
(8F10.5)	(GP(K), K = 1, NG)
(8F10.5)	(GV(K), K = 1, NG)

For the ground acceleration, read the following:

Format	Variables	
(8F10.5)	(ATTX(K), ATX(K), K = 1, 4)	If NFILE = 0, these values are on cards.
		If NFILE is not 0, these values are on file 1.
(8F10.5)	(ATTZ(K), ATZ(K), K = 1, 4)	If NFILE is 4, 5, or 6, these values are on file 2.
	Successive sets of four time–acceleration coordinates are read by each read statement from each card or file record.	

Computer program listing

```
            COMPONENT MODE COMPUTER PROGRAM
C
C
      COMMON SP(75,4),SW(75,4),SB(75,4),SJ(75,4),SM(75,4),SY(3,75,4),
     1GP(75),GW(75),GB(75),GM(75),GZ(75,4),GR(75,4),GY(3,75),GYACC(75),
     2  SZACC(4),GZACC,NG,SC(3,75,4),GC(3,75),QS(75,4),QG(75),SRACC(4)
     3  ,ATTX(4),ATX(4),ATTZ(4),ATZ(4),SYACC(75,4),SV(75,4),GX(75,4)
     4  ,ZOLD(4),ZNOW(4),ZNEW(4),ROLD(4),RNOW(4),RNEW(4),QR(4),QZ(4)
     5  ,XOLD(4),XNOW(4),XNEW(4),NS(4),QX(4),SXACC(4),GXACC,GV(75)
     6  ,SMO(4),JXO(4),JZO(4),SIO(4),XACC(4),ZACC(4),RACC(4),XOLDR(4),
     7  ZOLDR(4),ROLDR(4),XZ(4),XX(4),ZZ(4),XR(4),ZR(4),RR(4)
      REAL JXO,JZO
C
C
C  SM,GM   STRUCTURAL AND GROUND MODAL MASSES(75 MODES EACH)
C  SP,SV,CP,CV  STRUCTURAL AND GROUND MODAL PARTICIPATION FACTORS
C  SW,GW   STRUCTURAL AND GROUND MODAL FREQUENCIES, HERTZ
C  SB,GB   STRUCTURAL AMD GROUND MODAL DAMPING, BETA
C  SJ  STRUCTURAL ROTATIONAL PARTICIPATION FACTOR OF MODES
C  GX,GZ,GR  STRUCTURE BASE DISPLACEMENTS AND ROTATION IN GROUND MODES
C  SY,GY   STRUCTURAL AND GROUND MODAL DISPLACEMENTS
C  SYACC,GYACC  STRUCTURAL AND GROUND MODAL ACCELERATIONS
C  SXACC,SZACC  STRUCTURE BASE ACCELERATIONS
C  SRACC   STRUCTURE BASE ROTATIONAL ACCELERATION
C  GXACC,GZACC  GROUND BASE ACCELERATIONS,  G'S
C  NSS  NUMBER OF STRUCTURES
C  NS,NG  NUMBER OF MODES IN STRUCTURES AND GROUND
C  ATIX,ATTZ  STORES TIME OF GROUND ACCELERATION
C  ATX,ATZ  STORES ACCESLERATIONS OF GROUND ACCELERATION
C  SMO  THE STRUCTURE MASS
C  SIO  THE STRUCTURE MOMENT OF INERTIA IN ROTATION ABOUT THE ORIGIN
C  JXO  THE FIRST MOMENT, INTEGRAL OF X TIMES THE MASS
C  JZO  THE FIRST MOMENT, INTEGRAL OF Z TIMES THE MASS
C
C
 1001 FORMAT(1H0,12X,31HCOMPONENT MODE COMPUTER PROGRAM///)
      WRITE(6,1001)
      GRAV=386.
C
C  GRAV IS THE ACCELERATION OF GRAVITY IN THE POUND-INCH-SECOND SYSTEM
C  IN THE INTERNATIONAL NEWTON-METRE-SECOND SYSTEM USE GRAV=9.807.
C
 1002 FORMAT(8F10.5)
 1003 FORMAT(10E12.4)
C
C  WE WILL NOW READ THE MODAL DATA
C
 1004 FORMAT(10I4)
      READ(5,1004)NSS,(NS(K),K=1,NSS),NG,NFILE
C
C  NFILE=0 DATA FOR GROUND ACCELERATION IS ON CARDS--OTHERWISE ON FILE 1
C  NFILE=1 READS DATA FROM FILE 1 -- PRINTS OUTPUT EVERY TSTP
C  NFILE=2 READS DATA FROM FILE 1 -- PRINTS OUTPUT AND PUTS IT ON FILE 3
C       EVERY TSTP
C  NFILE=3 READS DATA FROM FILE 1 -- PRINTS OUTPUT EVERY TSTP AND PUTS
C  STRUCTURE MODAL DISPLACEMENTS AND ACCELERATIONS ON FILE 3 EVERY DEL
C  NFILE=4 LIKE NFILE=1 EXCEPT ON FILES 1 AND 2 FOR X AND Z INPUT
C  NFILE=5 LIKE NFILE=2 EXCEPT ON FILES 1 AND 2 FOR X AND Z INPUT
C  NFILE=6 LIKE NFILE=3 EXCEPT ON FILES 1 AND 2 FOR X AND Z INPUT
C
```

```
      WRITE(6,1004)NSS,(NS(K),K=1,NSS),NG,NFILE
      READ(5,1002)DEL,TSTP,TTL
C
C   DEL IS THE INTEGRATION TIME STEP
C   TSTP IS THE TIME STEP FOR PRINTING OUTPUT
C   TTL IS THE TOTAL TIME OF THE COMPUTATION
C
      WRITE(6,1003)DEL,TSTP,TTL
      DO 310 J=1,NSS
      NJ=NS(J)
      READ(5,1002)SMO(J),JXO(J),JZO(J),SIO(J)
      READ(5,1002)(SM(K,J),K=1,NJ)
      READ(5,1002)(SW(K,J),K=1,NJ)
      READ(5,1002)(SB(K,J),K=1,NJ)
      READ(5,1002)(SP(K,J),K=1,NJ)
      READ(5,1002)(SV(K,J),K=1,NJ)
      READ(5,1002)(SJ(K,J),K=1,NJ)
      READ(5,1002)(GX(K,J),K=1,NG)
      READ(5,1002)(GZ(K,J),K=1,NG)
      READ(5,1002)(GR(K,J),K=1,NG)
  310 CONTINUE
      READ(5,1002)(GM(K),K=1,NG)
      READ(5,1002)(GW(K),K=1,NG)
      READ(5,1002)(GB(K),K=1,NG)
      READ(5,1002)(GP(K),K=1,NG)
      READ(5,1002)(GV(K),K=1,NG)
      DO 311 J=1,NSS
      NJ=NS(J)
      WRITE(6,1003)SMO(J),JXO(J),JZO(J),SIO(J)
      WRITE(6,1003)(SM(K,J),K=1,NJ)
      WRITE(6,1003)(SW(K,J),K=1,NJ)
      WRITE(6,1003)(SB(K,J),K=1,NJ)
      WRITE(6,1003)(SP(K,J),K=1,NJ)
      WRITE(6,1003)(SV(K,J),K=1,NJ)
      WRITE(6,1003)(SJ(K,J),K=1,NJ)
      WRITE(6,1003)(GX(K,J),K=1,NG)
      WRITE(6,1003)(GZ(K,J),K=1,NG)
      WRITE(6,1003)(GR(K,J),K=1,NG)
  311 CONTINUE
      WRITE(6,1003)(GM(K),K=1,NG)
      WRITE(6,1003)(GW(K),K=1,NG)
      WRITE(6,1003)(GB(K),K=1,NG)
      WRITE(6,1003)(GP(K),K=1,NG)
      WRITE(6,1003)(GV(K),K=1,NG)
C
C   FORM MULTIPLIERS FOR DAMPING AND SPRING FORCES IN MODES AND
C   INITIALIZE
C
      DO 5 J=1,NSS
      NJ=NS(J)
      XX(J)=-SMO(J)
      XZ(J)=0.0
      ZZ(J)=-SMO(J)
      XR(J)=-JZO(J)
      ZR(J)=JXO(J)
      RR(J)=-SIO(J)
      DO 5 K=1,NJ
      XX(J)=XX(J)+SP(K,J)**2/SM(K,J)
      XZ(J)=XZ(J)+SP(K,J)*SV(K,J)/SM(K,J)
      ZZ(J)=ZZ(J)+SV(K,J)**2/SM(K,J)
```

```
      XR(J)=XR(J)+SP(K,J)*SJ(K,J)/SM(K,J)
      ZR(J)=ZR(J)+SV(K,J)*SJ(K,J)/SM(K,J)
      RR(J)=RR(J)+SJ(K,J)**2/SM(K,J)
      SW(K,J)=6.2832*SW(K,J)
    5 CONTINUE
      DO 6 K=1,NG
      GW(K)=6.2832*GW(K)
    6 CONTINUE
      DELG=0.5/GW(NG)
      DO 312 J=1,NSS
      NJ=NS(J)
      DELS=0.5/SW(NJ,J)
      IF(DELS.LT.DEL)DEL=DELS
  312 CONTINUE
      IF(DELG.LT.DEL) DEL=DELG
      DELSQ=DEL**2
      DO 313 J=1,NSS
      NJ=NS(J)
      DO 10 K=1,NJ
      BWD=SB(K,J)*SW(K,J)/DEL
      SC(1,K,J)=-3.0*BWD-SW(K,J)**2
      SC(2,K,J)=4.0*BWD
      SC(3,K,J)=-BWD
      DO 10 JJ=1,3
   10 SY(JJ,K,J)=0.0
  313 CONTINUE
      DO 20 K=1,NG
      BWD=GB(K)*GW(K)/DEL
      GC(1,K)=-3.*BWD-GW(K)**2
      GC(2,K)=4.*BWD
      GC(3,K)=-BWD
      DO 20 J=1,3
   20 GY(J,K)=0.0
      TIMTWX=-0.00002
      ACCLX2=0.0
      NSEISX=5
      XXNEW=0.0
      XXNOW=0.0
      XXOLD=0.0
      TIMTWZ=-0.00002
      ACCLZ2=0.0
      NSEISZ=5
      ZZNEW=0.0
      ZZNOW=0.0
      ZZOLD=0.0
      GZACC=0.0
      DO 314 J=1,NSS
      XNEW(J)=0.0
      XNOW(J)=0.0
      XOLD(J)=0.0
      XOLDR(J)=0.0
      ZNEW(J)=0.0
      ZNOW(J)=0.0
      ZOLD(J)=0.0
      ZOLDR(J)=0.0
      RNEW(J)=0.
      RNOW(J)=0.
      ROLD(J)=0.
      ROLDR(J)=0.0
  314 CONTINUE
```

```
   45 CONTINUE
      TCOUNT=-0.000001
C
C  SET INITIAL VALUE OF TIME
C
      TM=-DEL
C
C  START THE NUMERICAL INTEGRATION OF THE EQUATIONS OF MOTION
C
   50 TM=TM+DEL
      XXOLD=XXNOW
      XXNOW=XXNEW
      ZZOLD=ZZNOW
      ZZNOW=ZZNEW
      DO 315 J=1,NSS
      XACC(J)=(2.*XNEW(J)-5.*XNOW(J)+4.*XOLD(J)-XOLDR(J))/DELSQ
      XOLDR(J)=XOLD(J)
      XOLD(J)=XNOW(J)
      XNOW(J)=XNEW(J)
      ZACC(J)=(2.*ZNEW(J)-5.*ZNOW(J)+4.*ZOLD(J)-ZOLDR(J))/DELSQ
      ZOLDR(J)=ZOLD(J)
      ZOLD(J)=ZNOW(J)
      ZNOW(J)=ZNEW(J)
      RACC(J)=(2.*RNEW(J)-5.*RNOW(J)+4.*ROLD(J)-ROLDR(J))/DELSQ
      ROLDR(J)=ROLD(J)
      ROLD(J)=RNOW(J)
      RNOW(J)=RNEW(J)
  315 CONTINUE
      IF(TM-TIMTWX-0.00001)1206,1206,1207
 1207 TIMONX=TIMTWX
      ACCLX1=ACCLX2
 1204 FORMAT(8F10.5)
      IF(NSEISX.EQ.5.AND.NFILE.GT.0)READ(1,1204)(ATTX(K),ATX(K),K=1,4)
      IF(NSEISX.EQ.5.AND.NFILE.EQ.0)READ(5,1204)(ATTX(K),ATX(K),K=1,4)
      IF(NSEISX.EQ.5)NSEISX=1
      TIMTWX=ATTX(NSEISX)
      ACCLX2=ATX(NSEISX)
      NSEISX=NSEISX+1
      IF(TM-TIMTWX-0.00001)1208,1208,1207
 1208 ZFAX=(ACCLX2-ACCLX1)/(TIMTWX-TIMONX)
 1206 CONTINUE
      GXACC=GRAV*(ACCLX1+(TM-TIMONX)*ZFAX)
C
C  GXACC IS THE CURRENT VALUE OF THE DISTANT GROUND ACCELERATION
C
      XXNEW=GXACC*DELSQ+XXNOW+XXNOW-XXOLD
      IF(TM.GT.0.0) GO TO 175
      CORR=0.5*(XXOLD+XXNEW)-XXNOW
      XXNEW=XXNEW-CORR
      XXOLD=XXOLD+CORR
  175 CONTINUE
      IF(NFILE.LE.3)GO TO 375
      IF(TM-TIMTWZ-0.00001)2206,2206,2207
 2207 TIMONZ=TIMTWZ
      ACCLZ1=ACCLZ2
      IF(NSEISZ.EQ.5.AND.NFILE.GT.0)READ(2,1204)(ATTZ(K),ATZ(K),K=1,4)
      IF(NSEISZ.EQ.5)NSEISZ=1
      TIMTWZ=ATTZ(NSEISZ)
      ACCLZ2=ATZ(NSEISZ)
      NSEISZ=NSEISZ+1
```

```
       IF(TM-TIMTWZ-0.00001)2208,2208,2207
 2208 ZFAZ=(ACCL72-ACCLZ1)/(TIMTWZ-TIMONZ)
 2206 CONTINUE
       GZACC=GRAV*(ACCLZ1+(TM-TIMONZ)*ZFAZ)
C
C   GZACC IS THE CURRENT VALUE OF THE DISTANT GROUND ACCELERATION
C
       ZZNEW=GZACC*DELSQ+ZZNOW+ZZNOW-ZZOLD
       IF(TM.GT.0.0) GO TO 375
       CORR=0.5*(ZZOLD+ZZNEW)-ZZNOW
       ZZNEW=ZZNEW-CORR
       ZZOLD=ZZOLD+CORR
  375 CONTINUE
C
C   COMPUTE FORCES STRUCTURE APPLIES TO GROUND
C
       DO 316 J=1,NSS
       NJ=NS(J)
       QX(J)=XACC(J)*XX(J)+ZACC(J)*XZ(J)+RACC(J)*XR(J)
       QZ(J)=XACC(J)*XZ(J)+ZACC(J)*ZZ(J)+RACC(J)*ZR(J)
       QR(J)=XACC(J)*XR(J)+ZACC(J)*ZR(J)+RACC(J)*RR(J)
       DO 60 K=1,NJ
       QS(K,J)=0.0
       DO 61 JJ=1,3
   61 QS(K,J)=QS(K,J)+SC(JJ,K,J)*SY(JJ,K,J)
       QX(J)=QX(J)-QS(K,J)*SP(K,J)
       QZ(J)=QZ(J)-QS(K,J)*SV(K,J)
       QR(J)=QR(J)-QS(K,J)*SJ(K,J)
   60 CONTINUE
  316 CONTINUE
C
C   WE NOW COMPUTE THE GROUND MODAL ACCELERATIONS
C
       DO 70 K=1,NG
       GYACC(K)=-(GP(K)*GXACC+GV(K)*GZACC)/GM(K)
       DO 71 J=1,3
   71 GYACC(K)=GYACC(K)+GC(J,K)*GY(J,K)
       DO 317 J=1,NSS
       GYACC(K)=GYACC(K)+
     1  (QX(J)*GX(K,J)+QZ(J)*GZ(K,J)+QR(J)*GR(K,J))/GM(K)
  317 CONTINUE
   70 CONTINUE
C
C   COMPUTE STRUCTURE BASE ACCELERATIONS
C
       DO 318 J=1,NSS
       NJ=NS(J)
       SXACC(J)=GXACC
       SZACC(J)=GZACC
       SRACC(J)=0.0
       DO 80 K=1,NG
       SXACC(J)=SXACC(J)+GYACC(K)*GX(K,J)
       SZACC(J)=SZACC(J)+GYACC(K)*GZ(K,J)
       SRACC(J)=SRACC(J)+GYACC(K)*GR(K,J)
   80 CONTINUE
C
C   COMPUTE NEW BASE DISPLACEMENT AND ROTATION
C
       XNEW(J)=SXACC(J)*DELSQ+XNOW(J)+XNOW(J)-XOLD(J)
       ZNEW(J)=SZACC(J)*DELSQ+ZNOW(J)+ZNOW(J)-ZOLD(J)
```

```
      RNEW(J)=SRACC(J)*DELSQ+RNOW(J)+RNOW(J)-ROLD(J)
      IF(TM.GT.0.0)GO TO 275
      CORR=0.5*(XOLD(J)+XNEW(J))-XNOW(J)
      XNEW(J)=XNEW(J)-CORR
      XOLD(J)=XOLD(J)+CORR
      XOLDR(J)=XOLDR(J)+4.*CORR
      CORR=0.5*(ZOLD(J)+ZNEW(J))-ZNOW(J)
      ZNEW(J)=ZNEW(J)-CORR
      ZOLD(J)=ZOLD(J)+CORR
      ZOLDR(J)=ZOLDR(J)+4.*CORR
      CORR=0.5*(RNEW(J)+ROLD(J))-RNOW(J)
      RNEW(J)=RNEW(J)-CORR
      ROLD(J)=ROLD(J)+CORR
      ROLDR(J)=ROLDR(J)+4.*CORR
  275 CONTINUE
C
C   COMPUTE STRUCTURE MODAL ACCELERATIONS
C
      DO 90 K=1,NJ
      SYACC(K,J)=-(SP(K,J)*SXACC(J)+SV(K,J)*SZACC(J)
     1  +SJ(K,J)*SRACC(J))/SM(K,J)+QS(K,J)
   90 CONTINUE
C
C   COMPUTE NEW VALUES OF AMPLITUDES
C
      DO 110 K=1,NJ
      SY(3,K,J)=SY(2,K,J)
      SY(2,K,J)=SY(1,K,J)
      SY(1,K,J)=2.*SY(2,K,J)-SY(3,K,J)+DELSQ*SYACC(K,J)
      IF(TM.GT.0.0)GO TO 110
      CORR=0.5*(SY(3,K,J)+SY(1,K,J))-SY(2,K,J)
      SY(1,K,J)=SY(1,K,J)-CORR
      SY(3,K,J)=SY(3,K,J)+CORR
  110 CONTINUE
  318 CONTINUE
      DO 120 K=1,NG
      GY(3,K)=GY(2,K)
      GY(2,K)=GY(1,K)
      GY(1,K)=2.*GY(2,K)-GY(3,K)+DELSQ*GYACC(K)
      IF(TM.GT.0.0)GO TO 120
      CORR=0.5*(GY(3,K)+GY(1,K))-GY(2,K)
      GY(1,K)=GY(1,K)-CORR
      GY(3,K)=GY(3,K)+CORR
  120 CONTINUE
C
C   PRINT OUTPUT
C
      DO 319 J=1,NSS
      NJ=NS(J)
      IF(NFILE.EQ.3.OR.NFILE.EQ.6)WRITE(3,1003)TM,(SYACC(K,J),K=1,NJ),
     1   SXACC(J),SZACC(J),SRACC(J),(SY(2,K,J),K=1,NJ),XNOW(J),ZNOW(J)
     2   ,RNOW(J)
  319 CONTINUE
C
C   PRINTS ON FILE 3 ACCELERATIONS AND DISPLACEMENTS FOR THE STRUCTURAL
C   MODES
C
      IF (TM.LT.TCOUNT) GO TO 1095
 1007 FORMAT(1H0,     17H**********       ,7HTIME = ,F12.5,17H      ****
     C******)
```

```
1008 FORMAT(1H0,4HSY= ,8F13.8/(5X,8F13.8))
1009 FORMAT(1H0,4HGY= ,8F13.8/(5X,8F13.8))
1011 FORMAT(1H0,17X,32HDISPLACEMENT OF STRUCTURAL MODES,I4)
1012 FORMAT(1H0,17X,28HDISPLACEMENT OF GROUND MODES)
1285 FORMAT(1H0,31HDISPLACEMENT OF GROUND AND BASE,I4)
1284 FORMAT(6F13.3)
1286 FORMAT(1H0,8HGROUND= ,F13.6,3X,F13.6,3X,
    1  19HBASE DISPLACEMENT= ,2(F13.6,3X),15HBASE ROTATION= ,F13.6)
     TCOUNT=TCOUNT+TSTP
     WRITE(6,1007)TM
     IF(NFILE.EQ.2.OR.NFILE.EQ.5)WRITE(3,1003)TM
C
C  MODAL STRUCTURAL DISPLACEMENTS
C
     DO 320 J=1,NSS
     NJ=NS(J)
     WRITE(6,1011)J
     WRITE(6,1008)(SY(2,K,J),K=1,NJ)
     IF(NFILE.EQ.2.OR.NFILE.EQ.5)WRITE(3,1003)(SY(2,K,J),K=1,NJ)
 320 CONTINUE
C
C  MODAL GROUND DISPLACEMENTS
C
     WRITE(6,1012)
     WRITE(6,1009)(GY(2,K),K=1,NG)
     IF(NFILE.EQ.2.OR.NFILE.EQ.5)WRITE(3,1003)(GY(2,K),K=1,NG)
C
C  DISPLACEMEMTS OF GROUND AND OF STRUCTURE BASE.
C
     DO 321 J=1,NSS
     WRITE(6,1285)J
     WRITE(6,1286)XXNOW,ZZNOW,XNOW(J),ZNOW(J),RNOW(J)
     IF(NFILE.EQ.2.OR.NFILE.EQ.5)
    1  WRITE(3,1003)XXNOW,ZZNOW,XNOW(J),ZNOW(J),RNOW(J)
 321 CONTINUE
C
C  IF TIME IS LESS THAN TTL GO BACK TO STATEMENT 50
C
1095 IF (TM.LT.TTL) GO TO 50
 100 STOP
     END
```

Some specific numerical examples of a very simple nature are given in the following paragraphs, which show the manner in which the computer program can be used.

Some numerical examples

We first consider a rudimentary model of a building and ground as shown in Fig. 8.15. The building model consists of a block of unit mass supported by a spring of unit stiffness. The base is connected to a ground block of unit mass and supported by a spring of unit stiffness. The building thus has a single degree of freedom whose mode shape is a unit deflection of the building mass. For the building, therefore, we have the mode shapes:

$$\xi_1^{(1)} = 1 \cdot 0 \quad \text{and} \quad \gamma_1^{(1)} = 0 \cdot 0$$

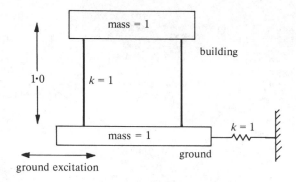

(a) Model of building and ground

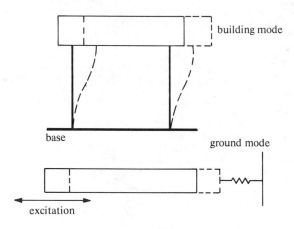

(b) Component modes of the building−ground system

FIGURE 8.15
Simple two-component model of building and ground

The building origin of coordinates is assumed to be such that the mass and inertial properties are given by the definitions that follow eqs. (8.19) as

$$M_0 = 1, \quad JX_0 = 0, \quad JZ_0 = 1, \quad I_0 = 1$$

The modal mass M_1 is given, according to the definition following eq. (8.16a), by $M_1 = 1$. Similarly, referring to the definitions following eq. (8.17a), the modal participation factors P_1, V_1, and J_1 are given by

$$P_1 = 1, \quad V_1 = 0, \quad J_1 = 1$$

The natural frequency of the building in Hz is given by $1/(2\pi)\sqrt{k/M_0}$, so that, in this case, $f_1 = 1/(2\pi)$. We assume that the structural damping in this mode is five per cent of critical, so that

$$\beta_1 = 0\cdot05$$

In the case of the ground, which is also modeled as a single degree-of-freedom system, the modal amplitudes at the base of the building are given by

$$\phi_{bx}^{(1)} = 1, \quad \phi_{bz}^{(1)} = 0, \quad \phi_{b\theta}^{(1)} = 0$$

The single natural frequency is $f_{G1} = 1/(2\pi)$ Hz, and we again assume that the damping is five per cent of critical, so that $\beta_{G1} = 0\cdot05$. The modal mass is $M_{G1} = 1$, and, according to the definitions following eq. (8.47), the modal participation factors are

$$P_{G1} = 1, \quad V_{G1} = 0$$

The excitation of the ground is assumed to be an x-direction acceleration increasing with time at a rate of $0\cdot001g/\text{sec}$.

The input data, constructed according to Table 8.3, is shown in Table 8.4. The output is plotted in Fig. 8.16. In that figure, SY is the building mode displacement and GY is the ground mode displacement.

Problem

8.1 Repeat this example, modeling it as a two degree-of-freedom system, and calculating its response by means of the computer program given in section 3.13. How do the results compare with those obtained by using the component mode analysis?

Table 8.4 INPUT DATA FOR THE MODEL OF FIG. 8.15

1	1	1	0
0.1		0.1	2.0
1.0		0.0	1.0	1.0
1.0			
0.15915			
0.05			
1.0			
0.0			
1.0			
1.0			
0.0			
0.0			
1.0			
0.15915			
0.05			
1.0			
0.0			
0.0		0.0	5.0	.005

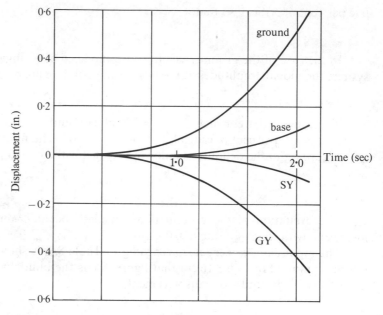

FIGURE 8.16
Response of simple system in Fig. 8.15

Table 8.5 INPUT DATA FOR THE TWO-MODE BUILDING MODEL OF FIG. 8.17

```
   1    2    2    0
0.01      0.1       2.0
2.0       0.0       30.0      500.0
3.618034  1.381966
0.098363  0.257518
0.0       0.0
2.618034  0.381966
0.0       0.0
42.3607   -2.36068
1.0       0.0
0.0       0.0
0.0       1.0
1.0       1.0
0.159154  0.159154
0.0       0.0
1.0       0.0
0.0       0.0
0.0       0.001     1.0       0.001     3.0       0.001
```

Now consider the building and ground system shown in Fig. 8.17. In this case, the building is modeled as a two mass system having two degrees of freedom, and the ground is allowed two degrees of freedom: rotational and translational. The excitation is a sudden acceleration of $0.001g$. The appropriate mass and modal parameters that describe the building system and the ground system are shown in Fig. 8.17. The input required for the computer program is shown in Table 8.5. The time step interval used is 0.01 sec. The output is plotted in Fig. 8.18. It is seen that the lowest ground mode response GY_1 tends to cancel out the ground

FIGURE 8.17
Two-mode building model, situated on a two-mode ground

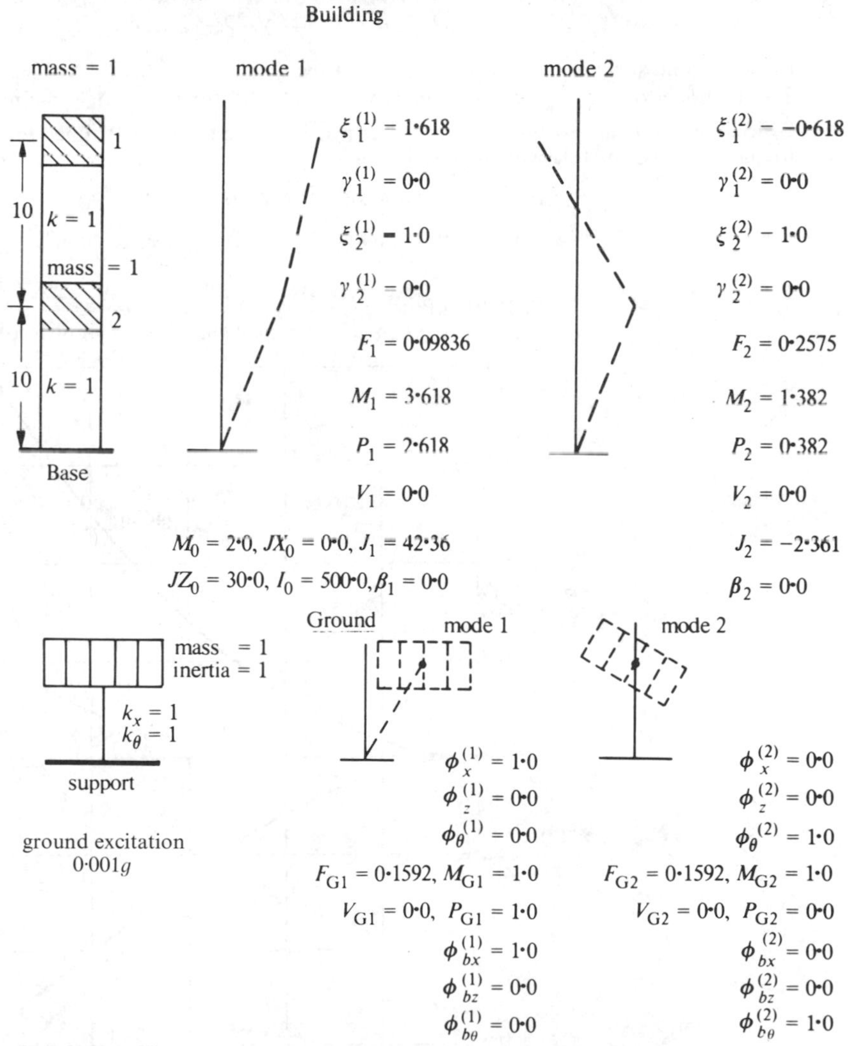

Building

mass = 1   mode 1   mode 2

$\xi_1^{(1)} = 1.618$ $\xi_1^{(2)} = -0.618$

$\gamma_1^{(1)} = 0.0$ $\gamma_1^{(2)} = 0.0$

$\xi_2^{(1)} = 1.0$ $\xi_2^{(2)} = 1.0$

$\gamma_2^{(1)} = 0.0$ $\gamma_2^{(2)} = 0.0$

$F_1 = 0.09836$ $F_2 = 0.2575$

$M_1 = 3.618$ $M_2 = 1.382$

$P_1 = 2.618$ $P_2 = 0.382$

$V_1 = 0.0$ $V_2 = 0.0$

$M_0 = 2.0, \ JX_0 = 0.0, \ J_1 = 42.36$ $J_2 = -2.361$

$JZ_0 = 30.0, \ I_0 = 500.0, \beta_1 = 0.0$ $\beta_2 = 0.0$

Ground   mode 1   mode 2

mass = 1, inertia = 1

$k_x = 1$
$k_\theta = 1$

support

ground excitation
$0.001g$

$\phi_x^{(1)} = 1.0$ $\phi_x^{(2)} = 0.0$

$\phi_z^{(1)} = 0.0$ $\phi_z^{(2)} = 0.0$

$\phi_\theta^{(1)} = 0.0$ $\phi_\theta^{(2)} = 1.0$

$F_{G1} = 0.1592, \ M_{G1} = 1.0$ $F_{G2} = 0.1592, \ M_{G2} = 1.0$

$V_{G1} = 0.0, \ P_{G1} = 1.0$ $V_{G2} = 0.0, \ P_{G2} = 0.0$

$\phi_{bx}^{(1)} = 1.0$ $\phi_{bx}^{(2)} = 0.0$

$\phi_{bz}^{(1)} = 0.0$ $\phi_{bz}^{(2)} = 0.0$

$\phi_{b\theta}^{(1)} = 0.0$ $\phi_{b\theta}^{(2)} = 1.0$

motion. Thus, for such a sudden shock, the ground acts as an isolator for the building. The difference between ground motion and GY_1 is the base motion. The responses in the structural modes SY_1 and SY_2 are much smaller by comparison. For this particular example, there is more response in SY_2 than SY_1. The base rocking permitted by the second ground mode is the probable cause of the situation.

Problems

8.2 Compute the response of the system shown in Fig. 8.19, in which there are two 'buildings' situated on a ground. Each structure is identical with that in Fig. 8.17. In this case,

$$\phi_{bx1}^{(1)} = 1, \quad \phi_{bz1}^{(1)} = 0, \quad \phi_{b\theta1}^{(1)} = 0, \quad \phi_{bx1}^{(2)} = 0, \quad \phi_{bz1}^{(2)} = 0\cdot1, \quad \phi_{b\theta1}^{(2)} = 1$$

for the first structure, while

$$\phi_{bx2}^{(1)} = 1, \quad \phi_{bz2}^{(1)} = 0, \quad \phi_{b\theta2}^{(1)} = 0, \quad \phi_{bx2}^{(2)} = 0, \quad \phi_{bz2}^{(2)} = -0\cdot1, \quad \phi_{b\theta2}^{(2)} = 1$$

for the second structure. Use a time step of 0·1 sec.

8.3 To the model of the two buildings in problem 8.2, add the effects of some modes in the z-direction (vertical), so that they have vertical frequencies of $\sqrt{10}$ times the horizontal frequencies. Use the following modes:

$$\xi_1^{(3)} = 0\cdot0 \qquad \gamma_1^{(3)} = 1\cdot618 \qquad \xi_2^{(3)} = 0\cdot0 \qquad \gamma_2^{(3)} = 1\cdot0$$

$$\xi_1^{(4)} = 0\cdot0 \qquad \gamma_1^{(4)} = -0\cdot618 \qquad \xi_2^{(4)} = 0\cdot0 \qquad \gamma_2^{(4)} = 1\cdot0$$

FIGURE 8.18
Response of system with two 'building' modes and two 'ground' modes

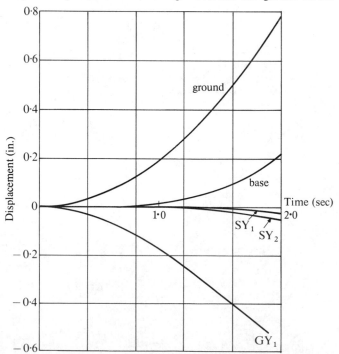

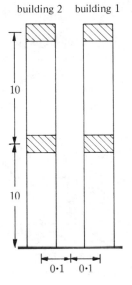

building 2   building 1

Each 'building' is
the same as in
Fig. 8.17.

10

10

'Buildings' offset
0·1 to either side
of ground centerline.

0·1   0·1

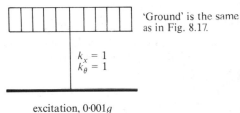

'Ground' is the same
as in Fig. 8.17.

$k_x = 1$
$k_\theta = 1$

FIGURE 8.19
Two 'buildings' with two modes each on
'ground' with two modes

excitation, 0·001g

giving

$$P_3 = 0·0 \qquad V_3 = 2·618 \qquad J_3 = 0·0$$
$$P_4 = 0·0 \qquad V_4 = 0·382 \qquad J_4 = 0·0$$

In addition, assume that

$$M_3 = 3·618 \qquad f_3 = 0·311$$
$$M_4 = 1·382 \qquad f_4 = 0·814$$

How much effect on the response have the vertical modes, when compared to the response
of the previous example, in which no vertical modes were used?

8.4   Compute the response, by the component mode method, of the three mass system shown
in Fig. 8.20, to a 0·25g oscillation increasing in frequency by 2·5 octaves per minute from
15 Hz to 60 Hz. Use the large cabinet as the ground system, and the two mass chassis
system as the building system. As an approximation, take the damping as 0·05 of critical
for the building modes. What is the maximum response and at what time does it occur?
Repeat this example using the three mass computer program of section 3.13. Are the
results the same? (*Answer:* Maximum response of 2 lb mass is 9·5g.)

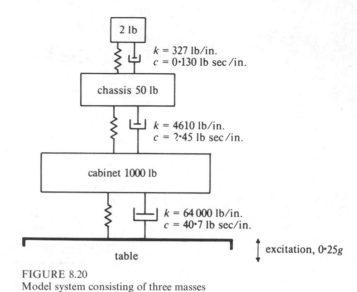

FIGURE 8.20
Model system consisting of three masses

8.10 SUMMARY

In this chapter we have discussed analysis of buildings subject to earthquake excitations. From a knowledge of earthquake records, design response spectra have evolved from which one can deduce a site design response spectrum. An upper bound to the building response can be estimated directly from these spectra. Finer analyses can be made from an artificial time history that is derived from the site design response spectra. The basic tools have been laid down for determining response spectra from given time histories and for the inverse (and more difficult) process of obtaining suitable time histories from a given response spectrum. The interaction of the soil and the building can be an important consideration, particularly if the building is very stiff or massive. Equivalent ground springs are often used to model this interaction, although recent investigations in soil–structure interaction have used the finite element method of analysis to aid in characterizing the numerous important effects. In that case, when a time history response is desired, a component mode analysis is used, and such a method has been developed. A computer program was described for this analysis, and a listing given. Some simple numerical examples using this program were worked out in detail to show how the program can be used.

8.11 REFERENCES

1. N. M. Newmark and E. Rosenblueth, *Fundamentals of Earthquake Engineering,* Prentice-Hall, Inc., Englewood Cliffs, New Jersey, 1971.
2. N. M. Newmark, J. A. Blume, and K. K. Kapur, 'Seismic design spectra for nuclear power plants,' *J. Power Division, Proc. ASCE,* **99**, No. PO2, 287–303, 1973.

3. M. Stoykovich, 'Seismic design and analysis of nuclear plant components,' in *Structural Design of Nuclear Plant Facilities,* vol. 1, pp. 1–28, ASCE, New York, 1973.
4. N.-C. Tsai, 'Spectrum-compatible motions for design purposes,' *J. Eng. Mechs. Div., Proc. ASCE,* **98**, 345–356, 1972.
5. P. C. Rizzo, D. E. Shaw, and S. J. Jarecki, 'Development of real/synthetic time histories to match smooth design spectra,' Paper K1/5, Second International Conference on Structural Mechanics in Reactor Technology, Berlin, W. Germany, 10–14 September 1973.
6. S. Levy and J. P. D. Wilkinson, 'Generation of artificial time histories, rich in all frequencies, from given response spectra,' Paper F2/5, Third International Conference on Structural Mechanics in Reactor Technology, London, UK, 1–5 September 1975.
7. L. K. Liu, C. L. Child, and B. Nowotny, 'The effects of parameter variations on floor response spectra,' paper presented at the Specialty Conference on Structural Design of Nuclear Plant Facilities, Chicago, Illinois, 17–18 December 1973.
8. H. Lamb, 'On the propagation of tremors over the surface of an elastic solid,' *Phil. Trans. Roy. Soc., London,* **203**, Ser. A, 1–42, 1904.
9. E. Reissner, 'Stationäre, Axialsymmetrische durch eine Schüttelnde Masse erregte Schwingungen eines homogenen elastischen Halbraumes,' *Ingenieur—Archiv,* **7**, 381–396, 1936.
10. E. Reissner, 'Freie und erzwungene Torsionschwingungen des elastischen Halbraumes,' *Ingenieur—Archiv,* **8**, 229–245, 1937.
11. G. N. Bycroft, 'Forced vibrations of a rigid circular plate on a semi-infinite elastic space and on elastic stratum,' *Phil. Trans. Roy. Soc., London,* **248**, Ser. A, 327–368, 1956.
12. F. E. Richart, J. R. Hall, Jr., and R. D. Woods, *Vibrations of Soils and Foundations,* Prentice-Hall, Inc., Englewood Cliffs, New Jersey, 1970.
13. R. V. Whitman, 'Soil–structure interaction,' in R. J. Hansen (Ed.), *Seismic Design in Nuclear Power Plants,* pp. 245–269, The MIT Press, Cambridge, Massachusetts, 1970.
14. N. M. Newmark and E. Rosenblueth, *op. cit.,* pp. 96–98.
15. H. Tajimi, 'Dynamic analysis of a structure embedded in an elastic stratum,' *Proc. Fourth World Conference on Earthquake Engineering,* Chile, 1969.
16. G. F. Weissmann, 'Torsional vibrations of circular foundations,' *Proc. ASCE, J. Soil Mechs. and Foundation Div.,* **97**, 1293–1316, 1971.
17. R. W. Clough, 'Analysis of structural vibrations and dynamic response,' in R. H. Gallagher, Y. Yamada, and J. T. Oden (Eds.), *Recent Advances in Matrix Methods of Structural Analysis and Design,* pp. 441–482 (particularly pp. 467, 472, 474), The University of Alabama Press, University, Alabama, 1970.
18. Y. Yamada, 'Dynamic analysis of civil engineering structures,' in R. H. Gallagher, Y. Yamada, and J. T. Oden (Eds.), *Recent Advances in Matrix Methods of Structural Analysis and Design,* pp. 487–512 (particularly pp. 488, 491–497), The University of Alabama Press, University, Alabama, 1970.
19. J. Lysmer, T. Udaka, H. B. Seed, and R. Hwang, 'LUSH—A computer program for complex response analysis of soil–structure systems,' Report No. EERC 74-4, College of Engineering, University of California, Berkeley, California, April 1974.
20. J. Isenberg and S. A. Adham, 'Interaction of soil and power plants in earthquakes,' *Proc. ASCE, J. Power Div.,* **98**, No. PO2, 273–291, 1972.
21. S. L. Chu, P. K. Agrawal, and S. Singh, 'Finite element treatment of soil–structure interaction problem for nuclear power plant under seismic excitation,' Paper K 2/4, Second International Conference on Structural Mechanics in Reactor Technology, Berlin, W. Germany, 10–14 September 1973.
22. H. B. Seed and I. M. Idris, 'Influence of soil conditions on ground motions during earthquakes,' *Proc. ASCE, J. Soil Mechs. and Foundation Div.,* **95**, No. SM1, 99–137, 1969.
23. J. Lysmer, H. B. Seed, T. Udaka, R. N. Hwang, and C.-F. Tsai, 'Efficient finite element analysis of seismic structure—soil–structure interaction,' paper presented at Second ASCE Specialty Conference on Structural Design of Nuclear Plant Facilities, New Orleans, Dec. 1975; also Report no. EERC 75-34, Earthquake Engineering Research Center, University of California, Berkeley, California, 1975.
24. K. Muto, T. Hayashi, K. Omatsuzawa, T. Ohta, K. Uchida, and Y. Kasai, 'Comparative forced vibration test of two BWR-type reactor buildings,' Paper K 5/3, Second International Conference on Structural Mechanics in Reactor Technology, Berlin, W. Germany, 10–14 September 1973.
25. K. Muto and K. Omatsuzawa, 'Earthquake response analysis for a BWR nuclear power plant using recorded data,' *Nuclear Engineering and Design,* **20**, 385–392, 1972.
26. T. Takemori and I. Hama, 'Seismic analysis of reactor containment facility for BWR plant (coupled vibration with attached buildings),' Paper K 4/6, Second International Conference on Structural Mechanics in Reactor Technology, Berlin, W. Germany, 10–14 September 1973.

27. W. C. Hurty, 'Dynamic analysis of structural systems using component modes,' *AIAA J.*, **3**, 678–685, 1965.
28. W. C. Hurty and M. F. Rubenstein, *Dynamics of Structures,* Prentice-Hall, Inc., Englewood Cliffs, New Jersey, 1964.
29. W. A. Benfield and R. F. Hruda, 'Vibration analysis of structures by component mode substitution,' *AIAA J.*, **9**, 1255–1261, 1971.

VIBRATIONS OF STRUCTURAL COMPONENTS SUBMERGED IN WATER

9.1 INTRODUCTION

When an object which is submerged in a fluid vibrates, its motion causes some of the fluid to move with it. The mass of the moving fluid has an important effect on the dynamics of the vibrating object, particularly on its natural frequencies. The dynamics of structural components in fluids are often studied by introducing the concept of the *added mass* (otherwise known as virtual mass, or hydrodynamic mass). This additional mass represents inertial effects that the component incurs due to the motion of the adjacent fluid. It is possible to estimate the amount of the added mass, and a method of so doing will be presented in this chapter.

When several structural components are submerged in the same fluid, the motion of the fluid during vibration causes an interaction between the components. Their vibrational responses become coupled. The added mass concept can be extended to this case by introducing an added mass matrix, which expresses the interactive influence of the fluid on the vibrating components.

This chapter will describe how the added mass matrix can be developed, and how, once the added mass is known, the response of submerged structural components can be computed by using the basic computer programs described in chapter 3. We shall limit the discussion to structural systems in which the flow of fluid is two-dimensional, although the basic concepts can be readily extended to three dimensions. Examples of such systems include boilers, heat exchangers, and other piping systems whose lengths are substantially greater than the cross-sectional dimensions. Thus, in what follows, we shall study, as examples of these structures, the vibrational response of long beams enclosed in a vessel of water and subject to external excitations like those developed during a seismic event.

9.2   POTENTIAL FLOW—A VARIATIONAL PRINCIPLE

For many purposes, the motion of the fluid may be adequately represented by the potential flow theory of an incompressible fluid. In the case of small amplitude motions, it can be shown[1,2] that the fluid pressure p is represented by Laplace's equation

$$\nabla^2 p = 0 \tag{9.1a}$$

subject to the boundary condition that, on the surfaces of the submerged structure,

$$\frac{\partial p}{\partial n} = -\rho a_n \tag{9.1b}$$

Here n is the normal from the boundary into the fluid, and a_n is the acceleration of the boundary in the normal direction. The fluid density is given by ρ.

In the two-dimensional case that we consider here, Laplace's equation is written as

$$\frac{\partial^2 p}{\partial x^2} + \frac{\partial^2 p}{\partial y^2} = 0 \tag{9.2a}$$

and the boundary condition can be stated as

$$l_x \frac{\partial p}{\partial x} + l_y \frac{\partial p}{\partial y} = -\rho a_n \tag{9.2b}$$

Here, x and y are the coordinates of the two-dimensional plane, while l_x and l_y are the direction cosines of the normal to the boundary directed into the fluid.

We shall first discuss a method for the solution of Laplace's equation in two dimensions, without specific reference to the added mass determination. To solve this equation, we shall use the finite element method. The development of the numerical analysis will be described in some detail because it will be approached in a slightly different (but entirely equivalent) manner from the development of the finite element method applied to an elastic continuum which was described in chapter 5. Having shown how the analysis proceeds in the general case, we shall adapt it to the specific problem of the added mass.

It is shown by Zienkiewicz and Nath[2] that the solution of eqs. (9.2) is equivalent to the variational problem of minimizing the functional

$$\Psi = 1/2 \int \int_A \left[\left(\frac{\partial p}{\partial x} \right)^2 + \left(\frac{\partial p}{\partial y} \right)^2 \right] dx \, dy + \int_C \rho a_n p \, ds \tag{9.3}$$

where A represents the area of the fluid section and C denotes the contour of the fluid boundaries. The integral around the boundaries is such that when the right hand faces the fluid, the direction of integration is forward. Thus, integration is carried out in an anticlockwise direction on interior boundaries and in a clockwise direction on the outer boundary. To demonstrate that the variational problem of minimizing Ψ in eq. (9.3) is equivalent to eqs. (9.2), we compute the variation $\delta\Psi$. Denoting $\partial p/\partial x$ by p_x and $\partial p/\partial y$ by p_y, we have

$$\delta\Psi = 1/2 \int \int_A \left[\frac{\partial}{\partial p_x} (p_x^2 + p_y^2) \delta p_x + \frac{\partial}{\partial p_y} (p_x^2 + p_y^2) \delta p_y \right] dx \, dy + \int_C \rho a_n \delta p \, ds \tag{9.4}$$

which reduces to

$$\delta\Psi = \int \int_A (p_x \delta p_x + p_y \delta p_y) \, dx \, dy + \int_C \rho a_n \delta p \, ds \tag{9.5}$$

Now the commutation rule applies between the operators δ and $\partial/\partial x$, so that

$$\delta p_x = \delta \left(\frac{\partial p}{\partial x} \right) = \frac{\partial}{\partial x} \delta p$$

and, similarly,

$$\delta p_y = \frac{\partial}{\partial y} \delta p$$

Thus,

$$\delta\Psi = \int \int_A \left[p_x \frac{\partial}{\partial x} \delta p + p_y \frac{\partial}{\partial y} \delta p \right] dx \, dy + \int_C \rho a_n \delta p \, ds \tag{9.6}$$

We now carry out an integration by parts. Observing that

$$\int \int_A p_x \frac{\partial}{\partial x} \delta p \, dx \, dy = \int_C p_x \delta p \, l_x \, ds - \int \int_A \frac{\partial}{\partial x} p_x \delta p \, dx \, dy \tag{9.7}$$

we have

$$\delta\Psi = \int \int_A \delta p \left[-\frac{\partial}{\partial x} p_x - \frac{\partial}{\partial y} p_y \right] dx \, dy + \int_C \delta p [\rho a_n + l_x p_x + l_y p_y] \, ds \tag{9.8}$$

In order to minimize the functional Ψ, we require that, for all variations δp,

$$\delta\Psi = 0 \tag{9.9}$$

This condition can only be satisfied if the expressions inside the integrals of eq. (9.8) vanish. Thus, we require

$$-\frac{\partial}{\partial x}p_x - \frac{\partial}{\partial y}p_y = 0 \text{ in } A \tag{9.10a}$$

$$\rho a_n + l_x p_x + l_y p_y = 0 \text{ on } C \tag{9.10b}$$

and these equations can be rewritten as

$$\frac{\partial^2 p}{\partial x^2} + \frac{\partial^2 p}{\partial y^2} = 0 \text{ in } A \tag{9.11a}$$

$$l_x \frac{\partial p}{\partial x} + l_y \frac{\partial p}{\partial y} = -\rho a_n \text{ on } C \tag{9.11b}$$

These equations, (9.11a) and (9.11b), are the same as those which we originally used to define the fluid pressures. Thus, we have demonstrated that the minimization of the functional Ψ of eq. (9.3) is equivalent to the solution of the field equation (9.2). The pressure field that minimizes eq. (9.3) is the same pressure field that satisfies Laplace's equation (9.2a), and the pressure gradient at the boundary that minimizes eq. (9.3) is the same pressure gradient at the boundary that satisfies the boundary condition eq. (9.2b).

9.3 THE FINITE ELEMENT METHOD APPLIED TO POTENTIAL FLOW

Having now shown that the pressure field that minimizes the functional Ψ of eq. (9.3) also satisfies the appropriate field equation and boundary conditions, we shall proceed to carry out the minimization by means of the finite element method.

Let the area A of the fluid surface be represented by a number of triangular elements, as shown in Fig. 9.1. There, we show an interior body surrounded by fluid and enclosed by an outer boundary. It is assumed that the accelerations of the boundaries are known. Now, attention is fixed on a single triangle, shown in Fig. 9.2(a). The nodes of the triangle are numbered by the letters k, l, n, as shown. We assume that within the triangular area, the pressure varies linearly in the coordinates x and y. Thus, we write

$$p = \alpha_1 + \alpha_2 x + \alpha_3 y \tag{9.12}$$

where α_1, α_2, and α_3 are constants to be determined. The pressure may also be written in terms of the pressures at the nodes, p_k, p_l, p_n. We observe that

$$\begin{Bmatrix} p_k \\ p_l \\ p_n \end{Bmatrix} = \begin{bmatrix} 1 & x_k & y_k \\ 1 & x_l & y_l \\ 1 & x_n & y_n \end{bmatrix} \begin{Bmatrix} \alpha_1 \\ \alpha_2 \\ \alpha_3 \end{Bmatrix} \tag{9.13}$$

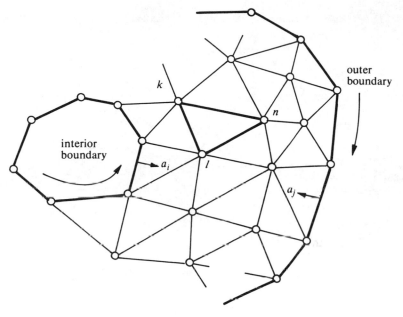

FIGURE 9.1
Representing the fluid field by triangular finite elements

so that, solving for α_i, we have

$$\begin{Bmatrix} \alpha_1 \\ \alpha_2 \\ \alpha_3 \end{Bmatrix} = \begin{bmatrix} 1 & x_k & y_k \\ 1 & x_l & y_l \\ 1 & x_n & y_n \end{bmatrix}^{-1} \begin{Bmatrix} p_k \\ p_l \\ p_n \end{Bmatrix}$$

$$= [d_{ij}] \begin{Bmatrix} p_k \\ p_l \\ p_n \end{Bmatrix} \tag{9.14}$$

In addition, we note from eq. (9.12) that

$$\frac{\partial p}{\partial x} = \alpha_2, \qquad \frac{\partial p}{\partial y} = \alpha_3$$

As a result,

$$\begin{Bmatrix} \dfrac{\partial p}{\partial x} \\ \dfrac{\partial p}{\partial y} \end{Bmatrix} = \begin{bmatrix} d_{21} & d_{22} & d_{23} \\ d_{31} & d_{32} & d_{33} \end{bmatrix} \begin{Bmatrix} p_k \\ p_l \\ p_n \end{Bmatrix}$$

$$= [D]\{p_e\} \tag{9.15}$$

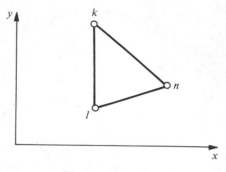

(a) A triangular element

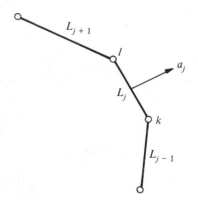

FIGURE 9.2
Element and boundary segment geometries

(b) Boundary segments

We have defined the pressures at the nodes of the element as $\{p_e\}$:

$$\{p_e\} = \begin{Bmatrix} p_k \\ p_l \\ p_n \end{Bmatrix} \tag{9.16}$$

Since

$$\begin{bmatrix} \dfrac{\partial p}{\partial x} & \dfrac{\partial p}{\partial y} \end{bmatrix} \begin{Bmatrix} \dfrac{\partial p}{\partial x} \\ \dfrac{\partial p}{\partial y} \end{Bmatrix} = \left(\dfrac{\partial p}{\partial x}\right)^2 + \left(\dfrac{\partial p}{\partial y}\right)^2 \tag{9.17}$$

we also have

$$\left(\frac{\partial p}{\partial x}\right)^2 + \left(\frac{\partial p}{\partial y}\right)^2 = \{p_e\}^T [D]^T [D] \{p_e\} \tag{9.18}$$

The first portion of the functional Ψ in this element is therefore given by

$$\Psi^e = \frac{1}{2} \int\!\!\int_{A_e} \left[\left(\frac{\partial p}{\partial x}\right)^2 + \left(\frac{\partial p}{\partial y}\right)^2 \right] dx \, dy$$

$$= \frac{1}{2} \{p_e\}^T [D]^T [D] \{p_e\} A_e \tag{9.19}$$

$$= \frac{1}{2} \{p_e\}^T [h_e] \{p_e\}$$

where A_e is the area of the element. The area is related to the coordinates of the element nodes, and it can be shown that

$$2A_e = \det \begin{bmatrix} 1 & x_k & y_k \\ 1 & x_l & y_l \\ 1 & x_n & y_n \end{bmatrix} \tag{9.20}$$

The remaining integral of the functional Ψ has a contribution only at the fluid boundary. Imagine that the jth segment of the boundary is formed by the line of length L_j between two nodes k and l, as shown by Fig. 9.2(b). The normal acceleration of the segment is denoted by a_j. Then the contribution Ψ^b of the segment is

$$\Psi^b = \int_{C_j} \rho a_j p \, ds = \tfrac{1}{2} \rho a_j L_j (p_k + p_l)$$

$$= \{g_b\}^T \{p_b\} \tag{9.21}$$

where

$$L_j = [(x_k - x_l)^2 + (y_k - y_l)^2]^{1/2}$$

and the elements of $\{g_b\}$ are made up of the coefficients of the segment boundary pressures $\{p_b\}$. The process of assembly follows by the addition of the element contributions. Thus, if we compute the sum of all element contributions, we have

$$\Psi^E = \sum_{\text{elements}} \Psi^e \tag{9.22}$$

and, similarly, the sum of all boundary segment contributions is

$$\Psi^B = \sum_{\text{segments}} \Psi^b \tag{9.23}$$

Now, the total functional Ψ is the sum of contributions from the elements and the boundaries,

$$\Psi = \Psi^E + \Psi^B \qquad (9.24)$$

and, as a result of these additions, we have

$$\Psi = \frac{1}{2}\{p\}^T[H]\{p\} + \{G\}^T\{p\} \qquad (9.25)$$

where $\{p\}$ is the vector of all nodal pressures. The matrices $[H]$ and $\{G\}$ are determined during the assembly process by the appropriate addition of the contributions $[h_e]$ and $\{g_b\}$, respectively.

In order to minimize the functional Ψ with respect to each unknown nodal pressure p, we require that

$$\frac{\partial \Psi}{\partial \{p\}} = 0 \qquad (9.26)$$

By carrying out the indicated differentiation, it will be found that

$$[H]\{p\} + \{G\} = 0 \qquad (9.27)$$

Equation (9.27) forms a set of simultaneous algebraic equations in the unknown nodal pressures $\{p\}$. Solution of this set of equations by Gaussian elimination yields the required pressure field in the fluid.

The similarity between the finite element process described here and that applied to an elastic continuum in chapter 5 is striking. Indeed, as shown by Zienkiewicz and Nath,[2] the processes are identical. The process of determining element matrices and subsequent assembly of these matrices is the same in both cases. For the elastic solid, the functional that is being minimized is the total work done by the internal and external forces. Thus, an alternative way to approach the finite element method in elastic continua is by way of the minimization of this functional. A related finite element approach to added mass calculations has been described by Hartz and Schmid.[3]

The use of triangular elements as opposed to quadrilateral elements (as in the elastic continuum) arises from a desire for simplicity, since it is adequate for the pressures to be linearly varying in the element. For the elastic continuum, on the other hand, we allowed the displacements to vary quadratically so that upon differentiation, the strains were found to be linear. Greater accuracy was required in that case, and the quadrilateral element was able to meet that demand.

9.4 THE FINITE ELEMENT METHOD APPLIED TO ADDED MASS CALCULATIONS

The determination of the added mass matrix involves a special case of the general field equation (9.2). In this case, we wish to determine the forces induced on all the submerged boundaries by a unit acceleration of each body in turn. The forces are a

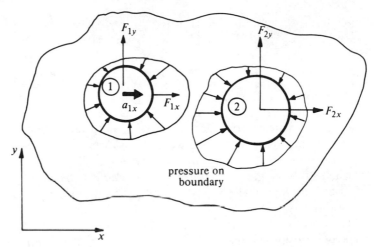

FIGURE 9.3

An acceleration a_{1x} of body 1 in the x-direction causes a pressure field in the fluid. The forces the bodies exert on the fluid in the x- and the y-directions can be computed by integrating the boundary pressures in those directions

direct expression of the added mass. Thus, consider two bodies immersed in a fluid, as shown in Fig. 9.3. Body 1 is given an acceleration a_{1x} in the x-direction. The resulting pressures in the fluid are calculated. The forces experienced by each body in the x- and y-directions are obtained by integration of the resultant boundary pressures in those directions. Thus, we have, because of the acceleration a_{1x}, the forces F:

$$\begin{Bmatrix} F_{1x,1x} \\ F_{1y,1x} \\ F_{2x,1x} \\ F_{2y,1x} \end{Bmatrix} = \begin{Bmatrix} M_{11} \\ M_{21} \\ M_{31} \\ M_{41} \end{Bmatrix} a_{1x} \tag{9.28}$$

Now, if we apply other accelerations a_{1y}, a_{2x}, and a_{2y}, in turn, we develop the forces

$$[F] = [M][a] \tag{9.29}$$

where

$$[a] = \begin{bmatrix} a_{1x} & 0 & 0 & 0 \\ 0 & a_{1y} & 0 & 0 \\ 0 & 0 & a_{2x} & 0 \\ 0 & 0 & 0 & a_{2y} \end{bmatrix}$$

$$[M] = \begin{bmatrix} M_{11} & M_{12} & M_{13} & M_{14} \\ M_{21} & M_{22} & M_{23} & M_{24} \\ M_{31} & M_{32} & M_{33} & M_{34} \\ M_{41} & M_{42} & M_{43} & M_{44} \end{bmatrix}$$

$$[F] = \begin{bmatrix} F_{1x,1x} & F_{1x,1y} & F_{1x,2x} & F_{1x,2y} \\ F_{1y,1x} & F_{1y,1y} & F_{1y,2x} & F_{1y,2y} \\ F_{2x,1x} & F_{2x,1y} & F_{2x,2x} & F_{2x,2y} \\ F_{2y,1x} & F_{2y,1y} & F_{2y,2x} & F_{2y,2y} \end{bmatrix}$$

In general, with n bodies, we can generate relationships like eq. (9.29), except that the matrices are $(2n \times 2n)$ in size. The elements of the matrix $[M]$ are the hydrodynamic mass coefficients, and represent the inertia coefficients that give the forces on each body when the accelerations of each body are specified.

To determine the added mass matrix, we apply individually a unit acceleration to each body in the two coordinate directions. The development of the element matrices and their assembly is done only once for all these variations in acceleration, since the matrix $[H]$ of eq. (9.27) is the same for all the boundary value problems—only the vector $\{G\}$ changes. The vector $\{G\}$ for a single acceleration condition is expanded to form a matrix $[G]$ having twice as many columns as the number of bodies. Successive pairs of columns of $[G]$ correspond to unit acceleration of a body in the x- and y-directions. Suppose we number consecutive nodes along a boundary, as shown in Fig. 9.4. We find that

$$\{g_b\}^T = \tfrac{1}{2}\rho a_n L_n [1, 1] \tag{9.30}$$

The body accelerations for the pair of columns in $[G]$ being considered is 1 for a_x and 0 for a_y for the first of the pair, and 0 for a_x and 1 for a_y for the second. For all other bodies, a_x and a_y are both zero. For the normal acceleration a_n of the segment, we have

$$a_n = a_x \cos \theta + a_y \sin \theta \tag{9.31}$$

FIGURE 9.4
Geometrical relationship between acceleration components on a triangle boundary

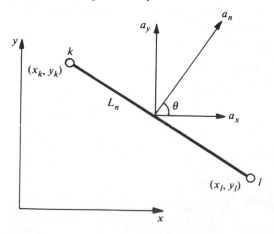

and since

$$\cos \theta = (y_k - y_l)/L_n$$

and

$$\sin \theta = (x_l - x_k)/L_n$$

then

$$\{g_b\}^T = \tfrac{1}{2}\rho\left[a_x(y_k - y_l) + a_y(x_l - x_n)\right]\begin{Bmatrix}1\\1\end{Bmatrix}^T \tag{9.32}$$

Having the segment matrices for each column in $[G]$ permits the assembly of $[G]$.
   A solution of the system

$$[H]\{p\} + [G] = 0 \tag{9.33}$$

gives columns of nodal pressures for each body acceleration condition. The forces that the body exerts on the fluid in the x- and y-directions are found by integrating the pressures around the boundaries. We have, for example, for the contribution f from a specific boundary segment between the two nodes k and l, the forces

$$\begin{Bmatrix}f_x\\f_y\end{Bmatrix} = \begin{bmatrix}1/2(y_k - y_l) & 1/2(y_k - y_l)\\1/2(x_l - x_k) & 1/2(x_l - x_k)\end{bmatrix}\begin{Bmatrix}p_l\\p_k\end{Bmatrix} \tag{9.34}$$

Addition of all contributions around the boundary of a body yields the total integrated forces on that body. These forces, in turn, make up a column of the added mass matrix.

   It is interesting to note that the same terms that enter into the assembly of $[G]$, eq. (9.32), are needed in going from pressure to virtual mass, except for the factor ρ. The added mass matrix is obtained from the computed pressure matrix $[p]$ and boundary matrix $[G]$ as

$$[M] = 1/\rho[G]^T[p] \tag{9.35}$$

Thus, the added mass matrix is found directly once the pressures are determined, without the specific need to actually perform the force integrations discussed previously.

9.5   EXAMPLE—CONCENTRIC CYLINDRICAL VESSELS

To illustrate the type of results that can be expected from the finite element approach described, we consider a 2 in. diameter cylinder situated concentrically in a 4 in. diameter cylinder, which is filled with a fluid of density 1·00 (for purposes of illustration). It has been shown[4] that the added mass matrix for concentric cyclinders of radius a and b is given by

$$\begin{Bmatrix}F_{1x}\\F_{1y}\\F_{2x}\\F_{2y}\end{Bmatrix} = \begin{bmatrix}m_{11} & 0 & m_{13} & 0\\0 & m_{22} & 0 & m_{24}\\m_{13} & 0 & m_{33} & 0\\0 & m_{24} & 0 & m_{44}\end{bmatrix}\begin{Bmatrix}a_{1x}\\a_{1y}\\a_{2x}\\a_{2y}\end{Bmatrix} \tag{9.36}$$

where

$$m_{11} = m_{22} = \rho \pi a^2 \left(\frac{b^2 + a^2}{b^2 - a^2} \right)$$

$$m_{13} = m_{24} = \frac{-2\rho \pi a^2 b^2}{(b^2 - a^2)}$$

$$m_{33} = m_{44} = \rho \pi b^2 \left(\frac{b^2 + a^2}{b^2 - a^2} \right)$$

(9.37)

For the case considered here, we get

$$m_{11} = m_{22} = 5 \cdot 23$$

$$m_{13} = m_{24} = -8 \cdot 38$$

$$m_{33} = m_{44} = 20 \cdot 95$$

Upon applying the finite element method with the mesh shown in Fig. 9.5, it will be found that

$$m_{11} = m_{22} = 5 \cdot 169$$

$$m_{13} = m_{24} = -8 \cdot 293$$

$$m_{33} = m_{44} = 20 \cdot 792$$

The difference in the exact and finite element solutions is about one per cent.

FIGURE 9.5
Finite element mesh used for two concentric cylinders

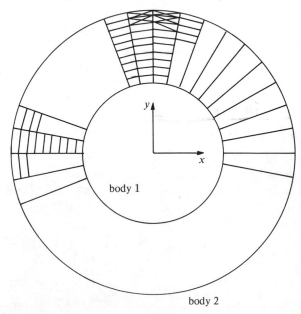

If the cylinders are placed eccentrically by moving the inner cylinder so its center lies 0·5 in. along the x-axis from the center of the outer cylinder, the finite element results become

$$
\begin{Bmatrix} F_{1x} \\ F_{1y} \\ F_{2x} \\ F_{2y} \end{Bmatrix} = \begin{bmatrix} 5\cdot605 & 0 & -8\cdot729 & 0 \\ 0 & 5\cdot598 & 0 & -8\cdot722 \\ -8\cdot720 & 0 & 21\cdot228 & 0 \\ 0 & -8\cdot722 & 0 & 21\cdot221 \end{bmatrix} \begin{Bmatrix} a_{1x} \\ a_{1y} \\ a_{2x} \\ a_{2y} \end{Bmatrix}
$$

It appears that there is therefore a moderate increase in all the coefficients.

Many other configurations of a simple geometric nature have been studied, and some exact solutions are available for some of them. Some have been compiled by Patton[5] and by Fritz.[6] Experimental analyses that show the effects of end-leakage and viscosity are discussed by Liu,[4] Fritz,[6] and Mavis.[7]

9.6 EXAMPLE—ARRAY OF SQUARE BEAMS IN A CYLINDRICAL CONTAINER

As an example of a more complex situation of irregular geometry, consider an array of four square beams, shown in Fig. 9.6. In this case, we shall obtain the added mass of the four beams when they are contained in a cylindrical vessel of water of radius 15 in.

The finite element mesh used for the calculation of the added masses is shown in Fig. 9.6. The resulting pressure fields corresponding to unit accelerations of one of the beams are shown in Fig. 9.7. Of interest are the relatively high pressures in the narrow passages between the beams. Such pressures are due to the rapid flow of water as it is squeezed out of the passageway due to the movement of the beams. The added mass of such passageways has been studied by Sharp and Wenzel.[8]

The added mass matrix is shown in Table 9.1 in the form that it is printed by the computer. This matrix is symmetric and the diagonal terms tend to dominate. The added masses will be used as a basic input when we come to study the dynamics of submerged beams of this cross-section.

9.7 FLOW IN NARROW PASSAGEWAYS—ADDED MASS APPROXIMATIONS

Another situation arises when two flat bodies, such as two beams, are situated close together, so that a relatively thin film or fluid separates them. In that case, an approximate solution for the added mass can be obtained which is a function of the distance between the beams. This expression therefore accounts for the closure between the beams. When used in a dynamic study, it causes a nonlinearity because the added mass varies with the geometry. In this development, we again consider the case of potential flow, valid generally for low Reynolds numbers. For higher Reynolds numbers (> 1000), viscous effects become progressively more important.

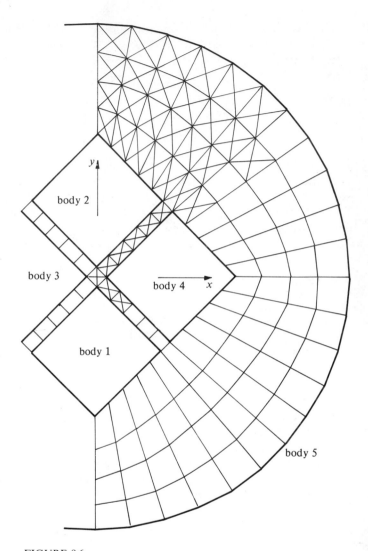

FIGURE 9.6
Section of beams and outer cylindrical container, showing finite element
mesh

Provision can be made for these viscous effects in an approximate way, but we shall
not pursue this point here.

Consider two long beams of length l, width h, and a distance s apart. The
cross-sectional geometry is shown in Fig. 9.8, and it is assumed that $s \ll h \ll l$. As
the two beams approach one another, water is forced out of the gap with a velocity

$$V = \frac{-\dot{s}x}{s} \qquad (9.38)$$

Table 9.1 HYDRODYNAMIC WATER MASS MATRIX FOR FOUR BEAMS AND CYLINDRICAL CONTAINER

1X BODY	1Y BODY	2X BODY	2Y BODY	3X BODY	3Y BODY	4X BODY	4Y BODY	5X BODY	5Y BODY
0.00654	0.00000	-0.00037	-0.00372	-0.00250	0.00000	-0.00037	0.00372	-0.00607	0.00000
0.00000	0.00524	-0.00182	-0.00037	0.00000	0.00065	0.00182	-0.00037	0.00000	-0.00792
-0.00037	-0.00182	0.00524	0.00000	-0.00037	0.00182	0.00065	0.00000	-0.00792	0.00000
-0.00372	-0.00037	0.00000	0.00654	0.00372	-0.00037	0.00000	-0.00250	0.00000	-0.00607
-0.00250	0.00000	-0.00037	0.00372	0.00654	0.00000	-0.00037	-0.00372	-0.00607	0.00000
0.00000	0.00065	0.00182	-0.00037	0.00000	0.00524	-0.00182	-0.00037	0.00000	-0.00792
-0.00037	0.00182	0.00065	0.00000	-0.00037	-0.00182	0.00524	0.00000	-0.00792	0.00000
0.00372	-0.00037	0.00000	-0.00250	-0.00372	-0.00037	0.00000	0.00654	0.00000	-0.00607
-0.00607	0.00000	-0.00792	0.00000	-0.00607	0.00000	-0.00792	0.00000	0.09391	0.00000
0.00000	-0.00792	0.00000	-0.00607	0.00000	-0.00792	0.00000	-0.00607	0.00000	0.09391

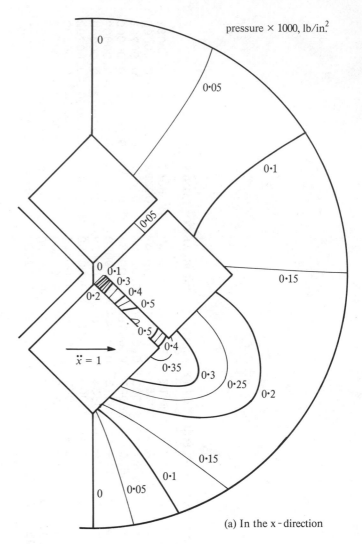

pressure × 1000, lb/in.^{2}

(a) In the x-direction

FIGURE 9.7
Pressure distribution for unit acceleration of body 1

The acceleration of the water is

$$\dot{V} = \left(\frac{x}{s}\right)\left(-\ddot{s} + \frac{\dot{s}^2}{s}\right) \tag{9.39}$$

Denoting by $\dot{V}_h$ the velocity at the end of the channel at $x = h/2$, we have, from eq. (9.39),

$$\dot{V} = \frac{2x}{h}\,\dot{V}_h$$

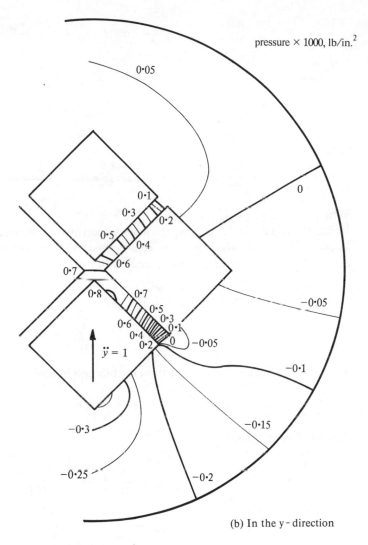

pressure × 1000, lb/in.^{2}

(b) In the y - direction

FIGURE 9.7—*continued*

Denoting the pressure by p and the density by ρ, we have, from considerations of equilibrium,

$$\frac{\mathrm{d}p}{\mathrm{d}x} = -\rho \dot{V} = -\rho \frac{2x}{h} \dot{V}_h \qquad (9.40)$$

Integration with respect to x gives

$$p = -\rho \dot{V}_h \frac{x^2}{h} + \text{constant} \qquad (9.41)$$

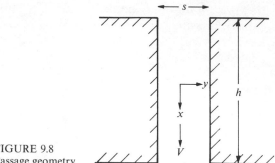

FIGURE 9.8
Passage geometry

When the gap opens, the pressure at $x = h/2$ is $-\rho V_h^2/2$, as a consequence of Bernoulli's theorem. When the gap closes, experiments with jets show that the pressure at $x = h/2$ is zero, since the flow separates at the exit. Using these conditions to evaluate the constant in eq. (9.41), we find that

$$
\left.
\begin{aligned}
p_{\text{opening}} &= \rho \left(\frac{h}{4} - \frac{x^2}{h} \right) \left(-\frac{\ddot{s}h}{2s} + \frac{\dot{s}^2 h}{2s^2} \right) - \rho \left(\frac{\dot{s}^2 h^2}{8s^2} \right) \\[2mm]
p_{\text{closing}} &= \rho \left(\frac{h}{4} - \frac{x^2}{h} \right) \left(-\frac{\ddot{s}h}{2s} + \frac{\dot{s}^2 h}{2s^2} \right)
\end{aligned}
\right\}
\tag{9.42}
$$

The force the beams exert on the fluid is obtained in each case by integration, so that

$$
F = l \int_{-h/2}^{h/2} p \, dx
\tag{9.43}
$$

Substituting eq. (9.42) gives

$$
\left.
\begin{aligned}
F_{\text{opening}} &= \frac{\rho l h^3}{12s} \left(-\ddot{s} - \frac{\dot{s}^2}{2s} \right) \\[2mm]
F_{\text{closing}} &= \frac{\rho l h^3}{12s} \left(-\ddot{s} + \frac{\dot{s}^2}{s} \right)
\end{aligned}
\right\}
\tag{9.44}
$$

We observe that the force on the beam depends on the gap size s, as well as on its velocity and acceleration of closure. The force component $-(\rho l h^3/12s)\ddot{s}$ can be thought of as an added mass term which varies with geometry, while the other component, which varies with the square of velocity, can be interpreted as a damping effect. Because of the divisor s, the forces become very large as the gap becomes small.

   In terms of the component element method, we can view the force expressions of eq. (9.44) as an element. Thus, we introduce into the computer programs another element whose force varies with gap size as in eq. (9.44). The acceleration term could be modeled in terms of time increment steps as

$$
\ddot{s} = \frac{(2s_0 - 5s_{-1} + 4s_{-2} - s_{-3})}{\Delta t^2}
\tag{9.45}
$$

However, when this expression is used, it can lead to divergence in the subsequent numerical integration of the equations of motion. For this reason, the $\ddot{s}$ term in eq. (9.44) is treated as an added mass of magnitude $l\rho h^3/12s$. However, the $\dot{s}^2$ term is treated as any other force would be, using the approximation

$$\dot{s} = \frac{(3s_0 - 4s_{-1} + s_{-2})}{2\Delta t} \tag{9.46}$$

to obtain the velocity of separation. When this element is used, the large forces that develop for very small s are considered unrealistic, since local roughness or lack of flatness is always present in actual configurations. To account for this effect, a stop element with an appropriate clearance should parallel the element modeling the force expressions of eq. (9.44).

To model the water-gap element in the computer programs of chapter 3, it is merely necessary to introduce them by using another INDX indicator, and by inserting the appropriate statements as described in section 4.1 on vehicle dynamics.

9.8 SEISMIC EXCITATION OF BEAMS ENCLOSED IN WATER

Consider a beam that is rigidly attached at one end to a rigid box full of water, as shown in Fig. 9.9(a). We shall study its response to a seismic input by applying a simple model having four degrees of freedom—two in translation and two in

FIGURE 9.9
Beam in water tank with seismic input

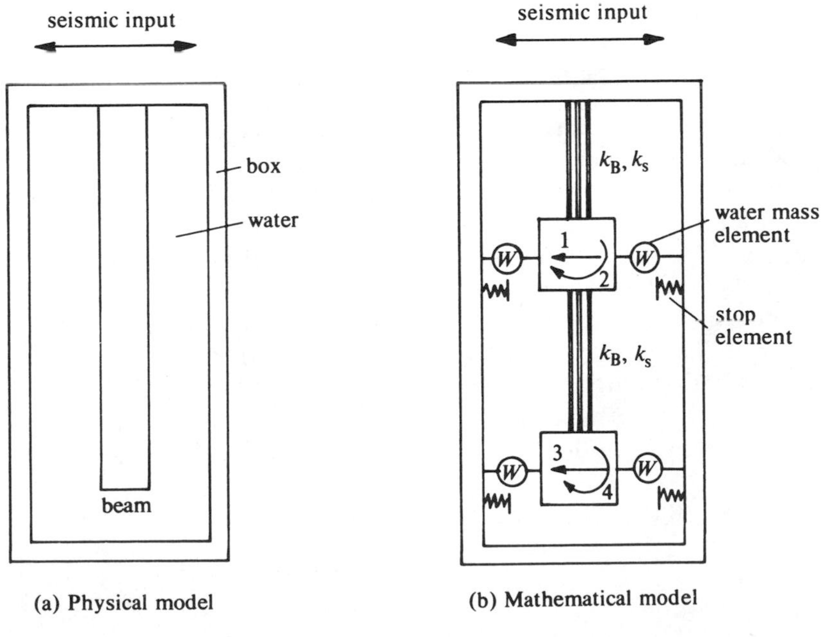

(a) Physical model   (b) Mathematical model

rotation, as shown in Fig. 9.9(b). The beam is 100 in. in total length and has an elastic modulus $E = 30 \times 10^6$ lb/in.^{2}, and a moment of inertia in bending of 10 in.^{4}. Consequently, using the formulas for the beam springs, eqs. (3.86) and (3.95), we find that

$$k_{\mathrm{B}} = 3 \times 10^6 \text{ lb/in.} \quad \text{and} \quad k_{\mathrm{S}} = 3600 \text{ lb/in.}$$

The beam is immersed in water and has 1/2 in. of clearance to the containing box.

During the vibration of the beam relative to the box, water is squeezed back and forth in the gap between it and the box, resulting in an added mass effect. The forces arising due to the water flow are modeled by the added mass elements discussed in the previous section and by stop elements that contacted at 0·02 in. from the wall. There are a total of four such pairs of elements, one pair to model the forces in each gap between the beam masses and the rigid box.

The coupling coefficients for the beam elements are given by eq. (3.102). The water mass elements on the left have coupling coefficients of -1 between the beam masses and the box, and those on the right have coefficients of $+1$.

The box is excited seismically. The excitation is given as a floor response spectrum, shown in Fig. 9.10(a). By using the method described in section 8.6, an acceleration time history was generated from this floor response spectrum. It is shown in Fig. 9.10(b). The duration of the excitation is 10 sec.

The results of the calculations are shown in Fig. 9.11, where the translational movement of the beam relative to the box is given for both masses. Because the water mass forces become very large as the gap closes, the side of the box acts as a stop and limits the travel of the beam. Note how the tip of the beam strikes the box a number of times during the excitation. The midpoint, however, does not move far enough to enter into contact.

9.9 EXCITATION OF BEAMS BY A SINUSOIDAL FREQUENCY SWEEP

The four beams whose added mass matrix was obtained in section 9.6 will be studied under a sinusoidal frequency sweep in the low frequency range. In this example, the beam 1 is transversely excited in the y-direction at one-third its length from the bottom. We introduce a limited number of coordinates to describe the motion of the right-hand half of the system. The four coordinates are denoted by A, B, C, D in Fig. 9.12(a). The actual motion of each beam is modeled as a single mode of a simply supported beam so that the displacement along the length l in the x- or y-direction (as the case may be) is $\sin(\pi z/l)$, where z is the axial coordinate. For such a deflection, the modal mass is given by

$$m = \int_0^l \rho y^2 \, \mathrm{d}z = \int_0^l \left(\frac{W}{lg}\right) \sin^2\left(\frac{\pi z}{l}\right) \mathrm{d}z = \frac{W}{2g} \tag{9.47}$$

where W is the weight of the beam. The modal stiffness of the beam is

$$k = \int_0^l EI \left(\frac{\mathrm{d}^2 y}{\mathrm{d}x^2}\right)^2 \mathrm{d}x = \frac{\pi^4 EI}{2l^3} \tag{9.48}$$

○ original spectrum
+ computed spectrum by section 8·6

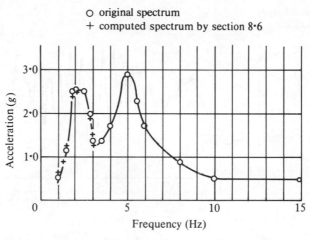

(a) Response spectrum of floor

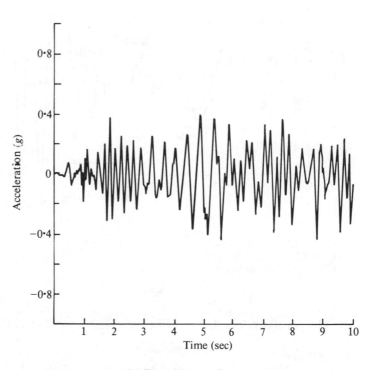

(b) Time history of acceleration

FIGURE 9.10
Seismic excitation

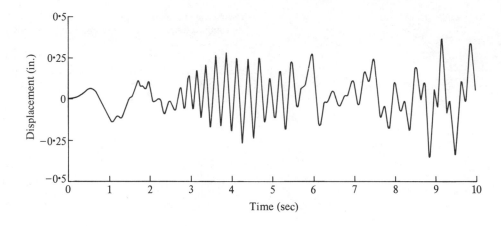

(a) Midpoint response

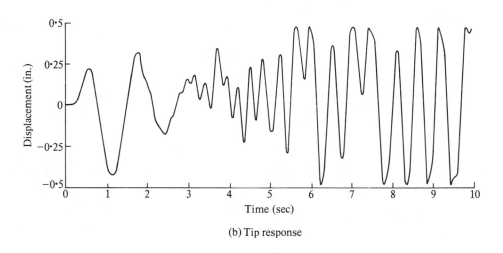

(b) Tip response

FIGURE 9.11
Response relative to box

The pertinent data that describe the mass and stiffness of the system are

beam weight = 615 lb

beam $EI = 108{\cdot}5 \times 10^6$ lb in.^{2}

density of water = 0·000 093 5 lb sec^2/in.^{4}

beam length = 165 in.

Thus, the modal masses in units of lb sec^2/in. are given by

$$m_A = 0\cdot398$$

$$m_B = 0\cdot398$$

$$m_C = 0\cdot796$$

$$m_D = 0\cdot796$$

and the modal stiffnesses in units of lb/in. are given by

$$k_A = 587\cdot5$$
$$k_B = 587\cdot5$$
$$k_C = 1175\cdot0$$
$$k_D = 1175\cdot0$$

We have already calculated the added mass matrix for this system in section 9.6. There, we included many more coordinates for the description of the motion, whereas, in the present example, we have limited ourselves to the four coordinates of Fig. 9.12. Because of the symmetry in Fig. 9.12(a), we can relate the mass effects of the co-ordinates A, B, C, D to those of the original system of Fig. 9.6(a) through the following transformations. The accelerations are related by

$$
\begin{Bmatrix} a_{1x} \\ a_{1y} \\ a_{2x} \\ a_{2y} \\ a_{3x} \\ a_{3y} \\ a_{4x} \\ a_{4y} \\ a_{5x} \\ a_{5y} \end{Bmatrix}
=
\begin{bmatrix}
0 & 0 & 0 & 0 \\
0 & 1 & 0 & 0 \\
0 & 0 & 0 & 0 \\
1 & 0 & 0 & 0 \\
0 & 0 & 0 & -1 \\
0 & 0 & 1 & 0 \\
0 & 0 & 0 & 1 \\
0 & 0 & 1 & 0 \\
0 & 0 & 0 & 0 \\
0 & 0 & 0 & 0
\end{bmatrix}
\begin{Bmatrix} a_A \\ a_B \\ a_C \\ a_D \end{Bmatrix}
$$

Denote this coupling matrix by $[MM_{\text{post}}]$. Also, the forces are related by

$$
\begin{Bmatrix} F_A \\ F_B \\ F_C \\ F_D \end{Bmatrix}
=
\begin{bmatrix}
0 & 0 & 0 & 1/2 & 0 & 0 & 0 & 0 \\
0 & 1/2 & 0 & 0 & 0 & 0 & 0 & 0 \\
0 & 0 & 0 & 0 & 0 & 0 & 0 & 1 \\
0 & 0 & 0 & 0 & 0 & 0 & 1 & 0
\end{bmatrix}
\begin{Bmatrix} F_{1x} \\ F_{1y} \\ F_{2x} \\ F_{2y} \\ F_{3x} \\ F_{3y} \\ F_{4x} \\ F_{4y} \end{Bmatrix}
$$

Denote this coupling matrix by $[MM_{\text{pre}}]$.

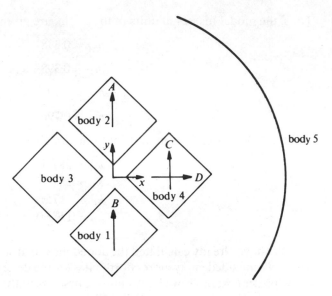

(a) Degrees of freedom in system

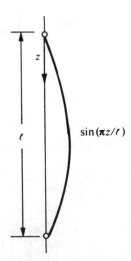

$\sin(\pi z/\ell)$

(b) Assumed mode shape in each degree of freedom

FIGURE 9.12
Degrees of freedom assumed in the system

By postmultiplying the full added mass matrix of Table 9.1 by $[MM_{post}]$ and premultiplying it by $[MM_{pre}]$, we obtain the following added mass matrix per unit length of beam:

$$\begin{bmatrix} 0\cdot003\ 27 & -0\cdot001\ 25 & -0\cdot000\ 37 & 0\cdot003\ 72 \\ -0\cdot001\ 25 & 0\cdot003\ 27 & -0\cdot000\ 37 & -0\cdot003\ 72 \\ -0\cdot000\ 37 & -0\cdot000\ 37 & 0\cdot005\ 89 & 0\cdot000\ 00 \\ 0\cdot003\ 72 & -0\cdot003\ 72 & 0\cdot000\ 00 & 0\cdot009\ 04 \end{bmatrix}$$

When this matrix is multiplied by $(l/2)$, we obtain the system added mass matrix:

$$[M_W] = \begin{bmatrix} 0\cdot270 & -0\cdot103 & -0\cdot031 & 0\cdot307 \\ -0\cdot103 & 0\cdot270 & -0\cdot031 & -0\cdot307 \\ -0\cdot031 & -0\cdot031 & 0\cdot489 & 0\cdot000 \\ 0\cdot307 & -0\cdot307 & 0\cdot000 & 0\cdot746 \end{bmatrix} \tag{9.49}$$

The system response to a sinusoidal force input on beam 1 in the y-direction at a position one-third the length up from the bottom can be found with the component element method. In this case, it is necessary to modify the computer program of section 3.16 so that the added mass matrix is read in as data, and so that this matrix is combined with the diagonal mass matrix of the coordinates. The equations of motion are then written as follows:

$$[M]_D \{\ddot{z}\} + [M_W] \{\ddot{z}\} = \{F\} \tag{9.50}$$

The accelerations $\ddot{z}$ at any time t_0 are determined by solving this equation, the forces $\{F\}$ being known at that particular time. Thus, symbolically,

$$\{\ddot{z}\} = [M + M_W]^{-1} \{F\} \tag{9.51}$$

The mass matrix couples the motions in the various coordinates. When numbering the coordinates, care should be taken to keep the numerical difference in coordinate number for water-coupled coordinates at a minimum so the bandwidth will be as small as possible for economy in computation.

Having determined $\{\ddot{z}\}$, the usual approximations for time derivatives were used to obtain the coordinates $\{z_1\}$ at time t_1:

$$\{z_1\} = 2\{z_0\} - \{z_{-1}\} + \{\ddot{z}_0\} (\Delta t)^2 \tag{9.52}$$

In this example, the water mass effects are based on using the added mass derived from the initial geometry of the system. Nonlinearities due to gap closure are neglected. However, the beams can strike one another if the amplitudes are large. This effect is accounted for by inserting stop elements (or bilinear spring elements) between the impacting beams, as was described in the previous section, 9.8. The effect of such stops comes into play through the force F in eq. (9.51).

The exciting force of 400 lb magnitude is sinusoidal, and its frequency increases from 3 Hz to 15 Hz in 36 sec. Its application causes the vibration of the system, some details of which are shown in Fig. 9.13 and Fig. 9.14. Figure 9.13 shows the movements of each beam. We observe that there are two natural frequencies: at

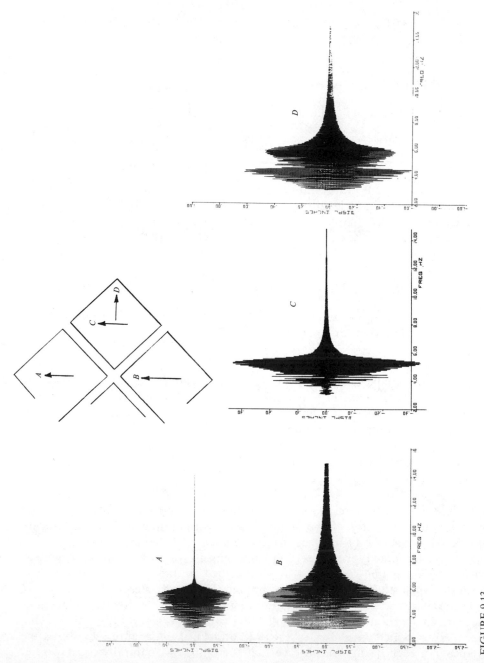

FIGURE 9.13
Beam displacements for a 400 lb force swept from 3 Hz to 15 Hz in 36 sec (note change in vertical scales)

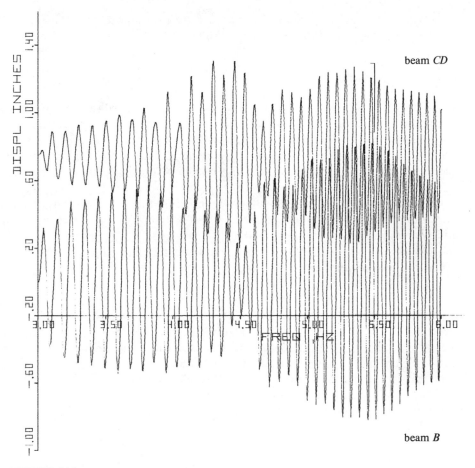

FIGURE 9.14
Motion of beams B and CD in a direction perpendicular to the gap between them (note the manner in which the gap periodically closes and opens)

approximately 3·8 Hz and 5·4 Hz. (The precise frequency varies slightly from beam to beam because of slightly differing added mass effects for each body.)

Some details in the dynamic interaction between beams may be observed in Fig. 9.14. There, we show the relative movements of beams B and CD in a direction perpendicular to the gap between them. The amplitudes of motion increase with frequency until the beams strike one another, first at near 3·6 Hz. At still higher frequencies (4·7 Hz to 6 Hz), the beams move more in unison, with their positions intertwining on the frequency axis.

The motions of this system are governed in part by the energy interchanges of the system having low-amplitude natural frequencies and mode shapes given by the equation

$$\left[[k]_D - \omega^2 [M]_D - \omega^2 [M_W] \right] \{y\} = 0$$

However, at higher amplitudes, the motions are also affected significantly by the nonlinear phenomena of the bumping of the beams.

9.10   SUMMARY

This chapter has been devoted to some aspects of the vibrations of structural components submerged in fluid. We developed the concept of the added mass due to the motion of the fluid, and showed how the added mass could be calculated by using potential flow theory for some general cases. The finite element method was used to make such calculations, and the results of the finite element calculations were compared to analytical results for the case of concentric cylinders, where good agreement was obtained. The method was also used to generate the added mass matrix for a complex configuration of beams in a container of water. Other specialized added mass elements were developed for flat beams in close proximity, and in this case it was shown how the added mass and associated damping depend on the size of the gap between the beams.

Two illustrative examples were discussed to show how the added mass effects are accounted for. In the first, the response was analyzed of a single beam in a water container, subject to seismic excitation. In this case, water mass elements were used which were based on the special derivation for the flow in a narrow gap. In the second example, an added mass matrix was used which had been derived from the finite element method as applied to potential flow. In this development, the added mass matrix served to couple the vibrational coordinates of the system. The modifications necessary in the basic component element method were discussed, so that the basic method developed in chapter 3 could be used to make the response calculations.

9.11   REFERENCES

1. O. C. Zienkiewicz, *The Finite Element Method in Engineering Science*, pp. 295–321, McGraw-Hill Book Company (UK) Ltd., London, 1971.
2. O. C. Zienkiewicz and B. Nath, 'Earthquake hydrodynamic pressures on arch dams—An electric analogue solution,' *Proc. Institution of Civil Engineers*, **25**, 165–176, 1963.
3. B. J. Hartz and G. Schmid, 'Finite element solutions for field equations with application to virtual mass coefficients in two and three dimensions,' *Quarterly Journal of the University of Washington*, pp. 20–26, College of Engineering, 1969.
4. L. K. Liu, 'Seismic analysis of the boiling water reactor,' *Proc. First Nat. Congress on Pressure Vessels and Piping*, pp. 90–102, ASME, San Francisco, 1971.
5. K. T. Patton, 'Tables of hydrodynamic mass factors for translational motion,' ASME Paper No. 65-WA/UNT-2, 1965.
6. R. J. Fritz, 'The effects of liquids on the dynamic motions of immersed solids,' *J. Engng. for Industry, Trans. ASME*, **94**, 167–173, 1972.
7. F. T. Mavis, 'Virtual mass of cylinders in water,' *Proc. ASCE, J. Hydraulics Division*, **98**, No. HY1, 319–323, 1972.
8. G. R. Sharp and W. A. Wenzel, 'Hydrodynamic mass matrix for a multibodied system,' ASME Paper No. 73-DET-121, 1973.

INDEX